D1751230

Edited by
John A. Pojman and Qui Tran-Cong-Miyata

Nonlinear Dynamics with Polymers

Related Titles

Xanthos, M. (ed.)

Functional Fillers for Plastics

2010
ISBN: 978-3-527-32361-6

Pascault, J.-P., Williams, R. J. J. (eds.)

Epoxy Polymers

New Materials and Innovations

2010
ISBN: 978-3-527-32480-4

Elias, H.-G.

Macromolecules

2005
ISBN: 978-3-527-31171-2

Elias, H.-G.

Macromolecules

Volume 4: Applications of Polymers

2009
ISBN: 978-3-527-31175-0

Dubois, P., Coulembier, O., Raquez, J.-M. (eds.)

Handbook of Ring-Opening Polymerization

2009
ISBN: 978-3-527-31953-4

Andrady, A. L.

Science and Technology of Polymer Nanofibers

2008
ISBN: 978-0-471-79059-4

Seavey, K., Liu, Y. A.

Step-Growth Polymerization Process Modeling and Product Design

2008
ISBN: 978-0-470-23823-3

Chern, C.-S.

Principles and Applications of Emulsion Polymerization

2008
E-Book
ISBN: 978-0-470-37793-2

Bruneau, C., Dixneuf, P. (eds.)

Metal Vinylidenes and Allenylidenes in Catalysis

From Reactivity to Applications in Synthesis

2008
ISBN: 978-3-527-31892-6

Severn, J. R., Chadwick, J. C. (eds.)

Tailor-Made Polymers

Via Immobilization of Alpha-Olefin Polymerization Catalysts

2008
ISBN: 978-3-527-31782-0

Edited by John A. Pojman and Qui Tran-Cong-Miyata

Nonlinear Dynamics with Polymers

Fundamentals, Methods and Applications

WILEY-VCH

WILEY-VCH Verlag GmbH & Co. KGaA

The Editors

Prof. John A. Pojman
Louisiana State University
Department of Chemistry
232, Choppin Hall
Baton Rouge, LA 70803
USA

Prof. Qui Tran-Cong-Miyata
Kyoto Inst. of Technology
Dept. of Polym. Science & Eng.
Matsugasaki Sakyoku
Kyoto 606-8585
Japan

All books published by Wiley-VCH are carefully produced. Nevertheless, authors, editors, and publisher do not warrant the information contained in these books, including this book, to be free of errors. Readers are advised to keep in mind that statements, data, illustrations, procedural details or other items may inadvertently be inaccurate.

Library of Congress Card No.: applied for

British Library Cataloguing-in-Publication Data
A catalogue record for this book is available from the British Library.

Bibliographic information published by the Deutsche Nationalbibliothek
The Deutsche Nationalbibliothek lists this publication in the Deutsche Nationalbibliografie; detailed bibliographic data are available on the Internet at <http://dnb.d-nb.de>.

© 2010 WILEY-VCH Verlag GmbH & Co. KGaA, Boschstr. 12, Weinheim

All rights reserved (including those of translation into other languages). No part of this book may be reproduced in any form – by photoprinting, microfilm, or any other means – nor transmitted or translated into a machine language without written permission from the publishers. Registered names, trademarks, etc. used in this book, even when not specifically marked as such, are not to be considered unprotected by law.

Cover Grafik-Design Schulz, Fußgönheim
Typesetting Laserwords Private Ltd., Chennai, India
Printing and Binding Fabulous Printers Pte Ltd

Printed in Singapore
Printed on acid-free paper

ISBN: 978-3-527-32529-0

Contents

List of Contributors *XI*

1 Introduction *1*
John A. Pojman and Qui Tran-Cong-Miyata
1.1 Overview *1*
1.2 What Follows *2*
1.3 The Future *4*
References *4*

2 What Is Nonlinear Dynamics and How Does It Relate to Polymers? *5*
Irving R. Epstein, John A. Pojman, and Qui Tran-Cong-Miyata
2.1 Introduction *5*
2.2 Nonlinear Dynamics *5*
2.3 Some Key Ideas of Nonlinear Chemical Dynamics *6*
2.3.1 Chemical Oscillations *7*
2.3.2 Waves and Patterns *7*
2.3.3 More Complex Phenomena *8*
2.4 Polymeric Systems *9*
2.4.1 What Is Special about Polymers? *10*
2.4.2 Challenges *10*
2.4.3 Sources of Feedback *10*
2.4.4 Nonlinear Dynamics and Phase Separation of Reacting Systems *12*
2.4.5 Spatial Structures in Polymeric Systems *13*
2.4.6 Approaches to Nonlinear Dynamics in Polymeric Systems *13*
2.4.6.1 Oscillations in a CSTR *16*
2.5 Conclusions *16*
References *16*

3 Evolution of Nonlinear Rheology and Network Formation during Thermoplastic Polyurethane Polymerization and Its Relationship to Reaction Kinetics, Phase Separation, and Mixing *21*
I. Sedat Gunes, Changdo Jung, and Sadhan C. Jana
3.1 Introduction *21*

Nonlinear Dynamics with Polymers: Fundamentals, Methods and Applications.
Edited by John A. Pojman and Qui Tran-Cong-Miyata
Copyright © 2010 WILEY-VCH Verlag GmbH & Co. KGaA, Weinheim
ISBN: 978-3-527-32529-0

3.2	Brief Overview of Evolution of Nonlinear Rheological Properties during Polymerization *22*
3.2.1	The Relationship between Nonlinear Rheology and the Extent of Polymerization during the Growth of Linear Chains *22*
3.2.2	Relationship between Nonlinear Rheology and the Extent of Polymerization during the Growth of Nonlinear Chains *24*
3.2.3	Chemical Structure of the Monomers and Polymerization Mechanism in Polyurethane Polymerization *25*
3.2.4	Evolution of Nonlinear Rheology during Polyurethane Polymerization *26*
3.2.5	Basic Reactions and Phase Separation Kinetics in Synthesis of Polyurethanes and Their Relationship to the Evolution of Nonlinear Rheology *27*
3.3	Evolution of Nonlinear Rheology and Network Formation during Thermoplastic Polyurethane Polymerization: Effects of Mixer Design, Mixing Protocol, Catalyst Concentration, and Timescales *27*
3.3.1	Effects of Mixing *29*
3.3.1.1	Mechanism of Mixing *29*
3.3.1.2	Laminar Mixing under Shear and Extensional Flow with Constant Shear and Elongation Rates *30*
3.3.1.3	Dispersive and Distributive Mixing *30*
3.3.1.4	Chaotic Mixing *31*
3.3.1.5	Effect of Mixing on Systems Undergoing Chemical Reactions *32*
3.3.2	Analysis of Timescale of Mixing and Chemical Reactions during TPU Polymerization *32*
3.3.3	Simultaneous Effects of Mixing, Chemical Reaction, and Molecular Diffusion on the Evolution of Nonlinear Rheological Properties *36*
3.4	Conclusions *38*
	References *40*
4	**Frontal Polymerization** *45*
	John A. Pojman
4.1	Introduction *45*
4.1.1	Requirements for Frontal Polymerizations *45*
4.1.2	Types of Systems *46*
4.1.3	Characteristics of Frontal Polymerization *47*
4.2	Applications *49*
4.2.1	Cure-On-Demand Putty *49*
4.2.2	Adhesive *50*
4.2.3	Coatings *51*
4.3	Motivation for Studying Nonlinear Dynamics with Frontal Polymerization *51*
4.4	Convective Instabilities *52*
4.4.1	Buoyancy-Driven Convection *52*
4.4.2	Effect of Surface-Tension-Driven Convection *55*

4.5	Thermal Instabilities	56
4.5.1	Effect of Complex Kinetics	57
4.5.2	Effect of Bubbles	58
4.5.3	Effect of Buoyancy	59
4.5.4	Other Factors	59
4.6	Snell's Law	59
4.7	Three-Dimensional Frontal Polymerization	60
4.8	Impact on Applications	61
4.9	Conclusions	62
	References	62
5	**Isothermal Frontal Polymerization**	**69**
	Lydia L. Lewis and Vladimir A. Volpert	
5.1	Introduction	69
5.1.1	A Comparison between TFP and IFP: Their Mechanisms and Front Properties	69
5.1.2	Background	71
5.2	Mathematical Models	74
5.3	Experimental IFP	79
5.4	Comparison of Experimental and Mathematical IFP	85
5.5	Conclusions	87
	Acknowledgments	88
	References	88
6	**Reaction-Induced Phase Separation of Polymeric Systems under Stationary Nonequilibrium Conditions**	**91**
	Hideyuki Nakanishi, Daisuke Fujiki, Dan-Thuy Van-Pham, and Qui Tran-Cong-Miyata	
6.1	Introduction	91
6.2	Overview of Theoretical Studies on Phase Separation Kinetics of Nonreactive and Reactive Binary Mixtures	92
6.2.1	Phase Separation of Nonreacting Mixtures	92
6.2.2	Phase Separation of Reacting Mixtures	94
6.3	Chemical Reactions in Polymeric Systems: the Non-Mean-Field Kinetics	97
6.3.1	Reaction Kinetics in the Bulk State of Polymer	97
6.3.2	Reaction Kinetics in the Liquid State of Polymer Mixtures	98
6.4	Reaction-Induced Elastic Strain and Its Relaxation Behavior	99
6.5	Phase Separation under Nonuniform Conditions in Polymeric Systems	101
6.5.1	Polymers with Spatially Graded Continuous Structures	101
6.5.2	Morphology with Arbitrary Symmetry and Distribution of Length Scales	105
6.5.2.1	The Computer-Assisted Irradiation Method	105

6.5.2.2	Polymers with an Arbitrary Distribution of Characteristic Length Scales *106*	
6.6	Conclusions *109*	
	Acknowledgments *110*	
	References *110*	

7	**Gels Coupled to Oscillatory Reactions** *115*	
	Ryo Yoshida	
7.1	Introduction *115*	
7.2	Design of Self-Oscillating Gel *116*	
7.3	Self-Oscillating Behaviors of the Gel *117*	
7.3.1	Self-Oscillation of the Miniature Bulk Gel *117*	
7.3.2	Control of Oscillating Behaviors *119*	
7.3.3	Peristaltic Motion of Gels with Propagation of Chemical Wave *119*	
7.3.4	Self-Oscillation with Structural Color Changes *121*	
7.4	Design of Biomimetic Micro-/Nanoactuator Using Self-Oscillating Polymer and Gel *122*	
7.4.1	Self-Walking Gel *122*	
7.4.2	Mass Transport Surface Utilizing Peristaltic Motion of Gel *124*	
7.4.3	Microfabrication of Self-Oscillating Gel for Microdevices *124*	
7.4.4	Control of Chemical Wave Propagation in Self-Oscillating Gel Array *126*	
7.4.5	Self-Oscillating Polymer Chains as "Nano-Oscillators" *127*	
7.4.6	Self-Flocculating/Dispersing Oscillation of Microgels *128*	
7.4.7	Fabrication of Microgel Beads Monolayer *129*	
7.4.8	Attempts of Self-Oscillation under Physiological Conditions *131*	
7.5	Conclusion *132*	
	References *132*	

8	**Self-Oscillating Gels as Biomimetic Soft Materials** *135*	
	Olga Kuksenok, Victor V. Yashin, Pratyush Dayal, and Anna C. Balazs	
8.1	Introduction *135*	
8.2	Methodology *137*	
8.2.1	Continuum Equations *137*	
8.2.2	Formulation of the Gel Lattice Spring Model (gLSM) *140*	
8.3	Sensitivity to Mechanical Deformation *143*	
8.3.1	Capturing Effects of Local Mechanical Impact on Homogeneous BZ Gels *143*	
8.3.2	Straining Heterogeneous BZ Gels *147*	
8.4	Sensitivity to Light *154*	
8.5	Conclusions *160*	
	Acknowledgments *160*	
	References *161*	

9	**Chemoelastodynamics of Responsive Gels** *163*	
	Jacques Boissonade, Pierre Borckmans, Patrick De Kepper, and Stéphane Métens	
9.1	Introduction *163*	
9.2	Elastodynamics of Responsive Gels: a Brief Survey *164*	
9.3	Oscillatory Gel Dynamics Using an Oscillating Chemical Reaction *166*	
9.3.1	The Approach *166*	
9.3.2	Coupling to the Oscillating Belousov–Zhabotinsky Reaction *169*	
9.3.3	Numerical Integration Results *171*	
9.4	Chemodynamic Oscillations Induced by Geometric Feedback *174*	
9.4.1	Spatial Bistability and Related Chemomechanical Instabilities *174*	
9.4.2	Simple Models *175*	
9.4.3	A More Realistic Model: The Polyelectrolyte Model *177*	
9.5	Experimental Observations *181*	
9.5.1	Experimental Results *182*	
9.5.1.1	Case of the Chlorite–Tetrathionate Reaction *182*	
9.5.1.2	Case of the Bromate–Sulfite Reaction *184*	
9.6	Conclusions and Perspectives *185*	
	References *186*	
10	**Oscillatory Systems Created with Polymer Membranes** *189*	
	Ronald A. Siegel	
10.1	Introduction *189*	
10.2	Survey of Synthetic Membrane Oscillators *191*	
10.2.1	Teorell Oscillator *191*	
10.2.2	Polyelectrolyte Membrane-Based Oscillators *193*	
10.2.3	Thermofluidic Oscillator *194*	
10.2.4	Lipid/Organic Membrane Analogs *196*	
10.2.5	Membrane/Enzyme Oscillators *196*	
10.2.6	General Discussion *198*	
10.3	Hydrogel–Enyzme Oscillator for Rhythmic Hormone Delivery *199*	
10.3.1	General Scheme *200*	
10.3.2	Bistability of Hydrogel Membrane Permeability *201*	
10.3.3	Oscillator Operation *204*	
10.3.4	Oscillator Prototype *205*	
10.3.5	Analysis of Factors Affecting Oscillations Over Time *207*	
10.3.6	Tuning pH Range of Oscillations *208*	
10.3.7	Discussion and Conclusion *211*	
	Acknowledgments *212*	
	References *212*	
	Further Reading *217*	

11	**Structure Formation in Inorganic Precipitation Systems** *219*	
	Oliver Steinbock and Jason Pagano	
11.1	Introduction *219*	
11.2	Permanent Patterns from Inorganic Precipitation and Deposition Processes *220*	
11.3	Tube Formation in Precipitation Systems and Silica Gardens *221*	
11.4	Historic and Cultural Links *222*	
11.5	Some Recent Developments *223*	
11.6	Experimental Methods *224*	
11.7	Growth Regimes *225*	
11.8	Wall Composition and Morphology *228*	
11.9	Relaxation Oscillations *230*	
11.10	Radius Selection *233*	
11.11	Bubbles as Templates *235*	
11.12	Toward Applications *237*	
11.13	Outlook and Conclusions *238*	
	Acknowledgments *239*	
	References *239*	

Index *243*

List of Contributors

Anna C. Balazs
University of Pittsburgh
Chemical Engineering
Department
Pittsburgh
PA 15261
USA

Jacques Boissonade
C.N.R.S
Centre De Recherche Paul Pascal
Avenue Schweitzer
Pessac 33600
France

Pierre Borckmans
Université Libre de Bruxelles
Service de Chimie Physique et
Biologie Théorique
CP 231 - Campus Plaine
1050 Brussels
Belgium

Pratyush Dayal
University of Pittsburgh
Chemical Engineering
Department
Pittsburgh
PA 15261
USA

Patrick De Kepper
C.N.R.S
Centre De Recherche Paul Pascal
Avenue Schweitzer
Pessac 33600
France

Irving R. Epstein
Brandeis University
Department of Chemistry
MS 015
Waltham
MA 02454-9110
USA

Daisuke Fujiki
Kyoto Institute of Technology
Graduate School of Science
and Technology
Department of Macromolecular
Science and Engineering
Matsugasaki
Kyoto 606-8585
Japan

I. Sedat Gunes
The University of Akron
Department of Polymer
Engineering
Akron
OH 44325-0301
USA

Sadhan C. Jana
The University of Akron
Department of Polymer
Engineering
Akron
OH 44325-0301
USA

Changdo Jung
Samsung Cheil Industries Inc.
Korea

Olga Kuksenok
University of Pittsburgh
Chemical Engineering
Department
Pittsburgh
PA 15261
USA

Lydia L. Lewis
Millsaps College
Department of Chemistry and
Biochemistry
1709 North State Street, Jackson
MS 39210
USA

Stéphane Métens
Université Paris 7-Denis Diderot
Matières et Systèmes Complexes
UMR CNRS 7057
10, rue Alice Domon et
Léonie Duquet
7525 Paris Cedex 13
France

Hideyuki Nakanishi
Kyoto Institute of Technology
Graduate School of Science
and Technology
Department of Macromolecular
Science and Engineering
Matsugasaki
Kyoto 606-8585
Japan

Jason Pagano
Saginaw Valley State University
Department of Chemistry
University Center
MI 48710
USA

John A. Pojman
Louisiana State University
Department of Chemistry
Baton Rouge
LA 70803
USA

Ronald A. Siegel
University of Minnesota
Twin cities Campus
Department of Pharmaceutics
and Biomedical Engineering
308 Harvard St. S.E, Minneapolis
MN 55455
USA

Oliver Steinbock
Florida State University
Department of Chemistry and
Biochemistry
Tallahassee
FL 32306-4390
USA

Qui Tran-Cong-Miyata
Kyoto Institute of Technology
Graduate School of Science
and Technology
Department of Macromolecular
Science and Engineering
Matsugasaki
Kyoto 606-8585
Japan

Dan-Thuy Van-Pham
Kyoto Institute of Technology
Graduate School of Science
and Technology
Department of Macromolecular
Science and Engineering
Matsugasaki
Kyoto 606-8585
Japan

Vladimir A. Volpert
Northwestern University
Department of Engineering
Sciences and Applied
Mathematics
2145 Sheridan Rd., Evanston
IL 60208-3145
USA

Victor V. Yashin
University of Pittsburgh
Chemical Engineering
Department
Pittsburgh
PA 15261
USA

Ryo Yoshida
The University of Tokyo
Graduate School of Engineering
Department of Materials
Engineering
7-3-1 Hongo
Bunkyo-ku
Tokyo 113-8656
Japan

1
Introduction
John A. Pojman and Qui Tran-Cong-Miyata

1.1
Overview

We have been friends for almost 20 years, originally drawn together by our seemingly quixotic mission of searching for advantages in studying nonlinear chemical dynamics with polymers. We are excited to have others join us in our quest but it has taken time. A focus issue appeared in *Chaos* in 1999 [1], and then a Conference Proceedings in 2000 [2].

This is the second book we have edited on the topic of nonlinear dynamics and polymers. Our first book appeared in 2003 [3], and it was a collection of symposium papers. Since that time, a great deal of progress has been made, and we chose to make a more focused volume containing invited chapters.

The two audiences for this book are nonlinear dynamicists who are interested in learning about polymers, and polymer researchers who are interested in learning about nonlinear dynamics. The problem of the former is they are unlikely to have a background in polymer chemistry. Polymer researchers are equally unlikely to have knowledge of dynamics.

So how do the ranchers and farmers become friends? We suggest collaboration, and we hope this book can aid such collaborations. A dynamicist can use the examples in the book to identify interesting phenomena and then ask a polymer scientist to help design the systems. Alternatively, a polymer scientist can seek the help of a dynamicist if he or she observes an unusual behavior.

Most of the "usual suspects" of nonlinear dynamics are presented in the following chapters, that is, temporal oscillations, chemical waves, propagating fronts, bifurcation analysis, spatial pattern formation, and the Belousov–Zhabotinsky (BZ) reaction. What distinguishes much of the work in this book from usual nonlinear dynamics is the goal of making some useful materials and devices. However, there is even more that distinguishes the work. New nonlinear phenomena arise in polymers, such as chemomechanical coupling in gel, phase separation induced by periodic forcing, and bistability in permeability.

Nonlinear Dynamics with Polymers: Fundamentals, Methods and Applications.
Edited by John A. Pojman and Qui Tran-Cong-Miyata
Copyright © 2010 WILEY-VCH Verlag GmbH & Co. KGaA, Weinheim
ISBN: 978-3-527-32529-0

1.2
What Follows

In Chapter 2, Epstein *et al.* provide an overview of nonlinear chemical dynamics and polymers. They distinguish between "self-assembly," which is an equilibrium phenomenon, and "self-organization," which only occurs far from equilibrium and on a much higher length scale. They continue with a discussion of oscillating reactions, chaos, chemical waves, and Turing patterns. They also consider the types of feedback that are known in polymers. Finally, they suggest approaches to creating dynamical systems with polymers.

Most polymeric fluids and solutions are non-Newtonian. Gunes *et al.* discuss the example of one such system in Chapter 3. They focus on the evolution of nonlinear rheological properties during polyurethane network formation in conjunction with the effects of reaction rate, extent of phase separation, diffusion limitations, and mixing protocols, although many of the general features can be readily applied to other polymerizing systems. They selected thermoplastic polyurethanes because of their complicated features; for example, they undergo phase separation and form hydrogen-bonded networks, and their rheological properties evolve during polymerization or even during processing. First, a brief overview of rheological properties of polymerizing systems is presented. This sets the stage for discussion of the rheological changes during polyurethane polymerization and helps in identifying the relationships between morphology and rheology of polyurethanes. Second, they discuss the rheological changes in polyurethanes during polymerization in detail. Third, they present some insight on the mutual relationship between rheology, extent and rate of polymerization, and the nature of the mixing process during polyurethane polymerizations.

This certainly is an important area that nonlinear dynamicists have overlooked, which we hope this contribution will help correct.

Pojman discusses thermal frontal polymerization in Chapter 4. He focuses on thermal frontal polymerization in which a localized reaction zone propagates through the coupling of thermal diffusion and the Arrhenius dependence of the kinetics of an exothermic polymerization. Frontal polymerization is close to commercial application for cure-on-demand applications and is also showing value as a way to make some materials that are superior to those prepared by traditional methods. It also manifests many types of instabilities, including buoyancy-driven convection, surface-tension-driven convection, and spin modes.

These different modes are worth studying because they can significantly interfere with the process. For example, buoyancy-driven convection can destroy a descending front when the polymer is a thermoplastic. Surface-tension-driven convection can quench a front in a thin layer unless the system has a critical viscosity. Nonplanar modes of propagation, called *"spin modes"* because in self-propagating high temperature synthesis (SHS) of inorganic materials, luminescent spots are observed spinning around the front, also appear. Buoyancy-driven convection can affect the appearance of these modes. These spin modes reduce the mechanical strength of the product.

Frontal polymerization can be used to study interesting modes not observable in other systems. For example, spherically propagating fronts can be studied.

Lewis and Volpert continue the discussion of the isothermal form of frontal polymerization in Chapter 5. Isothermal frontal polymerization is also a localized reaction zone that propagates but because of the autoacceleration of the rate of free-radical polymerization with conversion. A "seed" of poly(methyl methacrylate) is placed in contact with a solution of a peroxide or nitrile initiator, and a front propagates from the seed. The monomer diffuses into the seed, creating a viscous zone in which the rate of polymerization is faster than in the bulk solution. The result is a front that propagates but not with a constant velocity because the reaction is proceeding in the bulk solution at a slower rate. This process is used to create gradient refractive index materials by adding the appropriate dopant.

In Chapter 6, Nakanishi *et al.* review phase separation induced by combining photopolymerization with photo-cross-link reactions for a number of photoreactive binary polymer mixtures. The role of reaction inhomogeneity in the mode-selection process in phase separation is demonstrated for a number of binary polymer mixtures using laser-scanning confocal microscopy, light scattering, and Mach–Zehnder interferometry. From these experimental results, the triangular correlations among the reaction kinetics, the resulting transient elastic strain, and the morphological regularity are discussed for mixtures in both liquid and bulk states.

In Chapter 7, Yoshida reviews the work he and his coworkers have carried out to create "intelligent gels" by coupling the catalyst of the BZ reaction to hydrogels. They have created copolymer gels of *N*-isopropyl acrylamide in which ruthenium tris(2,2'-bipyridine) $(Ru(bpy)_3^{2+})$ is bound to the polymer chain. Gels expand or contract depending on the oxidation state of the ruthenium. They have created self-oscillating gels and self-walking gels. They have also created self-oscillating polymer chains.

Kuksenok *et al.* in Chapter 8 consider self-oscillating gels as biomimetic soft materials. They have modeled systems that exhibit irritability, "the ability to sense and respond to a potentially harmful stimulus." Specifically, they have modeled materials based on Yoshida's BZ gel systems that could emit a chemical "alarm signal" and directed motion in response to a mechanical deformation or impact. This could be a significant step to biomimetic materials with important applications.

Boissonade *et al.* consider the chemoelastodynamics of responsive gels in Chapter 9. This chapter is devoted to the spontaneous generation of mechanical oscillations by a responsive gel immersed in a reactive medium away from equilibrium. Two important cases are considered. In the first case, the chemomechanical instability is mainly driven by a kinetic instability leading to an oscillatory reaction. The approach is applied to the BZ reaction. The second case is a mechanical oscillatory instability that emerges from the cross-coupling of a reaction–diffusion process and the volume or size responsiveness of the supporting material. In this case, there is no need for an oscillatory reaction. Bistable reactions, namely, the chlorite-tetrathionate (CT) and the bromate-sulfite (BS) reactions, were chosen to support this approach. Several theories have been developed to account for gel-swelling mechanisms and their coupling to reactants.

In Chapter 10, Siegel reviews oscillatory systems created with polymer membranes, including membranes that support pressure-, electric-, and chemical-driven oscillations. The interaction of membrane transport with enzyme-catalyzed reactions is also reviewed. Within these systems, feedbacks of various kinds are central to the oscillation mechanisms. In the second part of the chapter, a hydrogel/enzyme system is described that can be used for rhythmic, pulsed delivery of drugs and hormones, driven by a constant external level of glucose, which serves as a free energy source. This system functions by negative, hysteretic feedback between the enzyme reaction and swelling/permeability of the hydrogel.

Steinbock reminds us that polymeric materials need not be organic. In Chapter 11, he examines self-organization in the silica garden system. This fun system is a common demonstration and actually dates back to the seventeenth century. In the conventional "chemical garden" experiment, small salt particles or crystals are seeded into aqueous solutions containing anions such as silicate, carbonate, borate, or phosphate. Such experiments are uncontrolled and cannot be made continuous. Steinbock explains how they replaced the salt crystals by a continuous flow of salt solution. He details the variety of instabilities that can occur and how bubbles can be used as templates for tube growth.

1.3
The Future

We are optimistic about the future of this field, because so much interesting work is present even though only a small fraction of polymer science is represented. We are confident that, as more polymeric systems are explored with the tools of nonlinear dynamics, more exciting and unusual phenomena will be discovered.

We hope this volume will inspire the reader to begin or continue with this new area of research.

References

1. Epstein, I.R. and Pojman, J.A. (1999) Overview: nonlinear dynamics related to polymeric systems. *Chaos*, **9**, 255–259.
2. Khokhlov, A.R., Tran-Cong-Miyata, Q., Davydov, V.A., Kuchanov S.I., and Yamaguchi T. (eds) (2000) *Nonlinear Dynamics in Polymer Science (Polynon '99)*, Wiley-VCH Verlag GmbH, Weinheim.
3. Pojman, J.A. and Tran-Cong-Miyata, Q. (ed.) (2003) *Nonlinear Dynamics in Polymeric Systems*, ACS Symposium Series 869, American Chemical Society, Washington, DC.

2
What Is Nonlinear Dynamics and How Does It Relate to Polymers?

Irving R. Epstein, John A. Pojman, and Qui Tran-Cong-Miyata

2.1
Introduction

Like its predecessor [1], which grew out of a symposium at the August 2002 National Meeting of the American Chemical Society, this volume is meant to address the connections and potential synergies between two fields that until recently have rarely intersected: nonlinear dynamics and polymer science. It seems fair to say that in the intervening 6 years some progress has been made in linking the two, but there is still a great deal of room for new ideas and collaborations. In this introduction which is largely pedagogical in nature, we attempt to lay out some of the key themes in nonlinear dynamics and suggest how they might be relevant to the study of polymer systems. A more detailed treatment of nonlinear dynamics in chemistry can be found in Ref. [2].

2.2
Nonlinear Dynamics

Generally speaking, dynamics is the study of systems that change with time. Its systematic study dates back to Newton, who was primarily concerned with the motion of objects subjected to forces, particularly gravity. Dynamical systems are typically described by differential equations, and can consist not only of falling apples or heavenly bodies but also, for example, of populations of animals, ocean currents, or, of particular interest here, chemically reacting species. When all the terms that describe the rate of change of the dependent variable (e.g., the position of an object or a chemical concentration) are constant or proportional to that variable, the differential equation is *linear* and relatively easy to solve, giving rise to such behaviors as exponential growth or decay, or the sinusoidal oscillation of a spring obeying Hooke's law. If these terms involve higher powers or more complex functional forms, the system becomes *nonlinear*, resulting in much less tractable mathematics, but a much richer array of phenomenology [3]. In chemistry, of course, systems that contain only zeroth- and first-order reactions

Nonlinear Dynamics with Polymers: Fundamentals, Methods and Applications.
Edited by John A. Pojman and Qui Tran-Cong-Miyata
Copyright © 2010 WILEY-VCH Verlag GmbH & Co. KGaA, Weinheim
ISBN: 978-3-527-32529-0

are rare, though sufficiently close to equilibrium, linear approximations work quite well. The subject of nonlinear chemical dynamics [2] deals with the behavior of chemical systems undergoing reaction, and possibly diffusion as well, under far-from-equilibrium conditions. We start off by presenting some of the key ideas of nonlinear chemical dynamics, commenting where appropriate on their relevance to polymer systems.

2.3
Some Key Ideas of Nonlinear Chemical Dynamics

The phenomena that we will be interested in have the common feature that they occur *far from equilibrium*. A useful way of thinking about this issue, suggested by Yamaguchi [4], is to distinguish between self-assembled structures and dissipative structures, which arise via self-organization. The former, crystals for example, arise under equilibrium or near-equilibrium conditions. The latter, the subject of our inquiry, are found far from equilibrium [5]. Table 2.1 summarizes the differences.

Self-assembled structures tend to be static and result from noncovalent intermolecular forces – hydrogen bonding, solvation, and so on. They require no external driving forces, and can be described by "ordinary" linear thermodynamics. Dissipative or self-organized structures are, to our biased way of thinking, much richer and more interesting. They can show variations in spatial as well as temporal dimensions and are relevant to living systems as well as to the inanimate world. Since they occur only away from equilibrium, they require fluxes of matter and/or energy to sustain them, and their thermodynamic description is correspondingly more complicated; for example, simple entropy minimization does not suffice to characterize them.

One feature that often characterizes systems of interest in nonlinear chemical dynamics is *feedback*, the influence of a species on the rate of its own production.

Table 2.1 Comparison of the properties of self-assembly and dissipative structures.

	Self-assembly	Dissipative structures
Periodicity	Spatial	Spatial and temporal
Wavelength[a]	$10^0 – 10^1$	$10^2 – 10^6$
Driving force[b]	10^1	10^2
Entropy production	Minimum	No universality
Potential function	Exists	Not known
Reversible	Yes	No
Characterized by	Phase transition	Instability and bifurcation

[a] With respect to the size of the components.
[b] With respect to the thermal energy (kT).
Adapted from Yamaguchi [4].

2.3 Some Key Ideas of Nonlinear Chemical Dynamics

The most commonly occurring form of feedback in these systems is *autocatalysis*, where the rate of production of species A increases with its concentration. In the simplest case, the rate is proportional to the concentration to some power:

$$d[A]/dt = k[A]^n \tag{2.1}$$

Autocatalysis, with $n = 1$, characterizes biological population growth, for example, since the number of offspring born is proportional to the number of individuals in the population. It leads to the Malthusian "population explosion." In chemical systems, where autocatalysis is less common, it can also result in explosion, since the solution to Eq. (2.1) is an exponentially growing concentration. Of relevance to polymer systems is the fact that any exothermic reaction is inherently autocatalytic, since an increase in the product concentration corresponds to production of heat, which leads to an increase in the rate constant of the reaction via the Arrhenius factor. If the reaction in question is lengthening a polymer chain by addition of the monomer, the rate should increase as the chain grows if the heat produced is not rapidly removed from the system.

We list below some of the most interesting and most thoroughly studied phenomena that arise in nonlinear chemical dynamics.

2.3.1
Chemical Oscillations

The study of nonlinear chemical dynamics began with chemical oscillators – systems in which the concentrations of one or more species increase and decrease periodically, or nearly periodically. While descriptions of chemical oscillators can be found at least as far back as the nineteenth century (and chemical oscillation is, of course, ubiquitous in living systems), systematic study of chemical periodicity began with two accidentally discovered systems by Bray [6] and by Belousov and Zhabotinsky (BZ) [7, 8]. These initial discoveries were met with skepticism by chemists who believed that such behavior violated the Second Law of Thermodynamics, but the development of a general theory of nonequilibrium thermodynamics [9] and of a detailed mechanism [10] for the BZ reaction brought credibility to the field by the mid-1970s. Oscillations in the prototypical BZ reaction are shown in Figure 2.1.

2.3.2
Waves and Patterns

One of the most striking features of the BZ reaction is that an apparently uniform solution unstirred in a thin layer spontaneously generates a striking pattern consisting of sets of concentric rings ("target patterns") or rotating spirals [11]. An example is shown in Figure 2.2. Similar patterns are seen in other chemical oscillators as well as in a variety of biological systems, including aggregating slime molds [12] and developing frog oocytes [13].

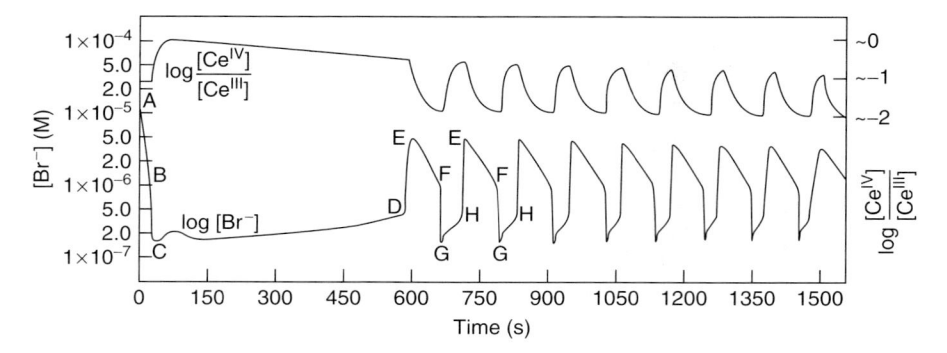

Figure 2.1 Oscillatory behavior at a platinum redox electrode and a bromide-sensitive electrode in a BZ reaction mixture containing an aqueous solution of bromate, malonic acid, a ceric salt, and sulfuric acid. Reproduced from Field et al. [10]. Copyright 1972 American Chemical Society.

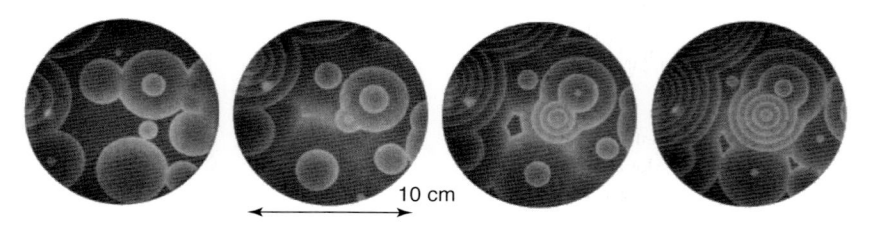

Figure 2.2 Target patterns in the BZ reaction.

Turing patterns [14] arise in systems in which an inhibitor species diffuses much more rapidly than an activator species. These patterns, which are often invoked as a mechanism for biological pattern formation, were first found experimentally in the chlorite–iodide–malonic acid reaction [15]. Figure 2.3 presents typical spot and stripe patterns.

2.3.3
More Complex Phenomena

"Simple" periodic chemical oscillations are now well understood. However, more complex behavior can arise when a single oscillator is subjected to new conditions or when it is coupled either to other oscillators or to external influences. Some chemical oscillators that are simply periodic under one set of conditions can exhibit complex, multipeaked, periodic, or even aperiodic (chaotic) [16] behavior at other concentrations and flow rates in an open reactor. Some examples of chaotic oscillations in the chlorite–thiosulfate system are shown in Figure 2.4. Coupling two or more reactions together can result in the generation of oscillations in a previously quiescent system or the cessation of oscillations in a pair of formerly

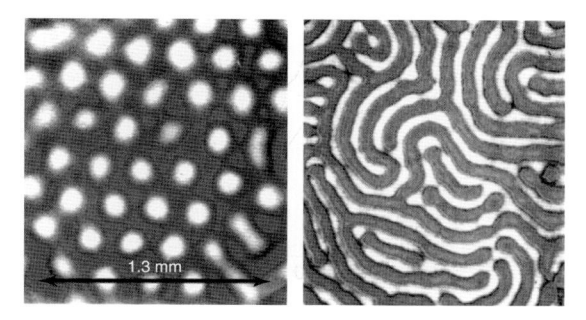

Figure 2.3 Turing patterns in the chlorite–iodide–malonic acid reaction. Dark areas show high concentrations of the starch–triiodide complex. Images courtesy of Patrick De Kepper.

Figure 2.4 Chaotic oscillation in the potential of a Pt redox electrode in the chlorite–thiosulfate reaction in a flow reactor. Adapted from Orbán and Epstein [16].

oscillating systems, or in a range of complex periodic and aperiodic oscillatory modes. Chemical oscillators containing charged or paramagnetic species are affected by electric [17, 18] or magnetic [19] fields and even, in the case of traveling waves, by gravitational fields [20, 21], as the chemical changes become coupled to convective motion.

2.4
Polymeric Systems

What is to be gained from applying nonlinear dynamics to polymer systems? Are there things that nonlinear dynamicists can learn or polymer scientists can make which would not be possible without bringing these two fields together? First, we briefly review some distinguishing characteristics of polymers. Next, we suggest three challenges that present themselves. We then examine sources of feedback in polymeric systems. Next, we propose several approaches to develop nonlinear dynamics with polymers.

We do not have the space to review polymers, so we refer the reader to several texts [22–24]. Instead, we seek to offer a brief overview of the most significant differences between polymeric systems and small-molecule systems, and review sources of feedback and approaches to nonlinear dynamics with polymers. We do not deal with biological systems, which certainly are polymeric systems. Biological systems are so complex and important that they deserve consideration separately; this was the approach of Goldbeter [25].

2.4.1
What Is Special about Polymers?

The distinguishing feature of polymers is their high molecular weight. The simplest synthetic polymer consists of hundreds to even millions of monomers that are connected end to end. However, a distribution of chain lengths always exists in a synthetic system. The molecular weight distribution can be quite broad, often spanning several orders of magnitude of molecular weight.

Linear polymers are often *thermoplastic*, meaning they can flow at some temperature depending on the molecular weight. Polymers need not be simple chains but can be branched or networked. Cross-linked polymers can be gels that swell in a solvent or *thermosets*, which form three-dimensional networks: for example, epoxy resins. This interconnectedness allows long-range coupling.

The physical properties of the reaction medium change dramatically during a reaction. For example, the viscosity almost always increases by orders of magnitude. These changes often will affect the kinetic parameters of the reaction and the transport coefficients of the medium.

Phase separation is ubiquitous with polymers because the entropy of mixing is low. Miscibility between polymers is the exception unless stronger intermolecular interactions occur, such as hydrogen bonding.

2.4.2
Challenges

In considering the possible advantages of applying nonlinear dynamics to polymeric systems, one might ask the following questions:

1) Are there new materials that can be made by deliberately exploiting the far-from-equilibrium behavior of processes in which polymers are generated?
2) Are there existing materials and/or processes that can be improved by applying the principles and methods of nonlinear dynamics?
3) Are there new nonlinear dynamical phenomena that arise because of the special properties of polymer systems?

2.4.3
Sources of Feedback

Synthetic polymer systems can exhibit feedback through several mechanisms. The simplest is thermal autocatalysis with an exothermic reaction, such as free-radical polymerizations. The reaction raises the temperature of the system, which increases the rate through the Arrhenius dependence of the rate constants. In a spatially distributed system, this mechanism allows propagation of thermal fronts.

Free-radical polymerizations of certain monomers exhibit autoacceleration at high conversion through the "gel effect" or "Norrish–Trommsdorff effect" [26, 27]. Free-radical polymerization occurs in three steps: initiation, propagation, and termination. A free radical is created by thermal decomposition or photochemical decomposition of an initiator. The initiating radical adds to a double bond of the monomer, creating a polymer chain, which can add to another monomer, propagating the chain. The chain growth terminates when two radicals encounter each other forming a chemical bond. As the polymerization proceeds, the viscosity increases. The diffusion-limited termination reactions are thereby slowed down, leading to an increase in the overall polymerization rate. The increase in the polymerization rate induced by the increase in viscosity builds a positive feedback loop into the polymerizing system. This is extremely important in the kinetics of multifunctional monomers, which polymerize much faster than monofunctional ones.

The reaction of dianhydrides with diamines can exhibit autocatalysis if performed in the proper solvent [28]. The reaction of the amine with the anhydride creates a carboxylic acid that catalyzes the reaction of the amine with an anhydride. Amine-cured epoxy systems exhibit autocatalysis because the attack on the epoxy group is catalyzed by OH, and one OH is produced for every epoxy group that reacts [29–31]. The synthesis of polyaniline by oxidation of aniline has been shown to be autocatalytic if performed electrochemically [32] or by the direct chemical oxidation [33, 34].

Because polymerization reactions are organic reactions, more studies should be made of autocatalysis in organic synthesis. For example, if an anhydride reacts with an amine, this reaction can be autocatalytic but will not produce a polymer. Difunctional molecules are required for this step-growth polymerization. (Free-radical polymerization is almost always a chain-growth reaction.) More autocatalytic organic reactions need to be identified so that more autocatalytic polymerization reactions can be created. We suspect that organic chemists do not worry about the kinetics unless they are specifically studying the mechanism of a reaction and so many interesting systems have yet to be identified.

Some polymer hydrogels exhibit "phase transitions" as the pH and/or temperature is varied [35, 36]. The gel can swell significantly as the conditions are changed and can also exhibit hysteresis [36, 37].

The necking phenomenon observed upon stretching a polymer film at a constant temperature is a consequence of a negative feedback loop driven by the interplay between the increase in temperature associated with the sample deformation and its glassification caused by the heat exchange with the environment [38]. Oscillatory behavior and period-doubling in the stress resulting from a constant strain rate have been experimentally observed.

Diffusion of small molecules, usually solvents, into glassy polymers exhibits "anomalous" or "non-Fickian" behavior [39]. As the solvent penetrates, the diffusion coefficient increases because the glass transition temperature is lowered. The solvent acts as a plasticizer, increasing the free volume and the mobility of the solvent. Thus we have an autocatalytic diffusion process.

Dissolving of some polymers in aqueous media can proceed by means of a front [40]. Water-dissolvable polymers are formed from esters, which create an acid upon hydrolysis that can catalyze further hydrolysis.

Finally, polymer melts and solutions are usually non-Newtonian fluids [41–43]. They often exhibit shear-thinning, which means the viscosity decreases as the shear is increased. This can lead to unusual phenomena. For example, when a polymer melt is extruded through a die, transient oscillations can occur [44, 45]. (Polymers can also exhibit shear-thickening.)

An unusual phenomenon is the Weissenberg effect, or the climbing of polymeric liquids up rotating shafts [43]. A Newtonian fluid, on the other hand, is depressed by rotation because of centrifugal forces.

2.4.4
Nonlinear Dynamics and Phase Separation of Reacting Systems

Nonlinear dynamics of polymers may arise from two types of instabilities depending on their nature: intrinsic or extrinsic. In the former, the spatiotemporal behavior arises from the intrinsic instabilities, which are generated as a consequence of some internal feedback processes existing inside the polymer system under consideration. Most of these instabilities originate from the thermomechanical couplings such as glass transition induced by a change in temperature associated with uniaxial extension of a polymer strip under a constant deformation rate [38, 46]. The spatiotemporal behavior of adhesives observed upon peeling of a polymer membrane loosely adhered on a surface also arises from this type of instability [47]. In these phenomena, viscoelasticity, which is a feature of chain-molecule systems, plays an indispensable role as a long-range effect. On the other hand, the rich dynamics of oscillatory chemical reactions such as BZ or pH-oscillatory reactions has led several research groups to attempt coupling the reaction kinetics to polymer dynamics [48]. Via this coupling, a piece of polyelectrolyte gel can exhibit oscillatory behavior in response to these periodic stimuli. In these experiments, the roles played by the difference between the characteristic timescales of the reaction and the relaxation times of polymer are not yet clearly understood. The kinetic mismatches between the two systems, that is, externally oscillatory reaction and relaxation of internal polymer segments, would play a key role in generating and controlling the spatiotemporal behavior of the polymer.

Irradiating a polymer mixture with light of high intensity can generate a gradient of quench depth along the pathway of light, leading to morphologies with a gradient of characteristic length scales [49]. This is an example of phase separation proceeding under stationary, nonequilibrium conditions imposed by a gradient of light intensity. Furthermore, by using the so-called computer-assisted irradiation (CAI) method, polymer mixtures can be maintained far from equilibrium under these stationary nonequilibrium conditions to generate and manipulate exotic morphologies as described in Chapter 6.

2.4.5
Spatial Structures in Polymeric Systems

A variety of spatial structures can emerge over a wide range of length scales, from nanometers to micrometers, as a result of the competitions between antagonistic interactions (hydrophobic vs hydrophilic, repulsion vs attraction in micellar systems, or suppression of phase separation at long range vs activation at short range in block copolymers). The existence of these competing interactions leads to the formation of universal ordered structures, such as hexagonal, cylindrical, gyroid or lamellar, and so on, observed in block copolymers, micellar systems, as well as liquid crystals. Because of this competing interaction mechanism, the morphology of block copolymers bears great resemblance to Turing structures [50, 51]. The difference in the length scales of these structures depends on the strength of the interactions between individual components in the system. Recently, it was found that typical structures resulting from these competing mechanisms, such as hexagonal phase with the length scales in the intermediate range (micrometers), can be also obtained by the competition between cross-link reaction and phase separation in polymer mixtures [52]. Finally, it is worth noting that by introducing some sort of anisotropy into these competing interaction processes, such as reaction anisotropy induced by linearly polarized light [53] or asymmetric interactions in phase separation of block copolymers by molecular design [54], a wide variety of ordered structures can be generated.

As a strong support for the roles of competing interactions in the emergence of structural regularity and orders, it has been recently shown by both theory [55] and experiments [55] that the competition between attractive and repulsive interactions in systems of charged micelles can lead to the formation of ordered structures with a rich variety of symmetry as well as regularity, such as icosahedral structures.

In summary, the competition between antagonistic interactions seems to be a general principle for the emergence of a variety of spatially ordered structures over a wide range of length scales, including the Turing structure. The competitions between antagonistic interactions along different orientations can also lead to breaking spatial symmetry, resulting in structures with various spatial symmetries. Competing interactions and breaking symmetry would be an efficient road to the design of functional materials.

2.4.6
Approaches to Nonlinear Dynamics in Polymeric Systems

We propose three approaches to creating nonlinear dynamical systems with polymers:

1) Couple polymers and polymer-forming reactions to other nonlinear systems (Type I).
2) Create a dynamical system using the inherent nonlinearities in polymeric systems (Type II).

3) Polymer systems are invariably characterized by polydispersity of the molecular weight distribution. One should be able to exploit the distribution of polymer lengths to amplify nonlinear effects in polymer systems, perhaps due to the molecular weight dependence of the diffusion coefficient. We know of no experimental work, but there has been a theoretical work considering such an effect on ester interchange reactions [56].

There have been several examples of the first approach. Given the importance of the BZ reaction in nonlinear chemical dynamics, it is not surprising that polymers and polymerizations would be coupled to it. Váradi and Beck have shown that adding acrylonitrile to the BZ reaction could inhibit oscillations, and a precipitate was produced that they assumed was polyacrylonitrile [57]. Pojman *et al.* [58] studied the BZ reaction to which acrylonitrile was added and showed that, after an inhibition period, the polyacrylonitrile was produced periodically in phase with the oscillations (Figure 2.5). Given that radicals are produced periodically from the oxidation of malonic acid by ceric ion, it seemed reasonable to assume that the periodic appearance of the polymer was caused by periodic initiation. However, Washington *et al.* showed that periodic termination by bromine dioxide caused the periodic polymerization [59]. Figure 2.6 shows that polymerization begins when the $BrO_2\bullet$ concentration drops.

Although this system is interesting and has provided unexpected results, nothing useful can be accomplished with it. If we were seeking periodic polymerization, we

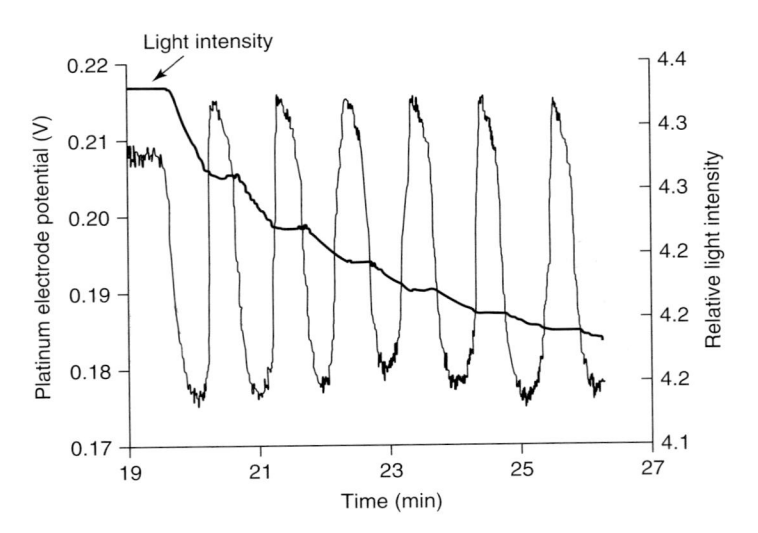

Figure 2.5 The evolution of a BZ reaction in which 1.0 ml acrylonitrile was present before the Ce(IV)/H_2SO_4 solution was added. [$NaBrO_3$]$_0$ = 0.077 M; [Malonic Acid]$_0$ = 0.10 M; [Ce(IV)]$_0$ = 0.0063 M; [H_2SO_4]$_0$ = 0.90 M. No oscillations occurred during the first 19 min. Adapted from Pojman *et al.* [58].

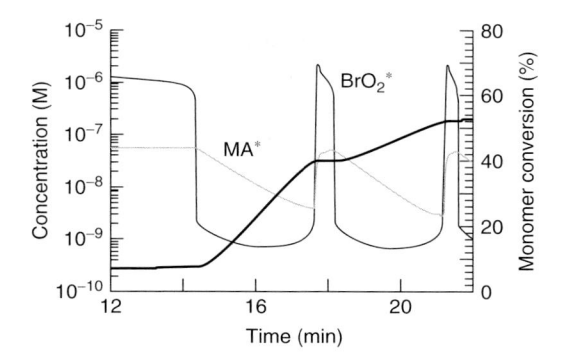

Figure 2.6 Simulations of the monomer conversion and the concentrations of malonyl radical and bromine dioxide as a function of time in the BZ reaction.

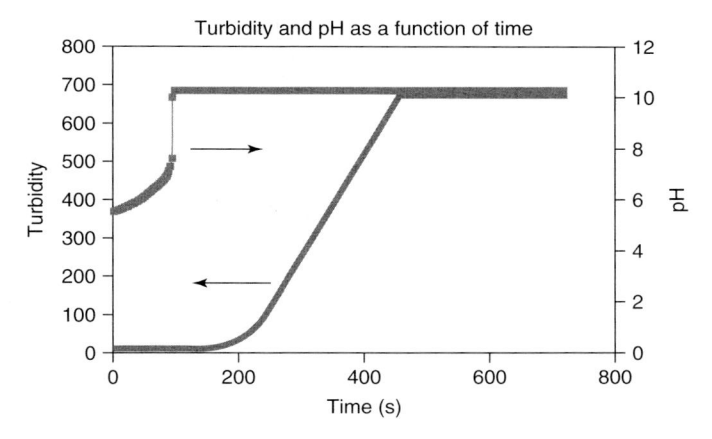

Figure 2.7 The turbidity of a dispersion of a trithiol and a triacrylate in the formaldehyde–bisulfite–sulfite clock reaction. An increase in turbidity corresponds to formation of the polymer.

could add a redox initiation system periodically or simply illuminate a photopolymerizable system periodically.

Hu et al. [60] used the formaldehyde–bisulfite–sulfite clock reaction [61] (Figure 2.7) as a trigger for the polymerization of a trithiol with a triacrylate. (The addition of a thiol to an acrylate is base-catalyzed.) The initial pH was 5, which then abruptly increased to 10 after 100 s. The polymerization was monitored from the turbidity of the solution, which increased after the pH increased. The system represents "time-lapse polymerization" in which the delay time can be programmed [62]. Unfortunately, this particular system is not practical because of the formaldehyde's toxicity and the short clock times; a system that could be tunable from minutes to hours would be desirable.

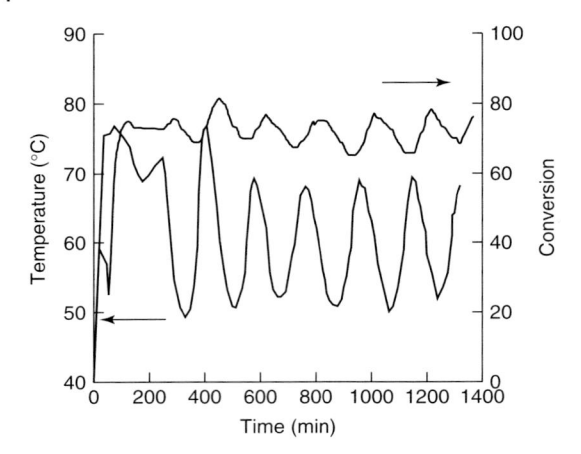

Figure 2.8 Oscillatory behavior of vinyl acetate polymerization in a CSTR. Adapted from Teymour and Ray [64].

2.4.6.1 Oscillations in a CSTR

With their combination of complex kinetics and thermal, convective, and viscosity effects, polymerizing systems would seem to be fertile ground for generating oscillatory behavior. Teymour and Ray reported both laboratory-scale Continuous Stirred-Tank Reactor (CSTR) experiments and modeling studies on vinyl acetate polymerization [63–66]. The period of oscillation was long, about 200 min, which is typical for polymerization in a CSTR (Figure 2.8). Papavasiliou and Teymour [67] reviewed nonlinear dynamics in CSTR polymerizations.

Emulsion polymerizations in a CSTR can exhibit oscillations in the extent of conversion and the interfacial tension [68].

2.5
Conclusions

The study of nonlinear dynamics has been limited for too long to inorganic systems. We hope this chapter will spark enough interest to make one read and observe how polymeric systems can exhibit many of the same temporal and spatial phenomena observed in systems such as the BZ reaction and some surprising new ones. Polymer systems may provide the first commercial applications of this field, and will certainly provide interesting results in the future.

References

1. Pojman, J.A. and Tran-Cong-Miyata, Q. (eds) (2003) *Nonlinear Dynamics in Polymeric Systems*, ACS Symposium Series No. 869, American Chemical Society, Washington, DC.

2. Epstein, I.R. and Pojman, J.A. (1998) *An Introduction to Nonlinear Chemical Dynamics: Oscillations, Waves, Patterns and Chaos*, Oxford University Press, New York.

3. Strogatz, S.H. (1994) *Nonlinear Dynamics and Chaos*, Addison-Wesley, Reading, MA.
4. Yoshida, R., Sakai, T., Ito, S., and Yamaguchi, T. (2003) in *Nonlinear Dynamics in Polymeric Systems*, ACS Symposium Series No. 869 (eds J.A. Pojman and Q. Tran-Cong-Miyata), American Chemical Society, Washington, DC, pp. 276–291.
5. Nicolis, G. and Prigogine, I. (1977) *Self-Organization in Nonequilibrium Systems*, John Wiley & Sons, Inc., New York.
6. Bray, W.C. (1921) A periodic reaction in homogeneous solution and its relation to catalysis. *J. Am. Chem. Soc.*, **43**, 1262–1267.
7. Belousov, B.P. (1958) A periodic reaction and its mechanism. *Sb. Ref. Radiats. Med.*, 145.
8. Belousov, B.P. (1985) in *Oscillations and Traveling Waves in Chemical Systems* (eds R.J. Field and M. Burger), John Wiley & Sons, Inc., New York, pp. 605–613.
9. Glansdorff, P. and Prigogine, I. (1971) *Thermodynamics of Structure, Stability and Fluctuations*, John Wiley & Sons, Inc., New York.
10. Field, R.J., Körös, E., and Noyes, R. (1972) Oscillations in chemical systems. II. Thorough analysis of temporal oscillation in the Bromate-Cerium-Malonic acid system. *J. Am. Chem. Soc.*, **94**, 8649–8664.
11. Zaikin, A.N. and Zhabotinskii, A.M. (1970) Concentration wave propagation in two-dimensional liquid-phase self-oscillating system. *Nature*, **225**, 535–537.
12. Kessler, D.A. and Levine, H. (1993) Pattern formation in dictyostelium via the dynamics of cooperative biological entities. *Phys. Rev.*, **E48**, 4801–4804.
13. Lechleiter, J., Girard, S., Peralta, E., and Clapham, D. (1991) Spiral Calcium wave propagation and annihilation in *Xenopus Laevis* oocytes. *Science*, **252**, 123–126.
14. Turing, A.M. (1952) The chemical basis of morphogenesis. *Philos. Trans. R. Soc. London Ser. B*, **237**, 37–72.
15. Castets, V., Dulos, E., Boissonade, J., and De Kepper, P. (1990) Experimental evidence of a sustained standing turing-type nonequilibrium chemical pattern. *Phys. Rev. Lett.*, **64**, 2953–2956.
16. Orbán, M. and Epstein, I.R. (1982) Complex periodic and aperiodic oscillation in the chlorite-thiosulfate reaction. *J. Phys. Chem.*, **86**, 3907–3910.
17. Sevcikova, H. and Marek, M. (1983) Chemical waves in electric field. *Phys. D*, **9**, 140–156.
18. Sevcikova, H. and Marek, M. (1986) Chemical waves in electric field-modelling. *Phys. D*, **21**, 61–77.
19. Boga, E., Kádár, S., Peintler, G., and Nagypál, I. (1990) Effect of magnetic fields on a propagating reaction front. *Nature*, **347**, 749–751.
20. Nagypál, I., Bazsa, G., and Epstein, I.R. (1986) Gravity induced anisotropies in chemical waves. *J. Am. Chem. Soc.*, **108**, 3635–3640.
21. Pojman, J.A. and Epstein, I.R. (1990) Convective effects on chemical waves. 1. Mechanisms and stability criteria. *J. Phys. Chem.*, **94**, 4966–4972.
22. Odian, G. (2004) *Principles of Polymerization*, 4th edn, John Wiley & Sons, Inc., New York.
23. Allcock, H.R. and Lampe, F.W. (1981) *Contemporary Polymer Chemistry*, Prentice Hall, Englewood Cliffs.
24. Sperling, L.H. (1992) *Introduction to Physical Polymer Science*, John Wiley & Sons, Inc., New York.
25. Goldbeter, A. (1996) *Biochemical Oscillations and Cellular Rhythms: The Molecular Bases of Periodic and Chaotic Behaviour*, Cambridge University Press, Cambridge.
26. Norrish, R.G.W. and Smith, R.R. (1942) Catalyzed polymerization of methyl methacrylate in the liquid phase. *Nature*, **150**, 336–337.
27. Trommsdorff, E., Köhle, H., and Lagally, P. (1948) Zur polymerisation des methacrylsäuremethylesters. *Makromol. Chem.*, **1**, 169–198.
28. Kaas, R.L. (1981) Autocatalysis and equilibrium in polyimide synthesis. *J. Polym. Sci. Polym. Chem. Ed.*, **19**, 2255–2267.
29. Mijovic, J. and Wijaya, J. (1994) Reaction kinetics of epoxy/amine model systems. The effect of electrophilicity of amine molecule. *Macromolecules*, **27**, 7589–7600.

30. Aspin, I.P., Barton, J.M., Buist, G.J., Deazle, A.S., and Hammerton, I. (1994) Kinetic and simulation studies of linear epoxy systems. *J. Mater. Chem.*, **4**, 385–388.
31. Eloundou, J.P., Feve, M., Harran, D., and Pascault, J.P. (1995) Comparative studies of chemical kinetics of an epoxy-amine system. *Die Ang. Makrom. Chem.*, **230**, 13–46.
32. Mu, S. and Kan, J. (1996) Evidence for the autocatalytic polymerization of aniline. *Electrochim. Acta*, **41**, 1593–1599.
33. Mazeikiene, R. and Malinauskas, A. (2000) Deposition of polyaniline on glass and platinum by autocatalytic oxidation of aniline with dichromate. *Synth. Met.*, **108**, 9–14.
34. Karunakaran, C. and Palanisamy, P.N. (2001) Autocatalysis in the sodium perborate oxidation of anilines in acetic acid–ethylene glycol. *J. Mol. Catal. A: Chem.*, **172**, 9–17.
35. Tanaka, T. (1986) Kinetics of Phase Transition in Polymer Gels. *Physica A: Stat. Mech.*, **140**, 261–268.
36. Addad, J.P.C. (1996) *Physical Properties of Polymer Gels*, John Wiley & Sons, Ltd, Chichester.
37. Hirotsu, S., Hirokawa, Y., and Tanaka, T. (1987) Volume–Phase Transitions of Ionized N-Isopropylacrylamide Gels. *J. Chem. Phys.*, **87**, 1392–1395.
38. Andrianova, G.P., Kechkyan, A.S., and Kargin, V.A. (1971) Self-oscillating mechanism of extension of polymers. *J. Poly. Sci.*, **9**, 1919–1933.
39. Crank, J. (1975) *Mathematics of Diffusion*, Clarendon, Oxford.
40. von Burkersroda, F., Schedl, L., and Opferich, A.G. (2002) Why degradable polymers undergo surface erosion or bulk erosion. *Biomaterials*, **23**, 4221–4231.
41. Severs, E.T. (1962) *Rheology of Polymers*, Reinhold, New York.
42. Ferry, J.D. (1980) *Viscoelastic Properties of Polymers*, John Wiley & Sons, Inc., New York.
43. Gupta, R.K. (2000) *Polymer and Composite Rheology*, Marcel Dekker, New York.
44. Meissner, J. (1972) Modifications of the Weissenberg rheogoniometer for measurement of transient rheological properties of molten polyethylene under shear. Comparison with tensile data. *J. Appl. Polym. Sci.*, **16**, 2877–2899.
45. Bird, R.B., Armstrong, R.C., and Hassager, O. (1987) *Dynamics of Polymeric Liquids*, 2nd edn, vol. 1, John Wiley & Sons, Inc., New York.
46. Toda, A., Tomita, C., Hikosaka, M., Hibino, Y., Miyaji, H., Nonomura, C., Suzuki, T., and Ishihara, H. (2002) Thermo-mechanical coupling and self-excited oscillation in the neck propagation of PET films. *Polymer*, **43**, 947–951.
47. Yamazaki, Y. and Toda, A. (2006) Pattern formation and spatiotemporal behavior of adhesive in peeling. *Phys. D*, **214**, 120–131.
48. Yoshida, R., Takahashi T, Yamaguchi, T., and Ichijo, H. (1996) Self-oscillating gel. *J. Am. Chem. Soc.*, **118**, 5134–5135.
49. Nakanishi, H., Namikawa, N., Norisuye, T., and Tran-Cong-Miyata, Q. (2006) Interpenetrating polymer networks with spatially graded morphology controlled by UV radiation curing. *Macromol. Symp.*, **242**, 157–164.
50. Ohta, T., Ito, A., and Tetsuka, A. (1990) Self-organization in an excitable reaction-diffusion system: synchronization of oscillatory domains in one dimension. *Phys. Rev. A*, **42**, 3225–3232.
51. Seul, M. and Andelman, D. (1995) Domain shapes and patterns: the phenomenology of modulated phases. *Science*, **267**, 476–483.
52. Nakanishi, H., Satoh, M., and Tran-Cong-Miyata, Q. (2008) Hexagonal phase induced by a reversible cross-link reaction in a polymer mixture. *Phys. Rev. E*, **77**, 020801(R).
53. Kataoka, K., Urakawa, O., Nishioka, H., and Tran-Cong, Q. (1998) Directional phase separation of binary polymer blends driven by photo-cross-linking with linearly polarized light. *Macromolecules*, **31**, 8809–8816.
54. Matsushita, Y. (2007) Creation of hierarchically ordered nanophase structures in block polymers having various competing interactions. *Macromolecules*, **40**, 772–776.

55. Vernizzi, G. and Olvera de la Cruz, M. (2007) Faceting ionic shells into icosahedra via electrostatics. *Proc. Natl. Acad. Sci.*, **104**, 18382–18386.
56. Pojman, J.A., Garcia, A.L., Kondepudi, D.K., and Van den Broeck, C. (1991) Nonequilibrium processes in polymers undergoing interchange reactions. 2. Reaction-diffusion processes. *J. Phys. Chem.*, **95**, 5655–5660.
57. Váradi, Z. and Beck, M.T. (1973) Inhibition of a homogeneous periodic reaction by radical scavengers. *J. Chem. Soc. Chem. Commun.*, 30–31.
58. Pojman, J.A., Leard, D.C., and West, W. (1992) The periodic polymerization of acrylonitrile in the cerium-catalyzed Belousov–Zhabotinskii reaction. *J. Am. Chem. Soc.*, **114**, 8298–8299.
59. Washington, R.P., Misra, G.P., West, W.W., and Pojman, J.A. (1999) Polymerization Coupled to Oscillating Reactions: I. A Mechanistic Investigation and Numerical Simulation of Acrylonitrile Polymerization in the Belousov-Zhabotinsky Reaction in a Batch Reactor. *J. Am. Chem. Soc.*, **121**, 7373–7380.
60. Hu, G., Pojman, J.A., Bounds, C., and Taylor, A.F. (2010) Time-lapse thiol-acrylate polymerization using a pH clock reaction. *J. Polym. Sci. Part A: Polym. Chem.*, **48**, 2955–2959.
61. Kovacs, K., McIlwaine, R., Gannon, K., Taylor, A.F., and Scott, S.K. (2004) Complex behavior in the formaldehyde-sulfite reaction. *J. Phys. Chem. A*, **109**, 283–288.
62. Norling, P.M. (1977) Time-lapse free-radical polymerizable composition. US Patent 4,000,150.
63. Teymour, F. and Ray, W.H. (1992) The dynamic behavior of continuous polymerization reactors – VI. Complex dynamics in full-scale reactors. *Chem. Eng. Sci.*, **47**, 4133–4140.
64. Teymour, F. and Ray, W.H. (1992) The dynamic behavior of continuous polymerization reactors – V. Experimental investigation of limit-cycle behavior for vinyl acetate polymerization. *Chem. Eng. Sci.*, **47**, 4121–4132.
65. Teymour, F. and Ray, W.H. (1991) Chaos, intermittency and hysteresis in the dynamic model of a polymerization reactor. *Chaos, Soliton Fractals*, **1**, 295–315.
66. Teymour, F. and Ray, W.H. (1989) The dynamic behavior of continuous polymerization reactors-IV. Dynamic stability and bifurcation analysis of an experimental reactor. *Chem. Eng. Sci.*, **44**, 1967–1982.
67. Papavasiliou, G. and Teymour, F. (2003) in *Nonlinear Dynamics in Polymeric Systems*, ACS Symposium Series No. 869 (eds J.A. Pojman and Q. Tran-Cong-Miyata), American Chemical Society, Washington, DC, pp. 309–323.
68. Schork, F.J. and Ray, W.H. (1987) The dynamics of the continuous emulsion polymerization of methylmethacrylate. *J. Appl. Poly. Sci. Chem.*, **34**, 1259–1276.

3
Evolution of Nonlinear Rheology and Network Formation during Thermoplastic Polyurethane Polymerization and Its Relationship to Reaction Kinetics, Phase Separation, and Mixing

I. Sedat Gunes, Changdo Jung, and Sadhan C. Jana

3.1
Introduction

Understanding the evolution of nonlinear rheological properties during polymerization has been of significant industrial and academic interest [1]. The industrial interest originated from the design, operation, and control requirements of polymerization processes [2, 3]. The industrial polymerization processes cover a broad spectrum from bulk production of thermoplastics, such as polyolefins, to production of thermosets, such as tires, comprising many simultaneous steps, such as chemical reactions and mixing, followed by compounding, molding, and cross-linking reactions [4, 5]. The most relevant and widely documented rheological property for success of industrial processes is the viscosity of the polymerizing medium, which directly influences mass, heat, and momentum transfer processes. The academic interest, on the other hand, has its roots in utilizing rheology as a valuable tool to understand the properties of polymer chains whether in bulk, in solution, or in a polymerizing medium [6]. Monitoring the changes in rheological properties, such as normal stresses, relaxation times, storage, and loss moduli, which are usually nonlinear functions of time, temperature, and morphology, has undeniable significance in providing insights into structure and properties of polymers as they develop during polymerization or during shape-forming operations such as of extrusion, injection molding, casting, and film blowing, to name a few.

The nonlinear rheological properties of growing polymer chains, irrespective of the growth mechanisms – such as free-radical or step-growth – are governed by similar parameters. Chemical structure, flexibility, and mobility of chains, the available free volume during polymerization, the degree of conversion, the extent of chain entanglements, the degree of phase separation, and the presence of chemical or physical cross-links are a few of the morphological features and physical phenomena [7] with strong influence on polymerization.

Nonlinear Dynamics with Polymers: Fundamentals, Methods and Applications.
Edited by John A. Pojman and Qui Tran-Cong-Miyata
Copyright © 2010 WILEY-VCH Verlag GmbH & Co. KGaA, Weinheim
ISBN: 978-3-527-32529-0

In this chapter, we focus on the evolution of nonlinear rheological properties during polyurethane network formation in conjunction with the effects of reaction rate, extent of phase separation, diffusion limitations, and mixing protocols, although many of the general features can be readily applied to other polymerizing systems. We selected thermoplastic polyurethanes (TPUs) because of their intricate nature; for example, they undergo phase separation, form hydrogen-bonded networks, and their rheological properties evolve during polymerization or even during processing. First, a brief overview of the rheological properties of polymerizing systems is presented. This will set the stage for discussion of the rheological changes during polyurethane polymerization and can help in identifying the relationships between the morphology and rheology of polyurethanes. Second, we discuss the rheological changes in polyurethanes during polymerization in detail. Third, we present some insight on the mutual relationship between rheology, extent and rate of polymerization, and the nature of mixing process during polyurethane polymerizations.

3.2
Brief Overview of Evolution of Nonlinear Rheological Properties during Polymerization

In this section, we first present an overview of the evolution of rheological properties during polymerization of linear chains by free-radical and step-growth mechanisms. Then, we turn our attention to thermosetting polymers that form permanently cross-linked polymer networks.

3.2.1
The Relationship between Nonlinear Rheology and the Extent of Polymerization during the Growth of Linear Chains

The first and foremost crucial influence of the extent of polymerization on the rheological properties is the increase of viscosity with the increase of molecular weight (M_w). The molecular weight at any given instant is the most influential parameter that determines the rheological properties during linear polymerization. For example, viscosity is a linear function of M_w in the early stages of polymerization where the growing chains are relatively short. However, the relationship between viscosity and molecular weight becomes nonlinear beyond a certain critical molecular weight, which indicates a strong morphological change in the polymerizing medium (Figure 3.1). The onset of the nonlinear relationship between viscosity and molecular weight is usually correlated with the occurrence of chain entanglements that significantly diminish the mobility of polymer chains and hence increase the resistance against flow. The onset point of nonlinear viscosity versus M_w relationship is dependent on the individual polymers. For example, the value of M_w for the onset in polystyrene (PS) was observed to be about an order higher (~40 000) compared to that of polyethylene (PE) (~4000) [8]. Note that PE has a

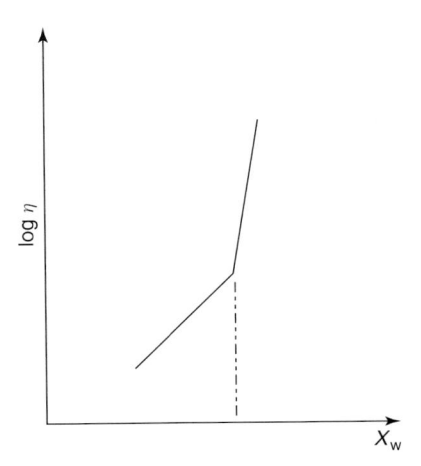

Figure 3.1 Typical dependence of viscosity (η) on molecular weight (M_w) for linear polymer melts. X_w is defined as $X_w = (s_o)^2 Z \varphi_1 / M_w \upsilon$, where $(s_o)^2$ is the mean square radius of gyration of an unperturbed molecule, Z is the number of chain atoms in the polymer, υ is the specific volume of polymer, and φ_1 is the volume fraction of polymer if the material is in solution. The slope of the curve is nearly unity below a critical value of X_w, and it assumes a value of about 3.4 above it. Note that the slopes below and above X_w are constant. The presence of a critical value of X_w is universal for polymers; however, its absolute value varies depending on the chemical structure of the polymer.

more flexible chemical structure than PS, which in turn can form entanglements at lower M_w.

Thus, it is evident that the degree of polymerization (DP) has a strong influence on the rheological properties of polymers. However, the existence of a mutual relationship between DP of linear polymers and rheology is based on a common a hypothesis. Purportedly, the best known and the most widely studied example in this context is the gel effect or Trommsdorff [9] effect. The gel effect has been identified with autoacceleration of the polymerization rate at intermediate to high monomer conversions during free-radical polymerization [10]. Note that autoacceleration and the gel effect have been considered as consequences of reduction in the rate of termination reactions in free-radical polymerization [11]. The gel effect can lead to catastrophic results, such as reactor explosion, in view of the highly exothermic nature of polymerization reactions [10]. The most common hypothesis on the origin of gel effect has been the formation of entanglements beyond a certain conversion level, which in turn diminish the mobility of growing chains and hence reduce the probability of termination reactions [12]. However, O'Neil et al. [13–15] also presented evidence for gel effect in the absence of entanglements during free-radical polymerization and established the role of the reduction in free volume as a potential cause for the onset of the gel effect. One may still argue that the formation of entanglements plays a role in the gel effect;

however, it is evident that the origin of the gel effect cannot be solely attributed to the formation of entanglements.

3.2.2
Relationship between Nonlinear Rheology and the Extent of Polymerization during the Growth of Nonlinear Chains

The influence of the extent of polymerization on rheological properties of thermosets is usually more dramatic compared to that of thermoplastics [16]. For example, the viscosity (η) of the polymerizing medium increases by orders of magnitude with conversion (α) at the onset of formation of cross-linked polymer networks (Figure 3.2). The conversion at the onset of formation of cross-linked polymer networks is defined as the gel point (α_c). The value of α_c is usually determined from rheological experiments [17]. Some investigators have determined the value of α_c from a plot of ($1/\eta$) versus α. The value of α_c is obtained by extrapolating the data to $1/\eta \rightarrow 0$ [18]. Several other investigators [19–21] have used the time at which the storage modulus (G') and loss modulus (G'') at a given oscillation frequency show a crossover as the gel time (t_{gel}). A few other researchers have found that the loss tangent ($\tan \delta = G''/G'$) becomes independent of the frequency at the gel point [22, 23]. At or beyond the gel point, the polymer is no longer a rheologically simple liquid. It turns into a viscoelastic solid with practically infinite viscosity. The crossover point in a loss and storage moduli versus reaction time curve is usually an accurate identification of the onset of gelation [24].

A more direct influence of polymerization on the rheological properties is derived from the thermal effects. Polymerization reactions are usually highly exothermic and hence the progress of polymerization can significantly increase

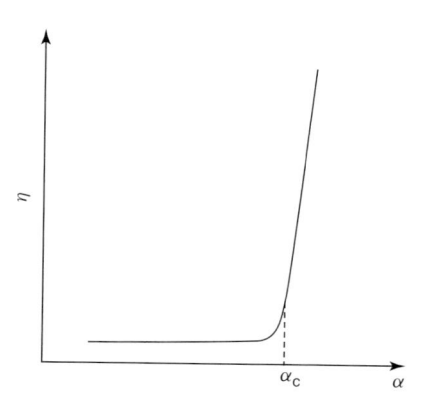

Figure 3.2 Typical dependence of viscosity (η) on the conversion factor (α) for thermosetting polymers. A continuous polymer network forms at a critical conversion (α_c), which is recognized as the gel point. Beyond α_c, the polymer exhibits the properties of a viscoelastic solid.

the temperature of the reaction mass, which in turn can reduce the viscosity significantly [25, 26].

3.2.3
Chemical Structure of the Monomers and Polymerization Mechanism in Polyurethane Polymerization

Polyurethanes can form thermoplastic or thermosetting polymers, depending upon the functionality of the monomers used in the formulation. The use of difunctional monomers yields a thermoplastic that flows upon heating, whereas higher functionality monomers result in thermosetting polymers. TPUs are synthesized by condensation of diisocyanates and dialcohols [27]. In usual practice, two different dialcohols, one with a relatively high molecular weight and the other with a low molecular weight, are utilized. The high molecular weight dialcohol is designated as the polyol – it forms the soft segments and is responsible for rubberlike elasticity of TPUs [28]. The low molecular weight alcohol – also called *short chain diols* – acts as the chain extender and, in conjunction with diisocyanate, forms the hard segments [29].

TPUs are capable of exhibiting physical properties of a thermosetting polymer network, such as the rubberlike elasticity upon deformation and instantaneous recovery from a state of stress. These properties originate from the phase separation of hard segments, which is promoted by strong hydrogen bonding between the hard segments and significant thermodynamic incompatibility between the highly polar hard segments and much less polar soft segments [30]. The phase-separated hard segments form the hard segment domains and act both as physical cross-link sites and as reinforcing entities in polyurethanes (Figure 3.3).

Figure 3.3 Morphology of thermoplastic polyurethanes (TPUs). The solid prisms illustrate the urethane linkages, which have been designated as the hard segments. The polyol soft segments are indicated as flexible strings. Note that a portion of hard segments undergoes phase separation and forms hard domains. The hard domains act both as physical cross-link sites and reinforcing fillers. Note that only a fraction of hard segments form hard segment domains upon phase separation.

Polyurethanes are condensation polymers formed by step-growth polymerization in which the chain length of the polymer increases steadily as the reaction progresses. Two major routes of polymerization of polyurethanes are one-step and two-step synthesis methods. In one-step synthesis, the diisocyanate, polyol, and a chain extender are mixed together and allowed to react. In the two-step route, an oligomer or prepolymer synthesized from diisocyanate and polyol and the chain extender are allowed to react. The two-step route offers some advantages over the one-step route, such as a reduced polydispersity index and a higher extent of phase separation.

3.2.4
Evolution of Nonlinear Rheology during Polyurethane Polymerization

Early studies on rheological analysis of polyurethane polymerizations mostly focused on synthesis of thermosetting polyurethanes [31]. The viscosity was observed to be a nearly linear function of molecular weight below the gel point: at or beyond the gel point it approached infinity [32–34]. This behavior is reminiscent of polymerization of other thermosetting polymers (Figure 3.2). Interestingly, TPUs had also been observed to exhibit a gel point where the viscosity increased sharply at a certain conversion [35, 36], which was attributed to the formation of physically cross-linked networks. Rheological studies indicated that phase-separated domains formed in the early stages of polymerization, which increased the viscosity. For example, the analysis of the rheological properties indicated a stronger relationship between the viscosity increase and molecular weight than that anticipated from an increase in molecular weight alone (Figure 3.4) [37].

The gel point and network formation are also identified from the frequency independence of the loss angle (tan $\delta = G''/G'$) at a critical reaction time or conversion [38]. Finally, it should also be mentioned that rheological properties of TPUs

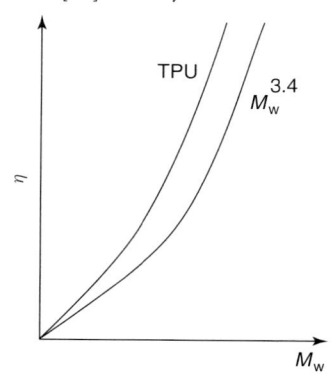

Figure 3.4 Typical dependence of viscosity (η) on molecular weight (M_w) for thermoplastic polyurethanes (TPUs). The commonly encountered relationship between η and M_w ($\eta \sim M_w^{3.4}$) is also shown, along with a sketch of the typical η versus M_w relationship observed during TPU polymerization. The stronger dependence on M_w in the case of TPU has been attributed to phase-separated domains that act as physical cross-links and in turn reduce the mobility of chains and augment the viscosity.

usually exhibit a strong dependence on thermal history [39]. Hence caution should be exercised and all experimental parameters, such as the temperature and the duration in preparation of specimens for rheological test, should be kept constant.

3.2.5
Basic Reactions and Phase Separation Kinetics in Synthesis of Polyurethanes and Their Relationship to the Evolution of Nonlinear Rheology

The kinetics of reactions between the monomers is a dominant parameter that impacts the rheological properties. For example, faster reactions lead to faster gelation and creation of a continuous polymer network. Another dominant parameter is the extent of phase separation. On one hand, it retards the rate of reactions and, on the other, promotes the formation of the polymer network and hence leads to an increase of viscosity. Accordingly, one may anticipate an intricate interplay between the reaction kinetics, phase separation, and network formation.

The condensation reactions between diisocyanates and polyols in bulk usually obey second-order kinetics [40–49]. However, the kinetic constants of polyurethane formation are strongly dependent on the chemical structure and the molecular weight of the reactants. Jung [50] performed a series of titration and Fourier-transform infrared (FT-IR) spectroscopic experiments to determine the kinetics of polyurethane polymerization reactions. A number of common diisocyanates, such as methylene diisocyanate (MDI), toluene diisocyanate (TDI), and H_{12}MDI (commonly known as *aliphatic MDI* or *diisocyanate–dicyclohexylmethane*), and various polyols, such as polypropylene glycols (PPGs) and polyester glycols of different molecular weights, were considered as the monomers. First, the effect of the chemical structure of the diisocyanate on reaction kinetics was determined, and the values of kinetic rate constants (k) of a series of condensation reactions between different diisocyanates and various polyols at different temperatures were compared. Note, in this case, that the reaction rate constants of both isocyanate groups were assumed to be the same. It was observed that the values of k of reactions of MDI and TDI with different polyols and at different temperatures were comparable with each other; however, that of H_{12}MDI was about an order of magnitude lower (Table 3.1). In addition, it was observed that the values of k were reduced in the case of lower molecular weight polyols. In this case, phase separation occurred rapidly, which retarded the progress of isocyanate–hydroxyl reactions by entrapping unreacted isocyanate groups and thereby reducing the rate of conversion (Figure 3.5).

3.3
Evolution of Nonlinear Rheology and Network Formation during Thermoplastic Polyurethane Polymerization: Effects of Mixer Design, Mixing Protocol, Catalyst Concentration, and Timescales

Several parameters influence the polymerization process in a reactor. Of these, the kinetic constant is a function of temperature. The type of mixing and the mixing

Table 3.1 Kinetic rate constants of several uncatalyzed diisocyanate–poly(ethylene glycol) reactions at several temperatures. The initial molar ratio of diisocyanate to polyol is 2 : 1.

Type of diisocyanate	$[NCO]_0$ (mol/L)	MW of PEG	$T(°C)$	k (L/mol·min)	E_a (kJ/mol)	A (L/mol·min)
MDI	3.17	1000	60	0.056	97.2	1.05×10^{14}
			70	0.186		
			80	0.408		
	1.90	2000	60	0.038	90.5	5.47×10^{12}
			70	0.079		
			80	0.243		
	1.29	3200	60	0.035	86.8	1.40×10^{12}
			70	0.082		
			80	0.207		
TDI	3.55	1000	60	0.040	112.5	1.75×10^{16}
			70	0.127		
			80	0.400		
	2.05	2000	60	0.024	100.3	1.18×10^{14}
			70	0.054		
			80	0.188		
	1.35	3200	60	0.023	91.6	5.31×10^{12}
			70	0.061		
			80	0.150		
H_{12}MDI	3.02	1000	60	0.0026	60.2	7.23×10^{6}
			70	0.0051		
			80	0.0089		
	1.84	2000	60	0.0054	52.2	8.56×10^{5}
			70	0.0101		
			80	0.0157		
	1.26	3200	60	0.0096	53.3	2.19×10^{6}
			70	0.0161		
			80	0.0286		

The abbreviations and symbols are as follows: MDI: methylene diisocyanate, TDI: toluene diisocyanate, H_{12}MDI: aliphatic MDI, $[NCO]_0$: initial isocyanate concentration, PEG: poly(ethylene glycol), MW: molecular weight, T: temperature, k: kinetic rate constant, E: The Arrhenius activation energy, A: pre-exponential factor. It is assumed that the reactions follow an Arrhenius relationship of $k = A \exp(-E_a/RT)$.

protocol as well as the shear rate in the mixer are among the crucial engineering parameters to affect polymerization. In the following section, we first present an analysis of the various parameters, such as the reaction kinetics and phase separation, encountered in various polyurethane systems. Next, we present the effects of engineering parameters, such as the mixing protocol and mixer design, on the evolution of nonlinear rheology of polyurethanes.

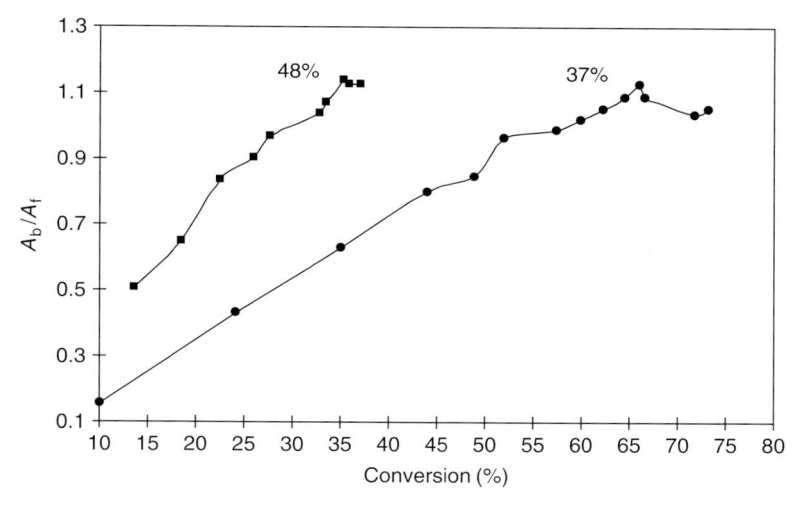

Figure 3.5 Variation of A_b/A_f ratio for C=O stretching absorption bands in FT-IR with conversion during polyurethane formation in a one-step route. MDI/polypropylene glycol of molecular weight of 1025 (PPG-1025)/butanediol (BD) A_b stands for the area under the hydrogen-bonded carbonyl peak, whereas A_f corresponds to that under the non-hydrogen bonded (free) peak. The percentages indicate hard segment content as the weight percent of the reaction mixture.

3.3.1
Effects of Mixing

3.3.1.1 Mechanism of Mixing

Reactants are mechanically mixed to reduce the nonuniformities or gradients in physical quantities, such as concentration, temperature, or size and shape, of the dispersed components. Mixing can be analyzed under two broad categories – macro- and micromixing. Macromixing reduces large-scale inhomogeneities in a material, such as the size of dispersed phase, whereas micromixing operates at the molecular scale and increases the configurational entropy of the system by promoting random mixing of the reactants. Macromixing can be achieved by both turbulent and laminar motions. In high-viscosity liquids, such as polymer melts and polymerizing media, laminar motion is predominant, with the Reynolds number varying between 0.1 and on the order of 100. Micromixing by molecular diffusion is a significantly slow process and is effective only at later stages of laminar mixing, when the diffusion paths are reduced as a result of thinning of the reactant striations.

In the absence of significant micromixing, either due to slow diffusion or due to thick striations separating the reactants, chemical reactions occur primarily at the interfaces between the adjacent striations of the reactants. In view of this, two straightforward functions of mixing can be realized in the form of continuous removal of the reaction products from the interfaces and supply of fresh reactants to the reaction sites. The former is needed to prevent shielding of the reacting species at the interfaces.

Let us now turn our attention to a brief overview of the mixing processes in liquids. The focus is to identify the relationships between flow, kinematics, chemical reactions, and nonlinear rheology.

3.3.1.2 Laminar Mixing under Shear and Extensional Flow with Constant Shear and Elongation Rates

Laminar mixing under constant shear flow is commonly observed during the flow of high-viscosity fluids in extruders [51]. If one considers a mixture in which the dispersed phase is present as droplets in a continuous matrix, the dispersed phase undergoes deformation and alignment along the flow direction under the action of laminar shear flow. In this case, the interfacial area in the system increases linearly with time. On the other hand, the interfacial area changes exponentially with time in elongational flow. However, elongational flow is rarely encountered in common processing equipment, although some extensional flow components have been considered in the design of certain extruder screws and mixers [52].

3.3.1.3 Dispersive and Distributive Mixing

In dispersive mixing, the dispersed phase undergoes size reduction and forms multiple smaller domains as a result of stresses acting at the interfaces [53, 54]. In this case, the viscous forces generated in the continuous phase overcome the restoring forces due to interfacial tension [55, 56]. Hence, the extent of deformation is determined by the ratio of the viscous and interfacial stresses, the relative magnitude of which can be expressed as a dimensionless number, the so-called capillary number [57, 58].

In distributive mixing, the flow field produces deformation in the dispersed phase, without inducing breakup. As a consequence, a lamellar morphology of alternating striations of dispersed phase and the matrix is obtained, which can be characterized by the striation thickness distribution. Note that the striation thickness (δ) is a function of the initial length scale (l_d), viscosity of the dispersed phase (η_d), total shear strain (γ) (product of shear rate and time), volume fraction of the dispersed phase (Φ_d), and viscosity of the matrix (η_m) [59] (Eq. (3.1)):

$$\delta = \left(\frac{l_d}{\gamma \Phi_d}\right)\left(\frac{\eta_d}{\eta_m}\right) \tag{3.1}$$

The extent of distributive mixing can be further enhanced by subjecting the fluid elements to frequent splitting, reorientation, and high levels of shear and elongational strains, and by facilitating uniform deformation history throughout the mixture [60]. These are achieved much more easily in the cases involving chaotic mixing.

3.3.1.4 Chaotic Mixing

The efficiency of mixing is significantly higher in chaotic flows due to the complex trajectories of fluid elements brought out by simple time-periodic or space-periodic velocity fields. The time series of fluid element locations undergoing chaotic advection is stochastic, and the Fourier spectra reveal multiple frequencies [61, 62]. The fluid–fluid interfaces periodically stretch and fold in chaotic flows. The periodic stretching and folding create new interfaces, which are essential for the rapid progress of chemical reactions [62].

A chaotic flow produces either transverse homoclinic or transverse heteroclinic intersections, and/or is able to stretch and fold material in such a way that it produces what is called a *horseshoe map*, and/or has positive Liapunov exponents. These definitions are not equivalent to each other, and their interrelations have been discussed by Doherty and Ottino [63]. The time-periodic perturbation of homoclinic and heteroclinic orbits can create chaotic flows. In bounded fluid flows, which are encountered in mixing tanks, the homoclinic and heteroclinic orbits are separate streamlines in an unperturbed system. These streamlines prevent fluid flux from one region of the domain to the other, thereby severely limiting mixing. These separate streamlines generate stable and unstable manifolds upon perturbation, which in turn dictate the mass and energy transports in the system [64–66].

Chaotic flows can also be identified by the presence of horseshoe maps (Figure 3.6). The formation of horseshoe maps is facilitated by consecutive folding of fluid elements. The horseshoe maps and their ramification on the interfacial area between the fluids are created by the low-order hyperbolic periodic points [66–70]. The consecutive folding of the interfaces also creates self-similar local microstructures, which are retained during the progress of mixing, and their length scales gradually become smaller. Note that this is contrary to turbulent mixing, which promotes randomization of local microstructures.

Chaotic flow can also be distinguished by the positive value of the Liapunov exponent (λ), which indicates that the nearby fluid element trajectories diverge exponentially with time because of stretching and folding. Note that λ is close to zero in nonchaotic laminar flows; in this case, the neighboring fluid element trajectories stay close to each other.

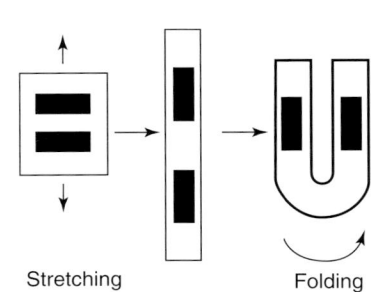

Figure 3.6 Schematic representation of a typical horseshoe map.

Chaotic mixing was produced in various mixers with different geometries, whereby the length and the area of the fluid elements increased exponentially with time as opposed to linear growth under steady shear flow conditions. Eccentric cylinder systems [70, 71], rectangular cavity [72], two rotating rods in tanks [73], stirred tanks [74], static mixers [75–77], and chaotic single-screw extruders [78–81] are among the examples of mixers with chaotic flows. The effectiveness of chaotic mixers in processing of polymer blends and polymer nanocomposites has also been demonstrated [81–89].

3.3.1.5 Effect of Mixing on Systems Undergoing Chemical Reactions

As presented in Section 3.3.1.3, macromixing plays a dominant role in creating fresh interfaces in both miscible and immiscible systems [90, 91]. It also reduces the striation thicknesses and hence the diffusion length scales, which can significantly accelerate the rate of chemical reactions [91]. The improved extents of mixing due to chaotic flows, which can be facilitated, for example, by back-and-forth application of shear [92, 93], were shown to be more efficient in augmenting the conversion than a one-way shear in an esterification reaction with immiscible reactants composed of maleic anhydride-grafted hydrogenated styrene–butadiene–styrene block copolymer and polycaprolactone diol. Similar observations were reported in the reactive blending of hydroxyl-functionalized PS with isocyanate-functionalized poly(methyl methacrylate). Feng and Hu [94] later found dramatic increases in the rates of reaction between –NCO and alcoholic –OH groups when the model reactants formed an immiscible system. These authors also observed that a miscible, homogeneous model system containing the same reacting groups produced a much lower rate of reaction. The rates of conversion of small molecular reactants have also been observed to increase under chaotic flows [95–104].

Two-component, catalyzed, interfacial polymerization of polyurethane has been previously modeled by means of a set of partial differential equations for the diffusion–reaction process, with concentration-dependent diffusivities [105–108]. Their main assumptions were that reactions were facilitated only toward the end of the mixing process, such as at the end of the reaction injection-molding process, wherein molecular diffusion was significant. It was also assumed that the diisocyanate and diols formed lamellar mixing structures, whereby alternative striations of diisocyanates and diols were present.

3.3.2
Analysis of Timescale of Mixing and Chemical Reactions during TPU Polymerization

Significant variations in material parameters with time and temperature make modeling of simultaneous chemical reaction, mixing, and molecular diffusion during polymerization unwieldy. The rheological properties, for example viscosity, change with conversion because of an increase in molecular weight, and also because of the exothermic heat of reaction. The change in rheological properties in turn leads to variations in the hydrodynamic conditions and stresses experienced by the fluid particles. The change in rheological properties further influences

3.3 Effects of Mixer Design, Mixing Protocol, Catalyst Concentration, and Timescales

macromixing and the molecular diffusion in a nonlinear manner. A difficulty specific to polyurethane polymerization should be mentioned here. Polyurethanes undergo phase separation during polymerization, which can exert significant impact on reaction, diffusion, and mixing conditions. The extent of phase separation at a given time is usually determined using FT-IR spectroscopy or light scattering methods [109, 110]. However, phase separation during polymerization is a dynamic process [111] and it is significantly difficult, if not impossible, to exactly identify the effect of phase separation on the local dynamics of polymerization. In view of these difficulties associated with a full-scale, reliable numerical simulation, a less detailed method based on timescales analysis can be used to gain some insight into the relative effects of simultaneous processes, such as reaction, mixing, and molecular diffusion, on polymerization.

The timescale of chemical reaction (t_{rxn}) between –NCO and –OH groups can be defined as follows:

$$t_{rxn} \sim \frac{1}{k_{cat}[NCO]_0 r} \tag{3.2}$$

where $[NCO]_0$ is the initial concentration of the isocyanate group coming from the prepolymer, $r \equiv [NCO]_0/[OH]_0$ is the stoichiometric ratio, with $[OH]_0$ being the initial concentration of –OH groups coming from the chain extender. The rate of urethane-forming reactions is considered to be second-order [42], as discussed previously:

$$-\frac{d[NCO]}{dt} = k[Cat]^a[NCO][OH] \tag{3.3}$$

In Eq. (3.3), [OH] and [NCO] are, respectively, the molar concentrations of the –OH and –NCO functional groups and [Cat] is the molar concentration of the catalyst. The apparent rate constant k_{cat} can be expressed as $k_{cat} \equiv k[Cat]^a$, where a is a constant and k is the reaction rate constant with Arrhenius-type dependence on temperature. Reflecting on the usual immiscibility of the prepolymer and the chain extender [50], the rate of polymerization should depend strongly on the degree of mixing of the reactants, especially if molecular diffusion is slow. Figure 3.7 presents a typical scenario involving three competing processes, such as molecular diffusion, reaction, and mixing by stretching. The chain extender phase, initially present as droplets, is stretched by the application of chaotic flow into elongated filaments. Note that, as a consequence, the concentration gradient inside the filament increases, which induces a higher diffusive flux of chain extender molecules to the interface (along the x_2 direction in Figure 3.7). In addition, fresh molecules of the chain extender are exposed to the interface via increased intermaterial area. Therefore, it is anticipated that the timescales of stretching and molecular diffusion both determine the concentration of the chain extender molecules at the interface.

Expressions for timescales associated with stretching and molecular diffusion for nonreacting systems are available in the literature [112]. A fluid element undergoes shear in two-dimensional chaotic flow with the principal stretching direction x_1 and compression direction x_2. Molecular diffusion becomes important after a time

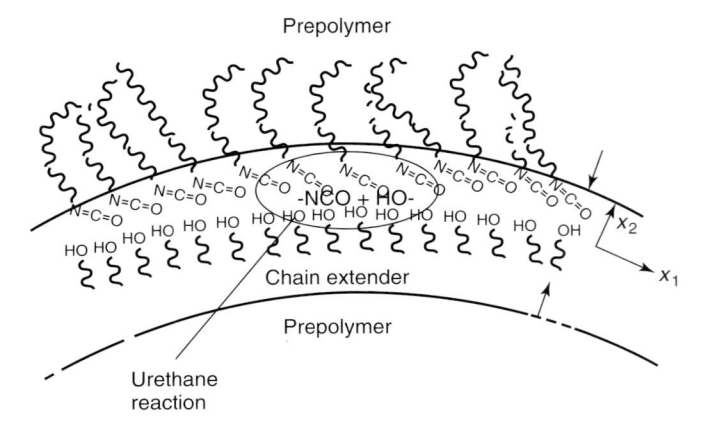

Figure 3.7 Stretching of a typical droplet of chain extender during mixing and the structure of the interface. Chemical reactions between the prepolymer and the chain extender occur at the interface. Local direction of stretching (x_1) and the direction of concentration variation across the thickness of the droplet (x_2) are shown.

when the concentration gradient remains constant over time. If the concentration of species A (e.g., −OH groups of the chain extender) is C_A and the concentration gradient of species A is $G_i \equiv \partial C_A/\partial x_i, i = 1, 2$, the mass balance of species A in the fluid element can be recast in the following form:

$$\frac{dG_i}{dt} = -G_j \frac{\partial v_j}{\partial x_i} + D_{AB} \Delta G_i \tag{3.4}$$

where v_j is the jth component of the velocity vector **v**, ΔG_i is the Laplacian of G_i, and D_{AB} is molecular diffusion coefficient. A reaction term similar to Eq. (3.3) can be incorporated in Eq. (3.4) to obtain a form suitable for the present work:

$$\frac{\partial}{\partial t}\left(\frac{\partial C_A}{\partial x_2}\right) + \frac{\partial}{\partial x_2}\left(v_2 \frac{\partial C_A}{\partial x_2}\right)$$
$$= D_{AB} \frac{\partial}{\partial x_2}\left(\frac{\partial^2 C_A}{\partial x_2^2}\right) - k_{cat} C_B \left(\frac{\partial C_A}{\partial x_2}\right) - k_{cat} C_A \left(\frac{\partial C_B}{\partial x_2}\right) \tag{3.5}$$

where C_A and C_B are, respectively, the concentrations of reactants A (e.g., −OH groups of the chain extender) and B (e.g., −NCO groups of prepolymer) at time t and D_{AB} is the molecular diffusion coefficient. It is assumed in Eq. (3.5) that the change in the concentration of A in the fluid element occurs only in the x_2 direction, that is, thickness direction (see also Figure 3.7). For initial concentrations of A and B, C_{A0} and C_{B0}, respectively, and $C_{A0} = rC_{B0}$, where r is the stoichiometric ratio, Eq. (3.5) can be rewritten as

$$\frac{d}{dt}\left(\frac{\partial C_A}{\partial x_2}\right) = D_{AB}\frac{\partial}{\partial x_2}\left(\frac{\partial^2 C_A}{\partial x_2^2}\right) - \frac{\partial v_2}{\partial x_2}\left(\frac{\partial C_A}{\partial x_2}\right)$$
$$- k_{cat}[C_{A0}(r-1) + 2C_A]\left(\frac{\partial C_A}{\partial x_2}\right) \quad (3.6)$$

The left-hand side of Eq. (3.6) represents the rate of increase of the concentration gradient in the fluid element (e.g., of the chain extender in Figure 3.7) as the fluid element is stretched in the chaotic mixer. The three terms on the right-hand side of Eq. (3.6) represent, respectively, the rate of decrease of concentration gradient in the stretched fluid element due to molecular diffusion, increase due to stretching, and decrease due to consumption of A in the bimolecular reaction with B. If the mixing is continued, a time is reached at which $(d/dt)(\partial C_A/\partial x_2) \sim 0$, and the cumulative changes in concentration of A in the fluid element due to molecular diffusion, chemical reaction, and stretching vanish. This time can be used to define a timescale of mixing t_{mix}, which can be interpreted as follows. The mechanical stirring of the ingredients should be continued for time t_{mix} by which time the domains containing reactant A become thin enough to trigger appreciable molecular diffusion of A and an appreciable rate of chemical reaction between A and B. At $t = t_{mix}$, Eq. (3.5) can be rewritten in terms of conversion of species A, $p_A \equiv (C_{A0} - C_A)/C_{A0}$, as

$$\frac{\partial v_2}{\partial x_2}\left(\frac{\partial p_A}{\partial x_2}\right) = D_{AB}\frac{\partial}{\partial x_2}\left(\frac{\partial^2 p_A}{\partial x_2^2}\right) - k_{cat}C_{A0}[(r-1) + 2(1-p_A)]\left(\frac{\partial p_A}{\partial x_2}\right) \quad (3.7)$$

The actual expression for t_{mix} depends on the relative values of the terms on the right-hand side of Eq. (3.6). Each term in Eq. (3.7) has the units of 1/(ms), and an appropriate scaling can be found such as

$$\frac{k_{cat}rC_{A0}L}{V} = f\left(\frac{D_{AB}}{VL}\right) \quad (3.8)$$

where two dimensionless groups $\prod_1 = k_{cat}rC_{A0}L/V$ and $\prod_2 = VL/D_{AB}$ are related by the function f, with L and V as characteristic length and velocity. To derive an appropriate expression of t_{mix}, a domain of reactant A (e.g., the chain extender) could be imagined, with an initial thickness d_0, undergoing stretching at time t:

$$d = d_0 e^{-\lambda t}, \lambda > 0 \quad (3.9)$$

In Eq. (3.9), λ is the Liapunov exponent, which can be computed from the knowledge of mixing conditions and the mixer geometry. The timescale of mixing t_{mix} for this case is given as [113]:

$$t_{mix} \sim \frac{1}{2\lambda}\ln\left[\frac{L^2/D_{AB}}{\lambda(\frac{1}{k_{cat}rC_{A0}})}\right] \quad (3.10)$$

where the length scale L is the same as the initial fluid element thickness d_0 or the initial diameter of the chain extender droplets.

3.3.3
Simultaneous Effects of Mixing, Chemical Reaction, and Molecular Diffusion on the Evolution of Nonlinear Rheological Properties

A reliable identification of the relationships between the effects of mixing conditions and the absolute values of rheological properties is difficult, if not impossible. The actual mixing conditions and geometry cannot be replicated exactly in a standard rheometer and the actual stress and strains are highly complex for accurate experimental determination due to the complex geometries of the mixers. However, the change in torque (Γ) values recorded during polymerization and mixing can be used as an approximation for the change of the viscosity of the system. The time at which maximum Γ was reached (t_{max}) is a reflection of the rate of polymerization – smaller values of t_{max} indicate higher rates of polymerization. Note that t_{max} also represents the reaction time needed to form a polymer network. The values of Γ can be correlated to the viscosity of the system by the following relationships [114]:

$$\Gamma = C\eta_0 N^n \tag{3.11}$$
$$\eta_0 = B(T) M_w^m \tag{3.12}$$

In Eqs (3.11) and (3.12), M_w is the weight-averaged molecular weight; η_0 is zero-shear viscosity, which is a function of temperature (T) and molecular weight (M_w); C is a characteristic constant that can be experimentally determined for a given mixer geometry; N is the rotor speed; n is the power law index in the equation between shear stress (τ) and rate of strain ($\dot{\gamma}$), $\tau = K(\dot{\gamma})^n$, with K as the consistency index; and $B(T)$ stands for the temperature dependency. The parameter m stands for the relationship between the molecular weight and viscosity. It has been known that its value is about unity for low molecular weight species, but increases to about 3.4 for high molecular weight, entangled chains. The value of m has been shown to be higher than 3.4, even right after the start of polymerization [37]. This high parametric dependence between the molecular weight and viscosity has been attributed to the formation of phase-separated domains, which act as physical cross-links and reduce the chain mobility. The values of t_{max}, that is, the time to reach a maximum of torque, can be related to the extent of conversion of functional groups and, therefore, to the effectiveness of mixing. Note that t_{max} is function of both M_w and the extent of phase separation, and the extent of phase separation is a function of M_w. Hence, it is possible to relate the change in values of torque to gradual evolution of M_w and conversion. The values of t_{max} are presented in Table 3.2 for polyurethane polymerizations in a chaotic mixer (mixer 1) (Figure 3.8) and in a conventional internal mixer (mixer 2) (Figure 3.9). The representative torque–time plots of both mixers are presented in Figure 3.10.

The data presented in Table 3.2 enables us to make several observations and comments. First, mixing appeared to become more effective at higher catalyst concentrations as a result of shorter timescales of chemical reactions. Note that the value of t_{max} was the same (11 min) for mixers 1 and 2 for a catalyst concentration

Table 3.2 Values of t_{rxn} and t_{mix} computed at 80 °C and values of t_{max} obtained from torque versus time curves presented in Figure 3.10. The initial concentration of isocyanate group (C_{A0}) was 0.75 mol l^{-1} and the ratio of the initial concentrations of isocyanate and hydroxyl groups (r) was 1.05. The timescale of mixing (t_{mix}) was obtained using a typical droplet diameter of 40 μm as a representative length scale L in Eq. (3.11). The value of molecular diffusivity D_{AB} of prepolymer based on PPG-2000 was taken to be 3.3×10^{-8} cm^2 s^{-1}. The mean values of the product of the Liapunov exponent and time period (λT) were found to be 1.3, whereby T was 6 s. The value of t_{mix} was not computed for mixer 2 and is specified with an entry N/A.

Polyol	Catalyst concentration ($\times 10^{-4}$ mol l^{-1})	Mixer	t_{rxn} (min)	t_{mix} (min)	t_{max} (min)
PPG-2000	0.56	1	0.51	0.03	11.0
	0.56	2	0.51	N/A	11.0
	1.13	1	0.28	0.08	6.3
	1.13	2	0.28	N/A	7.0
	2.26	1	0.13	0.14	3.5
	2.26	2	0.13	N/A	5.0
PPG-1025	1.13	1	0.55	0.03	1.7
	1.13	2	0.55	N/A	2.3

(a)

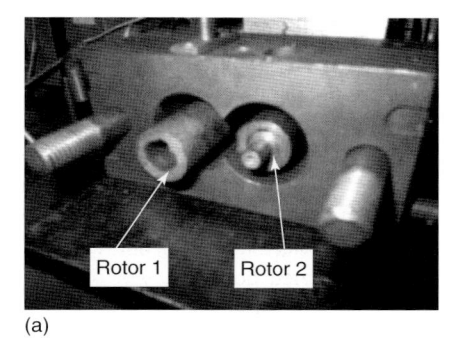

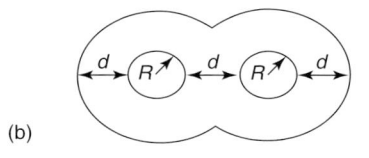

(b)

Figure 3.8 Mixing chamber and rotors of chaotic mixer. (a) Note that one of the rotors is exposed to the camera in order to ensure easier recognition of cylindrical rotor geometry. (b) The schematic of chaotic mixing chamber. R (the rotor radius) is 12.5 mm and d (the mixing gap) is 12.7 mm. The total capacity of the mixing chamber is 70 cm^3. The maximum shear rate at 65 rpm is 4.8 s^{-1}, the minimum shear rate is 2.7 s^{-1}, and the mean shear rate is 3.8 s^{-1}. The overall dimensions of the mixer are 15.0 cm × 9.0 cm × 7.5 cm.

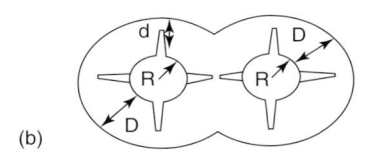

Figure 3.9 Mixing chamber and rotors of Brabender internal mixer. (a) One of the rotors exposed to the camera in order to ensure easier recognition of rotor geometry. (b) The schematic of mixing chamber. R (the rotor radius) is 10.8 mm, d (the minimum mixing gap) is 2.16 mm, and D (the maximum mixing gap) is 13 mm. The total capacity of the mixing chamber is 80 cm^3. The maximum shear rate at 60 rpm is 97 s^{-1}, and the minimum shear rate is 4.5 s^{-1}, and the mean shear rate is 4.6 s^{-1}. The overall dimensions of the mixer are 18.0 cm × 10.0 cm × 9.5 cm.

of 0.56×10^{-4} mol l^{-1}, indicating that chain extension reactions were kinetically controlled at this catalyst concentration and that mixing had almost no effect. Second, it is apparent that the chaotic mixer was significantly more effective for chain extension reactions than the conventional mixer used, especially at higher catalyst concentrations. For example, the time to reach maximum torque (t_{max}) in the chaotic mixer was 30% shorter than in the conventional mixer at a catalyst concentration of 2.26×10^{-4} mol l^{-1}, which corresponded to $t_{rxn} \sim 0.13$ min for chain extension of prepolymer based on PPG-2000. Note also that, in the light of the competing processes depicted in Figure 3.7, one may correctly identify that the expression of t_{mix} in Eq. (3.10) also includes t_{rxn} as defined in Eq. (3.1) and the diffusion timescale t_{diff} is on the order of L^2/D_{AB}. However, such dependence is logarithmic and, therefore, weak; the values of t_{mix} are primarily dictated by the value of the Liapunov exponent [112] and hence by the degree of chaotic mixing.

3.4
Conclusions

The rate and the extent of conversion in polymerization reactions, the evolution of nonlinear rheological properties, and the formation of polymer networks are closely related to the effectiveness and the type of mixing (chaotic vs nonchaotic)

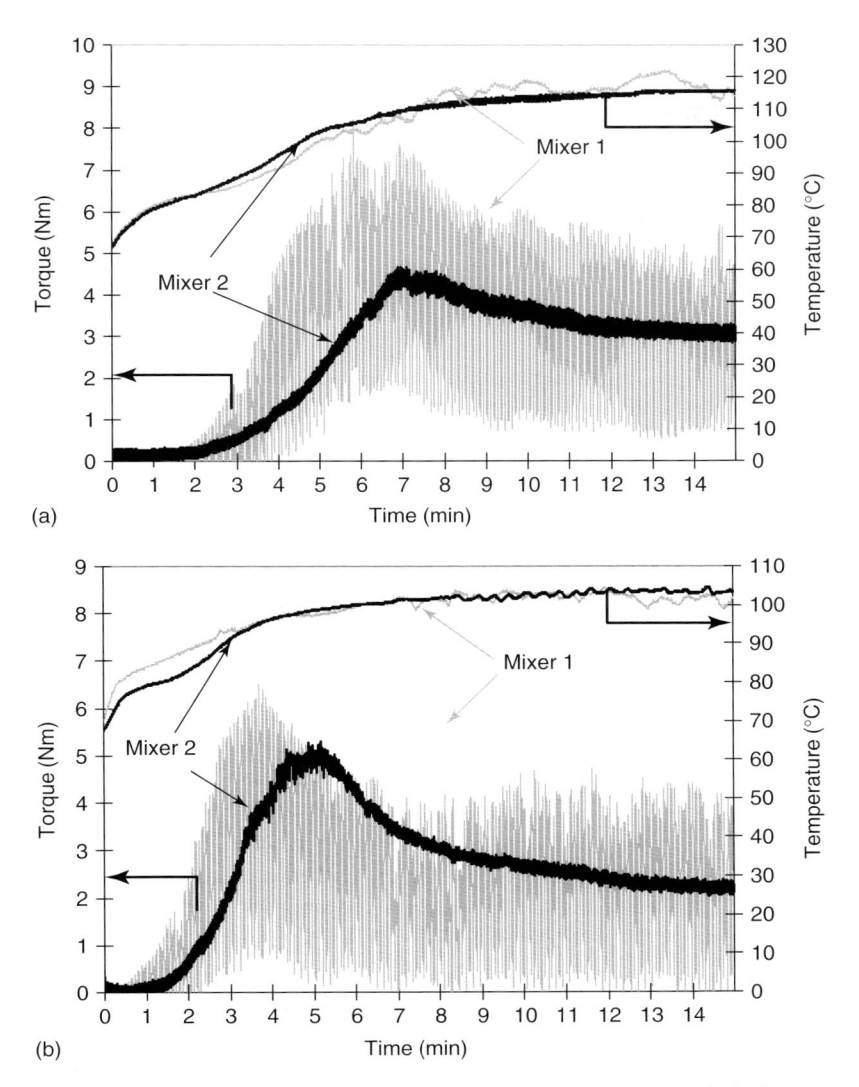

Figure 3.10 Representative plots of torque and temperature during chain extension reactions in mixers 1 and 2. The data for mixer 1 are designated by thin gray lines, whereas those for mixer 2 are presented in bold black lines. The upper curves in both images indicate the evolution of temperature in the mixer. Prepolymer was based on polypropylene glycol polyol of 2000 molecular weight. (a) Catalyst concentration 1.13×10^{-4} mol l^{-1} and (b) catalyst concentration 2.26×10^{-4} mol l^{-1}. The time that corresponds to the peak torque value is designated as t_{max}. Note the significantly longer t_{max} in mixer 2.

encountered. The mixing protocol and mixer design should be selected to obtain proper mixing of reactants. Other parameters, such as reaction temperature, chemical structure, and concentration of catalyst, should also be chosen in accordance with the mixing parameters. Polymerization reactions appear to be more easily and efficiently accomplished by inducing a globally chaotic mixing flow. The analysis presented in this chapter indicates that the timescale estimates present a relatively easy set of tools for the study and design of polymerization reactors. The data from laboratory experiments showed that the rate of polymerization reactions can be expedited when the magnitudes of the timescales of mixing and chemical reactions are made comparable to each other.

References

1. Denn, M.M. (2004) Anniversary article: fifty years of non-newtonian fluid dynamics. *AIChE J.*, **50**, 2335–2345.
2. Henderson, J.N. and Bouton, T.C. (eds) (1979) *Polymerization Reactors and Processes*, ACS Symposium Series, No. 104, American Chemical Society, Washington, DC.
3. Bouilloux, A., Macoscko, C.W., and Kotnour, T. (1991) Urethane polymerization in a counter-rotating twin-screw extruder. *Ind. Eng. Chem. Res.*, **30**, 2431–2436.
4. Morton, M. (ed.) (1987) *Rubber Technology*, Van Nostrand Reinhold, New York.
5. Cassagnau, P., Nietsch, T., and Michel, A. (1999) Bulk and dispersed phase polymerization of urethane in twin screw extruders. *Int. Polym. Proc.*, **14**, 144–151.
6. Morrison, F.A. (2001) *Understanding Rheology*, Oxford University Press, New York.
7. Middleman, S. (1968) *The Flow of High Polymers: Continuum and Molecular Rheology*, Interscience, New York.
8. Berry, G.C. and Fox, T.G. (1968) The viscosity of polymers and their concentrated solutions. *Adv. Polym. Sci.*, **5**, 262–357.
9. Trommsdorff, E., Kohle, H., and Lagally, P. (1948) Polymerization of methyl methacrylates. *Makromol. Chem.*, **1**, 169–198.
10. Norrish, R.G.W. and Smith, R.R. (1942) Catalyzed polymerization of methyl methacrylate in the liquid phase. *Nature*, **150**, 336–337.
11. Bamford, C.H., Barb, W.G., Jenkins, A.D., and Onyon, P.F. (1953) *The Kinetics of Vinyl Polymerizations by Radical Mechanisms*, Butterworth, New York.
12. O'Neil, G.A. and Torkelson, J.M. (1997) Recent advances in the understanding of the gel effect in free-radical polymerization. *Trends Polym. Sci.*, **5**, 349–355.
13. O'Neil, G.A., Wisnudel, M.B., and Torkelson, J.M. (1996) A critical experimental examination of the gel effect in free radical polymerization: do entanglements cause autoacceleration? *Macromolecules*, **29**, 7477–7490.
14. O'Neil, G.A., Wisnudel, M.B., and Torkelson, J.M. (1998) An evaluation of free volume approaches to describe the gel effect in free radical polymerization. *Macromolecules*, **31**, 4537–4545.
15. O'Neil, G.A. and Torkelson, J.M. (1999) Modeling insight into the diffusion-limited cause of the gel effect in free radical polymerization. *Macromolecules*, **32**, 411–422.
16. Malkin, A.Ya. (1980) Rheology in polymerization processes. *Polym. Eng. Sci.*, **20**, 1035–1044.
17. Kamal, M.R. (1974) Thermoset characterization for moldability analysis. *Polym. Eng. Sci.*, **14**, 231–239.
18. Bidstrup, S.A. and Macosko, C.W. (1990) Chemorheology relations for epoxy-amine crosslinking. *J. Polym. Sci. Part B: Polym. Phys.*, **28**, 691–709.

19. Tung, C.Y.M. and Dynes, P.J. (1982) Relationship between viscoelastic properties and gelation in thermosetting systems. *J. Appl. Polym. Sci.*, **27**, 569–574.
20. Chambon, F. and Winter, H.H. (1985) Stopping of crosslinking reaction in a PDMS polymer at the gel point. *Polym. Bull.*, **13**, 499–503.
21. Ng, H. and Manas-Zlocower, I. (1993) Chemorheology of unfilled and filled epoxy resins. *Polym. Eng. Sci.*, **33**, 211–216.
22. Winter, H.H. (1987) Can the gel point of a cross-linking polymer be detected by the $G' - G''$ Crossover? *Polym. Eng. Sci.*, **27**, 1698–1702.
23. Holy, E.E., Venktaraman, S.K., Chambon, F., and Winter, H.H. (1988) Fourier transform mechanical spectroscopy of viscoelastic materials with transient structure. *J. Non-Newton. Fluid Mech.*, **27**, 17–26.
24. Winter, H.H. (1987) Evolution of rheology during chemical gelation. *Progr. Colloid Polym. Sci.*, **75**, 104–110.
25. Ding, R. and Leonov, A.I. (1999) An approach to chemorheology of a filled SBR compound. *Rubber Chem. Technol.*, **72**, 361–383.
26. Joshi, P.G. and Leonov, A.I. (2001) Modeling of steady and time-dependent responses in filled, uncured, and crosslinked rubbers. *Rheol. Acta*, **40**, 350–365.
27. Odian, G. (2004) *Principles of Polymerization*, 4th edn, Interscience, New York.
28. Erman, B. and Mark, J.E. (1997) *Structures and Properties of Rubberlike Networks*, Oxford University Press, New York.
29. Woods, G. (1987) *The ICI Polyurethanes Book*, ICI Polyurethanes and John Wiley & Sons, Inc., The Netherlands.
30. Morbitzer, L. and Hespe, H. (1972) Correlations between chemical structure, stress-induced crystallization, and deformation behavior of polyurethane elastomers. *J. Appl. Polym. Sci.*, **16**, 2697–2708.
31. Lipshitz, S.D. and Macosko, C.W. (1976) Rheological changes during a urethane network polymerization. *Polym. Eng. Sci.*, **16**, 803–810.
32. Richter, E.B. and Macosko, C.W. (1980) Viscosity changes during isothermal and adiabatic urethane network polymerization. *Polym. Eng. Sci.*, **20**, 921–924.
33. Reboredo, M.M., Rojas, A.J., and Williams, R.J.J. (1983) Kinetic and viscosity relations for thermosetting polyurethanes. *Polym. J.*, **15**, 9–14.
34. John, R., Thachil, E.T., Neelkantan, N.R., and Subramanian, N. (1991) A viscometric approach to the study of the kinetics of polyurethane reactions. *Polym. Plast. Technol. Eng.*, **30**, 545–557.
35. Castro, J.M., Macosko, C.W., and Perry, S.J. (1984) Viscosity changes during urethane polymerization with phase separation. *Polym. Commun.*, **25**, 82–87.
36. Lee, Y.M. and Lee, L.J. (1987) Rheological changes for urethane polymerizations in bulk and in solution. *Polymer*, **28**, 2304–2309.
37. Castro, J.M., Lopez-Serrano, F., Camargo, R.E., Macosko, C.W., and Tirrell, M. (1981) Onset of phase separation in segmented urethane polymerization. *J. Appl. Polym. Sci.*, **26**, 2067–2076.
38. Prochazka, F., Nicolai, T., and Durand, D. (1996) Dynamic viscoelastic characterization of a polyurethane network formation. *Macromolecules*, **29**, 2260–2264.
39. Yoon, P.J. and Han, C.D. (2000) Effect of thermal history on the rheological behavior of thermoplastic polyurethanes. *Macromolecules*, **33**, 2171–2183.
40. Saunders, J.H. and Frisch, K.C. (1962) *Polyurethanes: Chemistry and Technology*, Interscience, New York.
41. Cummings, A.R.C. and Wright, P. (1976) *Solid Polyurethane Elastomers*, Maclaen, London.
42. Camargo, R.E., Macosko, C.W., Tirrell, M.V., and Wellinghoff, S.T. (1982) Experimental studies of phase separation in Reaction Injection-Molded (RIM) polyurethanes. *Polym. Eng. Sci.*, **22**, 719–728.

43. Hager, S.L., MacRury, T.B., Gerkin, R.M., and Critchfield, F.E. (1980) Urethane block polymers: kinetics of formation and phase development. *ACS Polym. Prep.*, **21**, 298–300.
44. Reegen, S.L. and Frisch, K.C. (1970) Isocyanate-catalyst and Hydroxyl-catalyst complex formation. *J. Polym. Sci. Part A: Polym. Chem.*, **8**, 2883–2891.
45. Lipshitz, S.D. and Macosko, C.W. (1977) Kinetics and energetics of a fast polyurethane cure. *J. Appl. Polym. Sci.*, **21**, 2029–2039.
46. Richter, E.B. and Macosko, C.W. (1978) Kinetics of fast (RIM) urethane polymerization. *Polym. Eng. Sci.*, **18**, 1012–1018.
47. Lee, Y.M., Kim, B.K., and Shin, Y.J. (1991) Kinetics of reactions between a Cycloaliphatic Diisocyanate (H_{12}MDI) and polyols. *Polymer (Korea)*, **15**, 447–452.
48. Camargo, R.E., Macosko, C.W., Tirrell, M., and Wellinghoff, S.T. (1985) Phase separation studies in RIM polyurethanes catalyst and hard segment crystallinity effects. *Polymer*, **26**, 1145–1154.
49. Broyer, E., Macosko, C.W., Critchfield, F.E., and Lawler, L.F. (1978) Curing and heat transfer in polyurethane reaction molding. *Polym. Eng. Sci.*, **18**, 382–387.
50. Jung, C. (2004) Synthesis of thermoplastic polyurethanes and polyurethane nanocomposites under chaotic mixing conditions. PhD Dissertation, The University of Akron, Akron, OH.
51. Brydson, J.A. (1970) *Flow Properties of Polymer Melts*, Iliffe Books, London.
52. Jones, J.C.R. (1994) Extensional flow mixer. US Patent 5,451,106.
53. Zumbrunnen, D.A. and Chhibber, C. (2002) Morphology development in polymer blends produced by chaotic mixing at various compositions. *Polymer*, **43**, 3267–3277.
54. Sau, M. and Jana, S.C. (2004) A study on the effects of chaotic mixer design and operating conditions on morphology development in immiscible polymer systems. *Polym. Eng. Sci.*, **44**, 407–422. Errata: (2004) *Polym. Eng. Sci.*, **44**, 1403.
55. Han, C.D. (1981) *Multiphase Flow in Polymer Processing*, Academic Press, New York.
56. Jana, S.C. and Sau, M. (2004) Effects of viscosity ratio and composition on development of morphology in chaotic mixing of polymers. *Polymer*, **45**, 1665–1678.
57. Taylor, G.I. (1932) The viscosity of a fluid containing small drops of another fluid. *Proc. Roy. Soc. (London)*, **A138**, 41–48.
58. Taylor, G.I. (1934) The formation of emulsions in definable fields of flow. *Proc. Roy. Soc. (London)*, **A146**, 501–523.
59. Mohr, W.D., Saxton, R.L., and Jepson, C.H. (1957) Theory of mixing in the single-screw extruder. *J. Ind. Eng. Chem.*, **49**, 1857–1862.
60. Erwin, L. (1978) Theory of laminar mixing. *Polym. Eng. Sci.*, **18**, 1044–1048.
61. Aref, H. (1984) Stirring by chaotic advection. *J. Fluid Mech.*, **143**, 1–21.
62. Ottino, J.M. (1989) *The Kinematics of Mixing: Stretching, Chaos, and Transport*, Cambridge University Press, New York.
63. Doherty, M.F. and Ottino, J.M. (1988) Chaos in deterministic systems: strange attractors, turbulence, and applications in chemical engineering. *Chem. Eng. Sci.*, **43**, 139–183.
64. Rom-Kedar, V., Leonard, A., and Wiggins, S. (1990) An analytical study of transport, mixing and chaos in an unsteady vortical flow. *J. Fluid Mech.*, **214**, 347–394.
65. Wiggins, S. (1997) *Introduction to Applied Non-linear Dynamical Systems and Chaos*, Springer, New York.
66. Wiggins, S. (1992) *Chaotic Transport in Dynamic Systems*, Springer, Berlin.
67. Ottino, J.M., Muzzio, F.J., Tjahjadi, M., Franjione, J.G., Jana, S.C., and Kusch, H.A. (1992) Chaos, symmetry, and self-similarity: exploiting order and disorder in mixing processes. *Science*, **257**, 754–760.

68. Jana, S.C., Metcalfe, G., and Ottino, J.M. (1994) Experimental and computational studies of mixing in complex stokes flows: the vortex mixing flow and multicellular cavity flows. *J. Fluid Mech.*, **269**, 199–246.
69. Ottino, J.M., Leong, C.W., Rising, H., and Swanson, P.D. (1988) Morphological structures produced by mixing in chaotic flows. *Nature*, **333**, 419–425.
70. Swanson, P.D. and Ottino, J.M. (1990) A comparative computational and experimental study of chaotic mixing of viscous fluids. *J. Fluid Mech.*, **213**, 227–249.
71. Chaiken, J., Chevray, R., Tabor, M., and Tan, Q.M. (1986) Experimental study of lagrangian turbulence in a stokes flow. *Proc. Roy. Soc. (London)*, **A408**, 165–174.
72. Leong, C.W. and Ottino, J.M. (1989) Experiments on mixing due to chaotic advection in a cavity. *J. Fluid Mech.*, **209**, 463–499.
73. Jana, S.C., Jahjadi, M., and Ottino, J.M. (1994) Chaotic mixing of viscous fluids by periodic changes in geometry: baffled cavity flow. *AIChE J.*, **40**, 1769–1781.
74. Fountain, G.O., Khakhar, D.V., and Ottino, J.M. (1998) Visualization of three-dimensional Chaos. *Science*, **281**, 683–686.
75. Li, H.Z., Fasol, C.F., and Choplin, L. (1996) Hydrodynamics and heat transfer of rheologically complex fluids in a Sulzer SMX static mixer. *Chem. Eng. Sci.*, **51**, 1947–1955.
76. Avalosse, T. and Crochet, M.J. (1997) Finite-element simulation of mixing. 1. Two-dimensional flow in periodic geometry. *AIChE J.*, **43**, 577–587.
77. Hobbs, D.M. and Muzzio, F.J. (1998) Optimization of a static mixer using dynamical systems techniques. *Chem. Eng. Sci.*, **53**, 3199–3213.
78. Kim, S.J. and Kwon, T.H. (1996) Enhancement of mixing performance of single-screw extrusion processes via chaotic flows. Part I. Basic concepts and experimental study. *Adv. Polym. Technol.*, **15**, 41–54.
79. Kim, S.J. and Kwon, T.H. (1996) Enhancement of mixing performance of single-screw extrusion processes via chaotic flows. Part II. Numerical study. *Adv. Polym. Technol.*, **15**, 55–69.
80. Jana, S.C., Scott, E.W. and Sundararaj, U. (1998) Single extruder screw for efficient blending of miscible and immiscible polymeric materials. US Patent 6,132,076.
81. Zumbrunnen, D.A. and Inamdar, S. (2001) Novel sub-micron highly multi-layered polymer films formed by continuous flow chaotic mixing. *Chem. Eng. Sci.*, **56**, 3893–3897.
82. Danescu, R.I. and Zumbrunnen, D.A. (1998) Creation of conducting networks among particles in polymer melts by chaotic mixing. *J. Thermoplast. Compos. Mater.*, **11**, 299–320.
83. Dharaiya, D. and Jana, S.C. (2005) Nanoclay-induced morphology development in chaotic mixing of immiscible polymers. *J. Polym. Sci. Part B: Polym. Phys.*, **43**, 3638–3651.
84. Perilla, J.E. and Jana, S.C. (2005) Coalescence of immiscible polymer blends in chaotic mixers. *AIChE J.*, **51**, 2675–2685.
85. Zumbrunnen, D.A., Miles, K.C., and Liu, Y.H. (1996) Auto-processing of very fine-scale composite materials by chaotic mixing of melts. *Compos. Part A: Appl. Sci. Manuf.*, **27**, 37–47.
86. Jimenez, G.A. and Jana, S.C. (2007) Electrically conductive polymer nanocomposites of polymethyl methacrylate and carbon nanofibers prepared by chaotic mixing. *Compos. Part A- Appl. Sci. Manuf.*, **38**, 983–993.
87. Jimenez, G.A. and Jana, S.C. (2007) Oxidized carbon nanofiber/polymer composites prepared by chaotic mixing. *Carbon*, **45**, 2079–2091.
88. Gunes, I.S., Jimenez, G.A., and Jana, S.C. (2009) Carbonaceous fillers for shape memory actuation of polyurethane composites by resistive heating. *Carbon*, **47**, 981–997.
89. Zumbrunnen, D.A., Inamdar, S., Kwon, O., and Verma, P. (2002) Chaotic advection as a means to develop nanoscale structures in viscous melts. *Nano Lett.*, **2**, 1143–1148.
90. Danckwerts, P.V. (1952) Temperature effects accompanying the absorption

of gases in liquids. *Appl. Sci. Res. A*, **3**, 385–390.
91. Ottino, J.M. (1980) Lamellar mixing models for structured chemical reactions and their relationship to statistical models: macro- and micromixing and the problem of averages. *Chem. Eng. Sci.*, **35**, 1377–1391.
92. Yang, I.K. and Lin, J.D. (2002) Effects of flow on polymeric reaction. *Polym. Eng. Sci.*, **42**, 753–759.
93. Hu, G.H. and Kadri, I. (1998) Modeling reactive blending: an experimental approach. *J. Polym. Sci. Part B: Polym. Phys.*, **36**, 2153–2163.
94. Feng, L.F. and Hu, G.H. (2004) Reaction kinetics of multiphase polymer systems under flow. *AIChE J.*, **50**, 2604–2612.
95. Muzzio, F.J. and Liu, M. (1996) Chemical reactions in chaotic flows. *Chem. Eng. J.*, **64**, 117–127.
96. Sawyers, D.R., Sen, M., and Chang, H.C. (1996) Effect of chaotic interfacial stretching on bimolecular chemical reaction in helical-coil reactors. *Chem. Eng. J.*, **64**, 129–139.
97. Bryden, M.D. and Brenner, H.A. (1996) Effect of laminar chaos on reaction and dispersion in eccentric annular flow. *J. Fluid Mech.*, **325**, 219–237.
98. Paireau, O. and Tabeling, P. (1997) Enhancement of the reactivity by chaotic mixing. *Phys. Rev. E*, **56**, 2287–2290.
99. Zalc, J.M. and Muzzio, F.J. (1999) Parallel-competitive reactions in a two-dimensional chaotic flow. *Chem. Eng. Sci.*, **54**, 1053–1069.
100. Metcalfe, G. and Ottino, J.M. (1994) Autocatalytic processes in mixing flows. *Phys. Rev. Lett.*, **72**, 2875–2878.
101. Chertkov, M. and Lebedev, V. (2003) Boundary effects on chaotic advection-diffusion chemical reactions. *Phys. Rev. Lett.*, **90**, 134501/1–134501/4.
102. Szalai, E.S., Kukura, J., Arratia, P.E., and Muzzio, F.J. (2003) Effect of hydrodynamics on reactive mixing in laminar flows. *AIChE J.*, **49**, 168–179.
103. Cox, S.M. (2004) Chaotic mixing of a competitive-consecutive reaction. *Phys. D*, **199**, 369–386.
104. Boesinger, C., Le Guer, Y., and Mory, M. (2005) Experimental study of reactive chaotic flows in tubular reactors. *AIChE J.*, **51**, 2122–2132.
105. Fields, S.D. and Ottino, J.M. (1987) Effect of stretching path on the course of polymerizations: applications to idealized unpremixed reactors. *Chem. Eng. Sci.*, **42**, 467–477.
106. Fields, S.D. and Ottino, J.M. (1987) Effect of segregation on the course of unpremixed polymerizations. *AIChE J.*, **33**, 959–975.
107. Fields, S.D. and Ottino, J.M. (1987) Effect of striation thickness distribution on the course of an unpremixed polymerization. *Chem. Eng. Sci.*, **42**, 459–465.
108. Chella, R. and Ottino, J.M. (1983) Modeling of rapidly-mixed fast-crosslinking exothermic polymerizations. Part I: adiabatic temperature rise. *AIChE J.*, **29**, 373–382.
109. Hepburn, C. (1999) *Polyurethane Elastomers*, 2nd edn, CRC Press, Boca Raton.
110. Coleman, M.M., Skrovanek, D.J., Hu, J., and Painter, P.C. (1988) Hydrogen bonding in polymer blends. 1. FTIR studies of urethane-ether blends. *Macromolecules*, **21**, 59–65.
111. Yang, W.P. and Macosko, C.W. (1989) Phase separation during fast (RIM) polyurethane polymerization. *Makromol. Chem. Makromol. Symp.*, **25**, 23–44.
112. Raynal, F. and Gence, J.N. (1997) Energy saving in chaotic laminar mixing. *Int. J. Heat Mass Trans.*, **40**, 3267–3273.
113. Jung, C., Jana, S.C., and Gunes, I.S. (2007) Analysis of polymerization in chaotic mixers using time scales of mixing and chemical reactions. *Ind. Eng. Chem. Res.*, **46**, 2413–2422.
114. Verhoeven, V.W.A., van Vondel, M.P.Y., Ganzeveld, K.J., and Janssen, L.P.B.M. (2004) Rheo-kinetic measurement of thermoplastic polyurethane polymerization in a measurement kneader. *Polym. Eng. Sci.*, **44**, 1648–1655.

4
Frontal Polymerization
John A. Pojman

4.1
Introduction

Frontal polymerization (FP) is a polymerization process in which polymerization occurs directionally. There are three types of FP. The first is isothermal FP, which is discussed in Chapter 5. The second is photofrontal polymerization in which the front is driven by the continuous flux of radiation, usually UV light [1–7]. The last type is thermal FP, which we will henceforth refer to as *frontal polymerization*, and it results from the coupling of thermal transport and the Arrhenius dependence of the reaction rate of an exothermic polymerization.

FP was discovered at the Institute of Chemical Physics in Chernogolovka, Russia. Chechilo and Enikolopyan studied methyl methacrylate polymerization under 3500 atm pressure [8–11]. The work from that Institute was reviewed in 1984 [12].

4.1.1
Requirements for Frontal Polymerizations

The essential criterion for FP is that the system must have an extremely low rate of reaction at the initial temperature but a high rate at the front temperature such that the rate of heat production exceeds the rate of heat loss. In other words, the system must react slowly or not at all at room temperature, have a large heat release, and have a high energy of activation. For free-radical polymerization, the peroxide or nitrile initiator provides the large activation energy. It is not possible to create a system that has a long pot life at room temperature and a rapid reaction at any arbitrary temperature if the system follows Arrhenius kinetics.

Frontal polyurethane polymerization [13–15], frontal atom transfer radical polymerization [16], and frontal ring-opening metathesis polymerization (ROMP) [17] all suffer from short pot lives: that is, bulk polymerization occurs in less than an hour. In some cases, the only way to avoid even faster bulk polymerization is to cool the reagents. For example, with frontal ring-opening metathesis polymerization of dicyclopentadiene with a Grubbs catalyst, the starting materials had to be frozen to prevent rapid bulk polymerization [17].

Nonlinear Dynamics with Polymers: Fundamentals, Methods and Applications.
Edited by John A. Pojman and Qui Tran-Cong-Miyata
Copyright © 2010 WILEY-VCH Verlag GmbH & Co. KGaA, Weinheim
ISBN: 978-3-527-32529-0

4.1.2
Types of Systems

Most work has been with free-radical systems but other chemistries can be used. Begishev *et al.* studied frontal anionic polymerization of ε-caprolactam [18, 19], and epoxy chemistry has been used as well [20–23]. Mariani *et al.* demonstrated frontal ring-opening metathesis polymerization [17]. Fiori *et al.* produced polyacrylate–poly(dicyclopentadiene) networks frontally [24], and Pojman *et al.* studied epoxy–acrylate binary systems [25]. Polyurethanes have been prepared frontally [13, 14, 26]. Frontal atom transfer radical polymerization has been achieved [16] as well as FP with thiol–ene systems [27]. Recent work has been done using FP to prepare microporous polymers [28–30], polyurethane–nanosilica hybrid nanocomposites [31], and segmented polyurethanes [32].

Epoxy resins can be cured in at least three ways: step-growth polymerization with amines, step-growth polymerization with thiols, or cationically via a chain-growth mechanism. The most common method for the preparation of composites with the greatest strength is with amine curing agents, although they are stoichiometric reactants with the epoxy groups. There has been work on the frontal curing of such systems [21–23, 33, 35]. Scognamillo *et al.* recently studied the cationic frontal curing of a tri-epoxy resin [36].

Crivello developed an interesting variation of FP [37–42]. Many alkyl glycidyl ethers form cations when illuminated by ultraviolet light. For some systems, the rate of cation initiation of ring-opening polymerization is low at room temperature, and systems will exhibit an induction period as the temperature rises from the slow reaction. If local heating is applied, a thermal front can propagate in very thin films (275 µm).

Given that FP is a thermally driven process, we would not expect it to work well at very low temperatures. Ilyashenko *et al.* did observe frontal acrylamide polymerization at 77 K but that was only with an extremely reactive monomer [43]. However, Barelko *et al.* discovered an amazing method for achieving front propagation at temperatures as low as 3 K! [44–46]. Acetaldehyde is a material that can be polymerized cryogenically [46]. Acetaldehyde is first frozen at 77 K and irradiated with a ^{60}Coγ source. The front is initiated by a brittle fracture caused by rapid heating of the surface. Front propagation occurs by a non-Arrhenius mechanism. Fronts would not propagate unless the irradiation dose exceeded 300 kGy. The front velocity was on the order of 1 cm s^{-1} and was proportional to the dosage. This is an amazingly high velocity when compared to the FP of multifunctional acrylates, which is on the order of 10 cm min^{-1} [47]. The front propagation is a result of "self-sustained brittle disruption (dispersion) of the solid matrix" [44].

Such cryogenic systems are interesting not only because of the novel propagation mechanism but also because they may provide insight into how complex molecules were generated in the universe. Simple molecules could have been irradiated with cosmic rays and then caused to react by collisions of meteoroids. Frontal cryopolymerization may also be a method to prepare composites on the Moon and/or Mars [46].

4.1.3
Characteristics of Frontal Polymerization

FP has two essential features – a sharp concentration gradient and a temperature gradient that propagates through the unstirred medium. Figure 4.1 shows a typical temperature profile.

We can also classify the systems on the basis of the states of the reactants and products. Monofunctional monomers such as benzyl acrylate are liquids and produce liquid polymers, and we call these "liquid/liquid" systems. Figure 4.2 shows a schematic of the changes in properties across a liquid/liquid front. Multifunctional monomers such as 1,6 hexanediol diacrylate (HDDA) are liquid but produce a thermoset, solid product, and these we call "liquid/solid" systems (Figure 4.3). Finally, solid monomers such as acrylamide [48, 49] and transition-metal nitrate complexes of acrylamide [50–52] can be polymerized frontally in "solid/solid" systems.

Fortenberry and Pojman studied FP of acrylamide without solvent using powdered acrylamide and persulfate [53]. They found that the initial or "green" density of the systems affected the front velocity. Such behavior is quite normal for self-propagating high-temperature synthesis (SHS) with inorganic components [54–56]. Solid monomers can be used in solvent as well [57].

We will focus henceforth on free-radical systems because they offer several control parameters. An unusual feature of free-radical systems is that they are nonstoichiometric. Adding more initiator usually does not affect the conversion

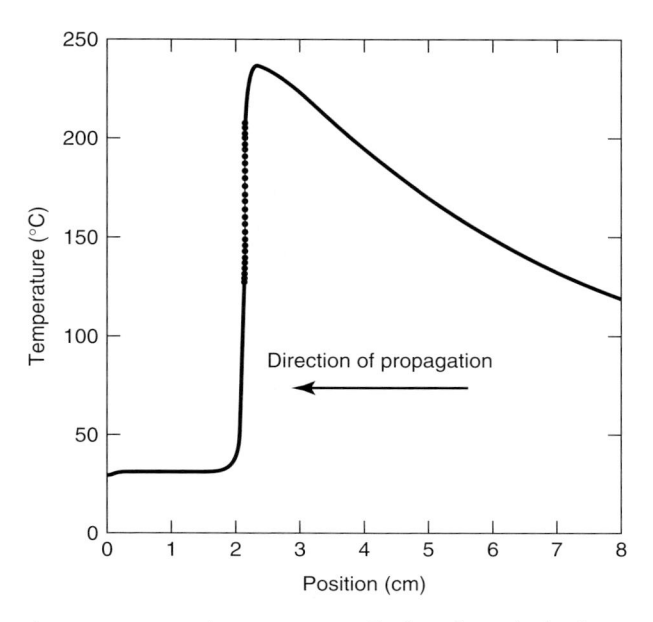

Figure 4.1 A typical temperature profile for a free-radical polymerization front.

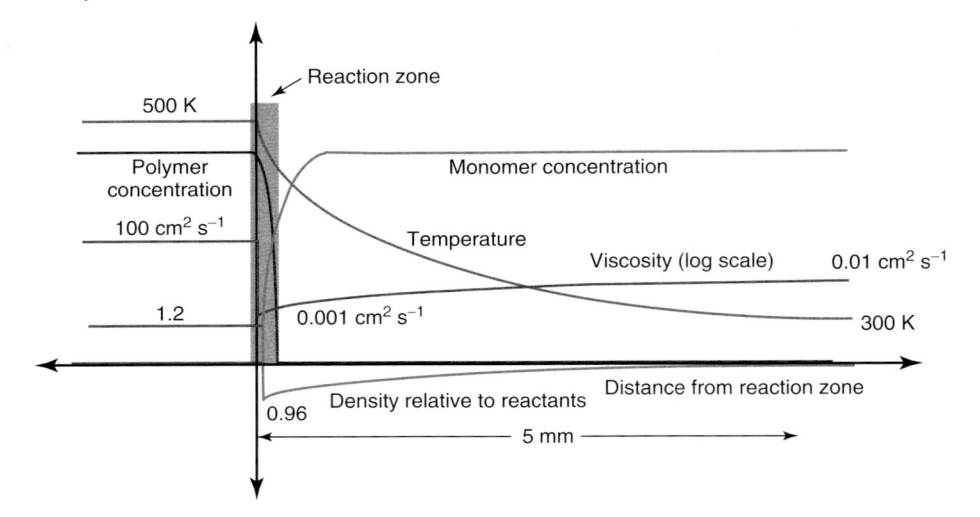

Figure 4.2 Schematic diagram showing changes in properties across a propagating polymerization front that produces a thermoplastic. Courtesy of Paul Ronney.

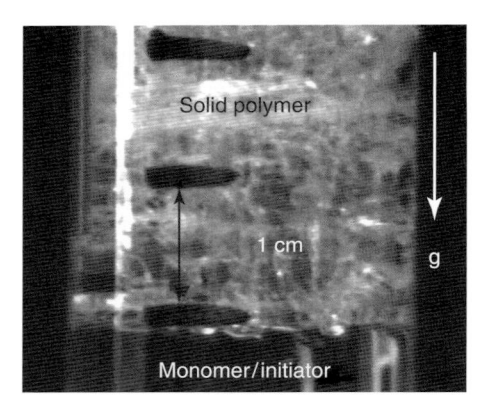

Figure 4.3 A descending case of frontal polymerization with triethylene glycol dimethacrylate and benzoyl peroxide as the initiator.

but only the front velocity. Also, by changing the functionality of the monomer, several aspects of the front can be altered.

Studies on the velocity dependence on temperature and initiator concentration have been performed [43, 58, 59]. FP in solution was performed [57], and initiators that do not produce gas were developed [60]. The velocity can be affected by the initiator type and concentration but is on the order of a few centimeters per minute for monofunctional acrylates and as high as 20 cm min^{-1} for multifunctional acrylates [60].

Free-radical FP has been discussed in detail by Pojman et al. [43] and by Washington and Steinbock [61]. The velocity dependence on the initiator concentration has been studied for several systems [58, 59, 62] and follows a power function dependence on the initiator concentration. Nason *et al.* investigated UV-induced FP of (meth)acrylates [47].

4.2 Applications

FP is not currently in commercial use, although there have been several patents issued related to it. The first is for an "In depth curing of resins induced by UV radiation," which used a combination of UV light to start free-radical FP [63]. In 2001, Gregory patented "Ultraviolet curable resin compositions having enhanced shadow cure properties" in which the frontal curing of cycloaliphatic epoxides was achieved with a cationic photoinitiator in tandem with a peroxide [64]. Pojman was awarded two patents on functionally gradient materials prepared by FP [65, 66]. The maker of chemical anchors, Hilti, has two patents on FP but has not commercialized either of them [67, 68].

FP is being pursued along three directions. The first direction is to use it to make new types of polymers or existing polymers more easily. For example, FP has been used for the preparation of amphilic gels [69], polymers for controlled release [70], and porous polyacrylamide hydrogels [71].

The second is to make unique polymeric devices. Chekanov *et al.* prepared functionally gradient materials by pumping monomer feed streams with varying compositions as a function of time on top of an ascending front [72]. Yan *et al.* prepared monolithic macroporous polymers via FP [73].

The third direction is to use FP as a process for accomplishing a task *in situ*. The Hilti patents propose FP as a means to rapidly cure a chemical anchor. Vicini *et al.* developed an FP method for the consolidation of stone [74, 75]. We have focused on cure-on-demand polymerization.

4.2.1
Cure-On-Demand Putty

FP can used to create a cure-on-demand putty for filling holes in wood, marble, and sheet rock. The putty has months to years of shelf life, is a one-pot formulation, can be applied leisurely, and then cured rapidly with a heat source. Figure 4.4 shows how holes in wood can be repaired. The curing was accomplished in a few seconds. The use of the ethoxylated trimethylol propane triacrylate (TMPTA) reduces the front temperature and thus smoking by the reaction, and it makes the product less brittle. In addition to kaolin, a variety of fillers can be used, including calcium carbonate, aluminum oxide, and sawdust. All formulations can be cured in dramatically less time than current air-dried/cured materials. Significantly, virtually any depth of hole can be filled because the reaction propagates from the surface. The cured materials can be stained or painted. Figure 4.5 illustrates hole repair in a stone floor. (NB: "phr" stands for "parts per hundred resin.")

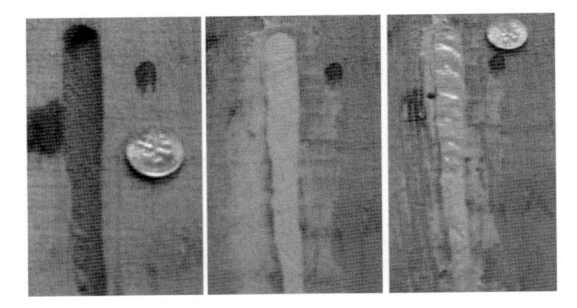

Figure 4.4 Repair of a hole in wood by frontal polymerization. The formulation was composed of 44% mass of a solution containing 3 phr Luperox 231 in trimethylolpropane ethoxylate triacrylate (1/1 EO/OH) ($M_n \sim 428$), 15% mass of a solution of 3 phr Luperox 231 in TMPTA-n, and 41% mass Polygloss 90.

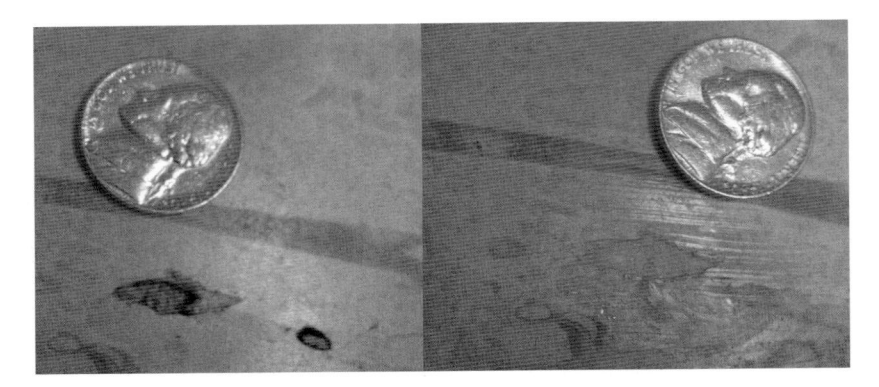

Figure 4.5 Repair of a hole in a stone floor. The hole was filled with the same formulation as in Figure 4.4. The frontal curing was started by heating through an aluminum foil with a laundry iron. The reaction was complete within seconds.

4.2.2
Adhesive

Wood and plastic–wood composites can be glued together using a frontally cured adhesive. Typical formulations use 1 phr Luperox 231 in TMPTA with 7 phr fumed silica. Figure 4.6 shows how such an adhesive can be applied to pieces of wood and then rapidly cured.

The shear strength is very high. (For reference, 1 MPa is equal to 145 psi.) Figure 4.7 presents the strength as a function of the amount of initiator added to a TMPTA–fumed silica system. We hypothesize that the voids formed by the volatile byproducts of the initiator decomposition reduce the strength.

Figure 4.6 Wood being glued with a frontally cured adhesive containing car tire particles.

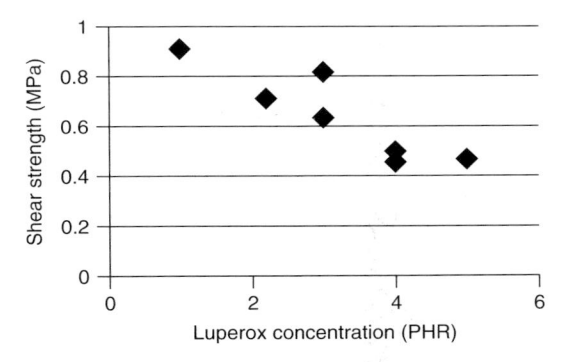

Figure 4.7 The shear strength of a frontal wood adhesive based on TMPTA as a function of the concentration of Luperox 231.

4.2.3
Coatings

Coatings with thickness of 1 – 0.1 mm can be created with TMPTA filled with kaolin clay. To create a smooth surface, a sheet of Kapton was placed on top and then pressed down with a piece of wood. Figure 4.8 shows the surface of such a material, half of which was stained with a standard wood stain.

4.3
Motivation for Studying Nonlinear Dynamics with Frontal Polymerization

There are at least three motives for studying nonlinear dynamics in FP systems. First, FP can be useful for finding interesting phenomena. (For studying nonlinear

Figure 4.8 A glossy layer on wood created by frontal polymerization. The right half was stained.

phenomena, free-radical FP is the most versatile of the types of polymerization because many parameters can be independently varied including front temperature, front velocity, initial viscosity, and energy of activation.) Secondly, FP can be a model for thermal fronts in condensed media, such as SHS [76]. The third reason is that some modes of dynamic behavior can adversely affect the FP process and/or the properties of the material produced.

4.4
Convective Instabilities

4.4.1
Buoyancy-Driven Convection

Because of the large thermal and concentration gradients, polymerization fronts are highly susceptible to buoyancy-induced convection. Garbey *et al.* performed the linear stability analysis for the liquid/liquid and liquid/solid cases [77–79]. The bifurcation parameter was a "frontal Rayleigh number":

$$R = g\beta q\kappa^2/\nu c^3 \qquad (4.1)$$

where g is the gravitational acceleration, β the thermal expansion coefficient, q the temperature increase at the front, κ the thermal diffusivity, ν the kinematic viscosity, and c the front velocity.

Let us first consider the liquid/solid case. Neglecting heat loss, the descending front is always stable because it corresponds to heating a fluid from above. The front is always flat. If the front is ascending, convection may occur depending on the parameters of the system.

Bowden *et al.* experimentally confirmed that the first mode is an antisymmetric one, followed by an axisymmetric one [80]. Figure 4.9 shows a flat descending front as well as axisymmetric and antisymmetric modes of ascending fronts. Figure 4.10 shows the stability diagram in the viscosity-front velocity plane. Most importantly, they confirmed that the stability of the fluid was a function not only of the viscosity but also of the front velocity. This means that the front dynamics affects the fluid dynamics.

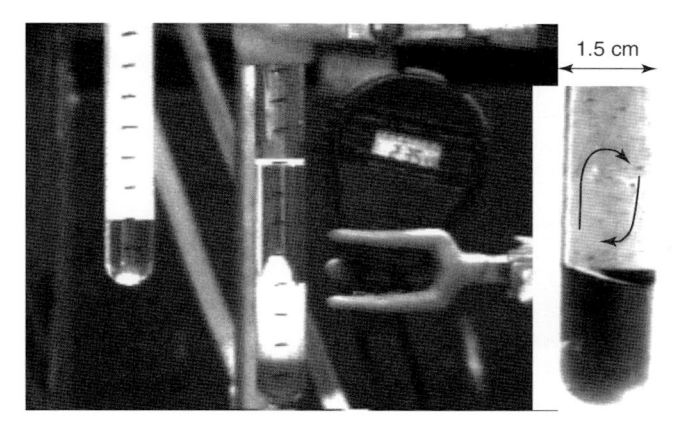

Figure 4.9 (a) The front on the left is descending and the one on the right ascending with an axisymmetric mode of convection. (b) An antisymmetric mode of an ascending front. The system is the acrylamide/bis-acrylamide polymerization in DMSO with persulfate initiator. Adapted from Bowden et al. [80].

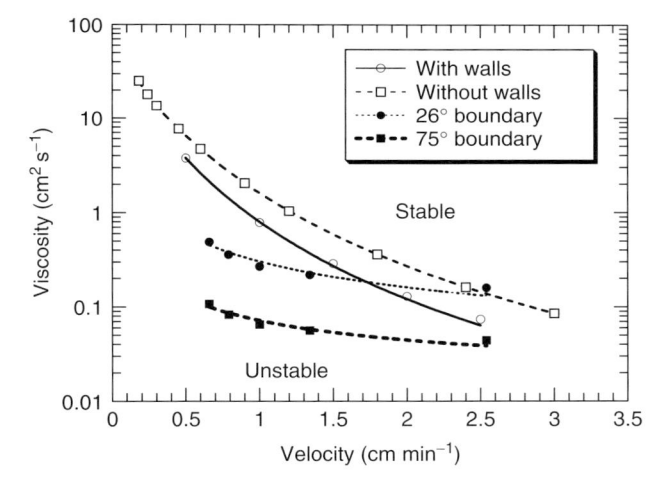

Figure 4.10 The stability diagram for the system in Figure 4.2. Adapted from Bowden et al. [80].

If the reactor is not vertical, there is no longer the question of stability–there is always convection. Bazile et al. studied descending fronts of acrylamide/bis-acrylamide polymerization in dimethyl sulfoxide (DMSO) as a function of tube orientation [81]. The fronts remained nearly perpendicular to the vertical but the velocity projected along the axis of the tube increased with the inverse of the cosine of the angle.

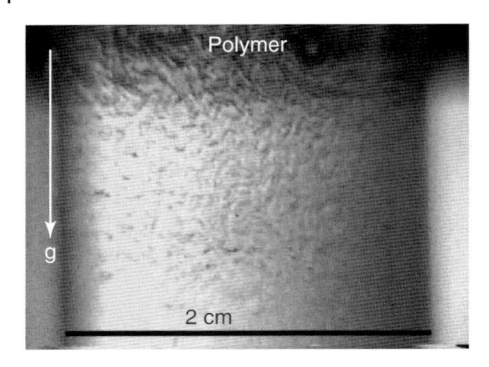

Figure 4.11 Rayleigh–Taylor instability with a descending front of butyl acrylate polymerization.

Liquid/liquid systems are more complicated than the previous case because a descending front can exhibit the Rayleigh–Taylor instability. Consider the schematic in Figure 4.2. The product is hotter than the reactant but is denser, and because the product is a liquid, fingering can occur. Such a front degeneration is shown in Figure 4.11. The Rayleigh–Taylor instability can be prevented by using high pressure [11], adding a filler [59], using a dispersion in salt water [82], or performing the fronts in vacuum [82].

McCaughey et al. tested the analysis of Garbey et al. and found the same bifurcation sequence of antisymmetric to axisymmetric convection in ascending fronts [83] as seen with the liquid/solid case.

Garbey et al. also predicted that, for a descending liquid/liquid front, instability could arise even though the configuration would be stable for unreactive fluids [77–79]. This prediction has yet to be verified experimentally because liquid/liquid FP exhibits the Rayleigh–Taylor instability. A thermal frontal system with a product that is less dense than the reactant is required.

Texier-Picard et al. analyzed a polymerization front in which the molten polymer was immiscible with the monomer and predicted that the front could exhibit the Marangoni instability even though comparable unreactive fluids would not exhibit the instability [84]. However, no liquid/liquid frontal system with an immiscible product has been identified. Even if such a system could be found, the experiment would have to be performed in weightlessness to prevent buoyancy-induced convection from interfering.

We note a significant difference between the liquid/liquid and the liquid/solid cases. For the liquid/solid case, convection in ascending fronts increases the front velocity but in the liquid/liquid case, convection slows the front. Convection increases the velocity of pH fronts and BZ waves. Why the difference between liquid/liquid FP and other frontal systems? In liquid/liquid systems, the convection also mixes cold monomer into the reaction zone, which lowers the front temperature. The front velocity depends more strongly on the front temperature than on the effective transport coefficient of the autocatalyst. Convection cannot mix the monomer into the reaction zone of a front with a solid product but only increases thermal transport so the velocity is increased.

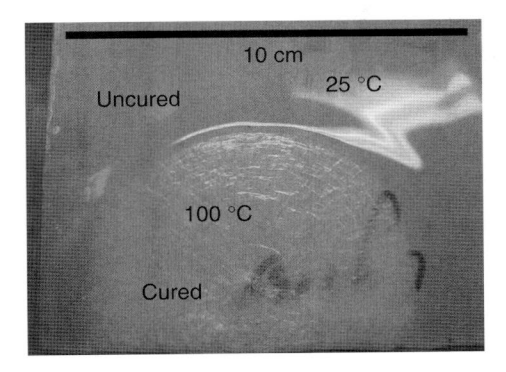

Figure 4.12 A front of pentaerythritol tetra-acrylate copolymerization with a trithiol on the surface of pine. The thickness of the liquid layers is approximately 1 mm. The initiator was Azobisisobutyronitrile (AIBN), and 3 phr of a surface active agent, BYK060N, was added to reduce bubble formation.

4.4.2
Effect of Surface-Tension-Driven Convection

If there is a free interface between fluids, gradients in concentration and/or temperature parallel to the interface cause gradients in the surface (interfacial) tension, which cause convection [85]. This convection, also known as *Marangoni convection*, is especially noticeable in thin layers (or weightlessness) in which buoyancy-driven convection is greatly reduced.

We had long thought that FP could not occur in thin layers. In fact, interfacial-tension-driven convection can cause so much heat loss that fronts are quenched. Figure 4.12 shows a front propagating in a thin layer (about 1 mm) of a tetra-acrylate in which fumed silica is dispersed. The large temperature gradient created by the reaction "pushes" monomer ahead but not enough to quench the front.

For a given surface, three variables affect whether a front will propagate: specifically, the viscosity (determined by the inherent viscosity of the monomer and the amount of fumed silica), the initiator concentration, and the thickness of the layer. For a fixed layer thickness, we determined that for trimethylolpropane triacrylate, with 2 phr fumed silica, no front would propagate with 1 phr Luperox 231 but would propagate with 1.1 phr. Fronts also can expand from the gas produced by the peroxide, pushing the monomer ahead. Figure 4.13 shows the patterns that can

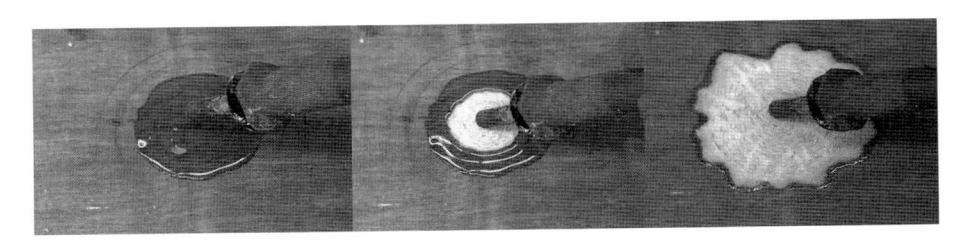

Figure 4.13 A front propagating in a thin layer of a triacrylate with 4 phr fumed silica and 2.1 phr Luperox 231 (a peroxide initiator). The rings on the wood are spaced 1 cm apart.

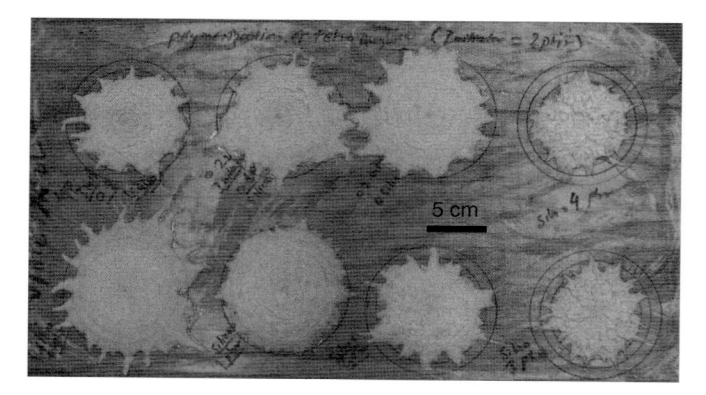

Figure 4.14 The variety of patterns that result in the frontal polymerization of a tetra-acrylate on wax paper.

develop as the monomer is pushed into unwetted regions of the wood. Figure 4.14 presents an array of patterns that develop during the FP of a tetra-acrylate on wax paper. The fronts propagate circularly until they reach the regions that are not covered with the monomer. We believe that the pattern is caused by a fingering instability of a wetting fluid [86–88].

4.5
Thermal Instabilities

Fronts do not have to propagate as planar fronts. Analogous to oscillating reactions, a steady state can lose its stability as a parameter is varied and exhibit periodic behavior, either as pulsations or "spin modes" in which a hot spot propagates around the reactor as the front propagates, leaving a helical pattern (Figure 4.15). This mode was first observed in SHS [89].

Linear stability analysis of the longitudinally propagating fronts in cylindrical adiabatic reactors with one reaction predicted that the expected frontal mode for

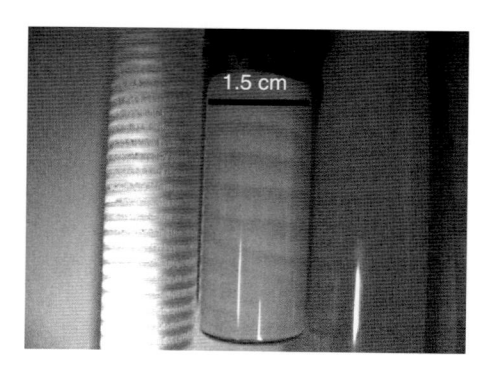

Figure 4.15 Helical patterns produced by "spin modes" in three different frontal polymerization systems.

the given reactive medium and diameter of reactor is governed by the Zeldovich number:

$$Z = \frac{T_m - T_o}{T_m} \frac{E_{eff}}{RT_m} \qquad (4.2)$$

For FP, lowering the initial temperature (T_0), increasing the front temperature (T_m), increasing the energy of activation (E_{eff}) all increase the Zeldovich number. The planar mode is stable if $Z < Z_{cr} = 8.4$. By varying the Zeldovich number beyond the stability threshold, subsequent bifurcations leading to higher spin mode instabilities can be observed. Secondly, for a cylindrical geometry the number of spin heads or hot spots is also a function of the tube diameter. We point out that polymerization is not a one-step reaction, so that the above form of the Zeldovich number does not directly apply. However, estimates of the effective Zeldovich number can be obtained from the overall energy of activation with the steady-state assumption for free-radical polymerization.

The most commonly observed case with FP is the spin mode in which a "hot spot" propagates around the front. A helical pattern is often observed in the sample. The first case observed was with the FP of ε-caprolactam [18, 19], and the next one was discovered by Pojman *et al.* in the methacrylic acid system in which the initial temperature was lowered [90].

Spin modes have also been observed in the FP of transition-metal nitrate/acrylamide complexes [50, 52], which are solid, but were not observed in the frontal acrylamide polymerization system [53].

The single-head spin mode was studied in detail by Ilyashenko and Pojman [91]. They were able to estimate the Zeldovich number using data on the initiator and methacrylic acid. The value at room temperature was about 7, which is less than the critical value for the appearance of spin modes. In fact, the fronts at room temperature were planar and spin modes appeared only by lowering the initial temperature. However, spin modes could be observed by increasing the heat loss from the reactor by immersing the tube in water or oil.

4.5.1
Effect of Complex Kinetics

Solovyov *et al.* performed two-dimensional numerical simulations using a standard three-step free-radical mechanism [92]. They calculated the Zeldovich number from the overall activation energy using the steady-state theory and determined the critical values for bifurcations to periodic modes and found that the complex kinetics stabilized the front.

Shult and Volpert performed the linear stability analysis for the same model and confirmed this result [93]. Spade and Volpert studied linear stability for nonadiabatic systems [94]. Gross and Volpert performed a nonlinear stability analysis for the one-dimensional case [95]. Commissiong *et al.* extended the nonlinear analysis to two dimensions [96]. They confirmed that, unlike in SHS [97], uniform pulsations are difficult to observe in FP. In fact, no such one-dimensional pulsating modes have been observed.

An interesting problem arises in the study of fronts with multifunctional acrylates. At room temperature, acrylates such as HDDA and triethylene glycol dimethacrylate (TGDMA) exhibit spin modes. In fact, if an inert diluent, such as DMSO is added, the spins modes are more apparent even though the front temperature is reduced. Masere and Pojman found spin modes in the FP of a diacrylate at ambient conditions [98]. Thus, although the mechanical quality of the resultant polymer material can be improved by using multifunctional acrylates, spin modes may appear and a nonuniform product may result. This observation implicates the role of polymer cross-linking in the front dynamics. In the same work, Masere and Pojman showed that pH indicators could be added to act as dyes that would be bleached by the free radicals, making the observation of the spin pattern readily apparent.

Tryson and Schultz studied the energy of activation of photopolymerized multifunctional acrylates and found that it increased with increasing conversion because of cross-linking [99]. Gray found that the energy of activation of HDDA increased exponentially during the reaction [100]. Applying the steady-state theory of polymerization to Gray's results, Masere et al. calculated the effective energy of activation for thermally initiated polymerization (photoinitiation has no energy of activation) by including the energy of activation (E_a) of a typical peroxide [101]. They calculated that E_a of HDDA polymerization increased from 80 kJ/mol at 0% conversion (which is the same as that of methacrylic acid) to 140 kJ mol^{-1} at 80% conversion. This can explain how spin modes appear at room temperature with diacrylates but not monoacrylates. The Zeldovich number for methacrylic acid polymerization at room temperature is 7.2, which is below the stability threshold. Using the activation energy at the highest conversion that can be obtained with HDDA, Masere et al. estimated a Zeldovich number of 12.

Masere et al. studied fronts with a peroxide initiator at room temperature and used two bifurcation parameters [101]. They added an inert diluent, DMSO, to change the front temperature and observed a variety of modes. More interestingly, they also varied the ratio of a monoacrylate, benzyl acrylate, to HDDA, keeping the front temperature constant. Changing the extent of cross-linking changed the effective energy of activation, which revealed a wide array interesting spin modes.

The three-dimensional nature of the helical pattern was studied by Manz et al. using magnetic resonance imaging (MRI) [102]. Pojman et al. observed zigzag modes in square reactors [103] and bistability in conical reactors [104].

4.5.2
Effect of Bubbles

Pojman et al. found an unusual mode of propagation when there are large amounts of very small bubbles that can occur when a linear polymer precipitates from its monomer [90]. In studying fronts of methacrylic acid polymerization, they observed convection that periodically occurred under the front at the same time as the front deformed and undulated. The period of convection was about 20 s and remained constant during the entire front propagation.

Volpert *et al.* analyzed the effect of the thermal expansion of the monomer on the thermal stability and concluded that the reaction front becomes less stable than without thermal expansion [105]. The effective thermal expansion can be increased because of the bubbles, and it can considerably affect the stability conditions.

4.5.3
Effect of Buoyancy

The first experimental confirmation that gravity plays a role on spin modes in a liquid/solid system came from the study of descending fronts in which the viscosity was significantly increased with silica gel. Masere *et al.* [101] found that silica gel significantly altered the spin behavior, as predicted by Garbey *et al.* [77]. Pojman *et al.* made a similar observation in square reactors [103]. Pojman *et al.* studied the dependence of spin modes on viscosity with the FP of HDDA with persulfate initiator [106]. They found that the number of spins was independent of the viscosity until a critical viscosity was reached, at which point the spins vanished.

The question arises why the analysis of Ilyashenko and Pojman worked so well for the methacrylic acid system even though they did not consider the effect of convection. They induced spin modes by reducing the initial temperature to 0 °C – below the melting point of methacrylic acid. Thus, the system was a solid/solid system and so hydrodynamics played only a small role.

4.5.4
Other Factors

McFarland *et al.* observed that spin modes did not occur when the initiator was microencapsulated [107, 108]. Not only were spin modes not observed but the material was 10 times stronger, which the authors attributed to the absence of spin modes. More studies of the relationship between spin modes and the size of the microcapsules are required. The front velocity as a function of microcapsules size is shown in Figure 4.16 for the FP of a diacrylate with encapsulated cumene hydroperoxide. Even for particles as large as 100 μm, the front velocity is 90% of the value with a dissolved initiator.

4.6
Snell's Law

Viner *et al.* found that Snell's law of refraction can be applied to quasi-two-dimensional fronts [109]. They prepared mixtures of a triacrylate with kaolin clay and peroxide and spread them into slabs. Figure 4.17 shows fronts propagating in such slabs. The upper slab contained less peroxide and so it propagated more slowly.

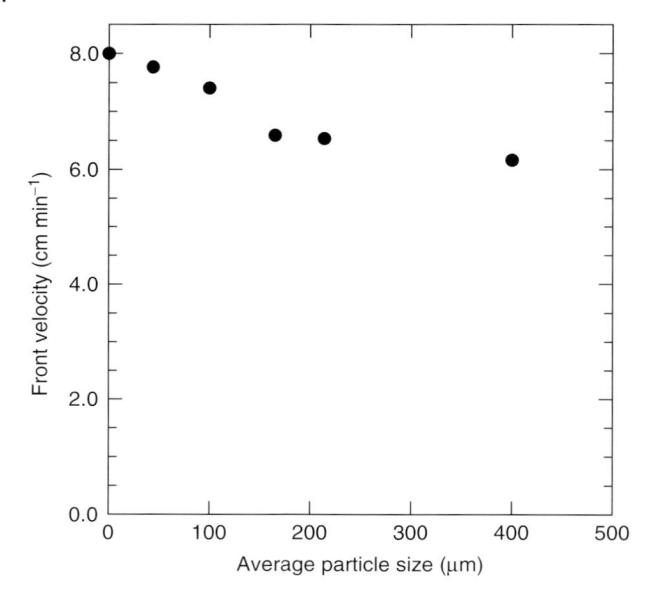

Figure 4.16 The front velocity for 1,6 hexanediol diacrylate polymerization as a function of the microcapsule (containing cumene hydroperoxide) size.

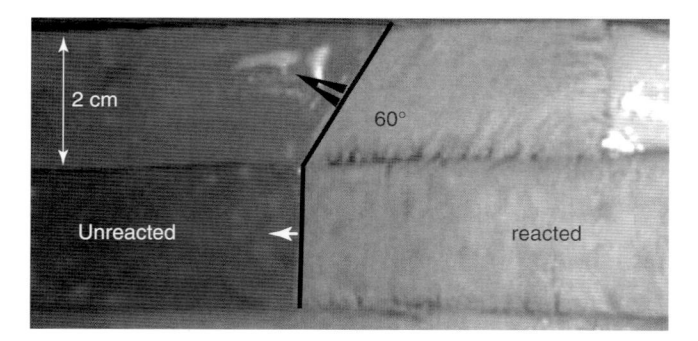

Figure 4.17 Fronts of triacrylate polymerization propagating side by side in two dimensions. The arrows indicate the direction of propagation.

4.7
Three-Dimensional Frontal Polymerization

FP allows the study of spherically propagating fronts. Binici *et al.* developed a system that was a gel created by the base-catalyzed reaction of a trithiol with a triacrylate, in which a peroxide was also dissolved [110]. The gel was necessary to suppress convection. The ratio of triacrylate to trithiol was such that the gel point was reached when all the thiol was consumed but enough triacrylate remained to

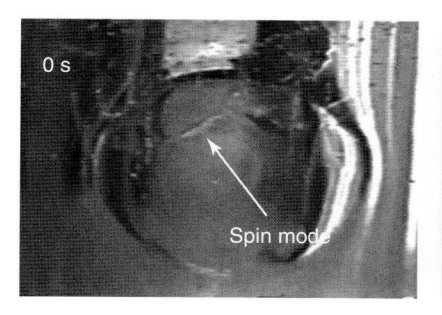

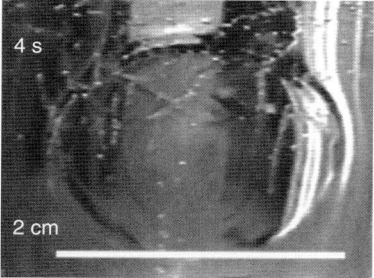

Figure 4.18 A spin mode on the surface of a spherically expanding front of trimethylolpropane triacrylate polymerization.

support FP via free-radical polymerization. (The specific system had been shown to support spin modes in a cylindrical reactor.) A solution of triacrylate and a photoinitiator was then injected into the center of the gel. The polymerization was started by illuminating the gel with 365 nm (UV) radiation. The solution with the photoinitiator polymerized, releasing heat that started an expanding front. Fronts propagated on the expanding front, as seen in Figure 4.18. No theoretical analysis exists for this system.

4.8
Impact on Applications

In addition to intrinsic interest in the different modes of polymerization, attempts to commercialize FP are affected. Consider Figure 4.19, which shows a putty prepared from trimethylolpropane triacrylate with 40% aluminum oxide. On the left, the unreacted material was extruded into a cylinder. The front was initiated at the end nearest to the bottom of the image. The product had the appearance of a fuzzy caterpillar. It is not clear whether the pattern results from only a pulsating

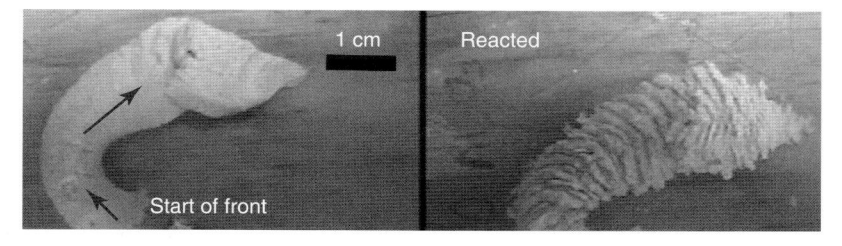

Figure 4.19 Trimethylolpropane triacrylate with 40% aluminum oxide filler and Luperox 231 (1,1-di(*tert*-butylperoxy)-3,3,5-trimethylcyclohexane). The arrows show the direction of front propagation.

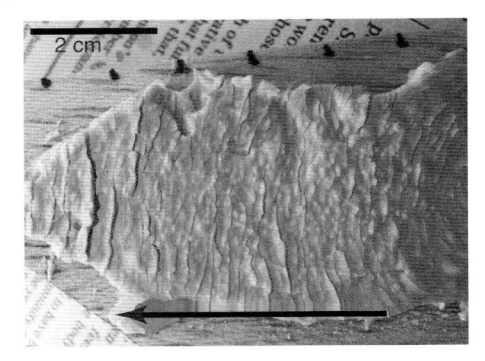

Figure 4.20 Image of a frontally polymerized triacrylate with 40% kaolin clay with 5 phr Luperox 231. The arrow indicates the direction of propagation.

front or from the stresses induced by the contractions upon polymerization. Unless someone will buy caterpillar decoys, this is not a useful product.

Figure 4.20 shows another pattern formed by a front of triacrylate polymerization with 40% kaolin clay, 4% dibutyl phthalate plasticizer, and 5 phr Luperox 231. Again, the mechanism of the pattern formation is unknown but it certainly affects the aesthetics of the final product.

4.9
Conclusions

Thermal FP is a powerful tool for studying the effects of convection on chemical reactions and for observing instabilities in thermal fronts. Buoyancy-driven convection and interfacial-tension-driven convection can occur, depending on the manner of performing the front. Convection can prevent front propagation, but it can be suppressed by adding a filler. Thermal instabilities appear as "spin modes" in which localized hot spots propagate along the front. These hot spots produce inhomogeneities in the product, which reduce the strength.

Not only is FP useful for observing interesting dynamics, but it is necessary to identify the conditions for the onset of each instability in order to avoid them for commercial applications.

References

1. Pearlstein, A.J. (1985) Criteria for the absence of thermal convection in photochemical systems. *J. Phys. Chem.*, **89**, 1054–1058.
2. Pearlstein, A.J., Harris, R.M., and Terronesq, G. (1989) The onset of convective instability in a triply diffusive fluid layer. *J. Fluid. Mech.*, **202**, 443–465.
3. Terrones, G. and Pearlstein, A.J. (1989) The onset of convection in a multicomponent fluid layer. *Phys. Fluids A*, **1**, 845–853.
4. Briskman, V.A. (2001) in *Polymer Research in Microgravity: Polymerization and Processing*, Acs Symposium Series No. 793 (eds J.P. Downey and J.A. Pojman), American Chemical Society, Washington, DC, pp. 97–110.

5. Terrones, G. and Pearlstein, A.J. (2001) Effects of optical attenuation and consumption of a photobleaching initiator on local initiation rates in photopolymerizations. *Macromolecules*, **34**, 3195–3204.
6. Terrones, G. and Pearlstein, A.J. (2004) Diffusion-induced nonuniformity of photoinitiation in a photobleaching medium. *Macromolecules*, **37**, 1565–1575.
7. Cabral, J.T., Hudson, S.D., Harrison, C., and Douglas, J.F. (2004) Frontal photopolymerization for microfluidic applications. *Langmuir*, **20**, 10020–10029.
8. Chechilo, N.M. and Enikolopyan, N.S. (1974) Structure of the polymerization wave front and propagation mechanism of the polymerization reaction. *Dokl. Phys. Chem.*, **214**, 174–176.
9. Chechilo, N.M. and Enikolopyan, N.S. (1975) Effect of the concentration and nature of initiators on the propagation process in polymerization. *Dokl. Phys. Chem.*, **221**, 392–394.
10. Chechilo, N.M. and Enikolopyan, N.S. (1976) Effect of pressure and initial temperature of the reaction mixture during propagation of a polymerization reaction. *Dokl. Phys. Chem.*, **230**, 840–843.
11. Chechilo, N.M., Khvilivitskii, R.J., and Enikolopyan, N.S. (1972) On the phenomenon of polymerization reaction spreading. *Dokl. Akad. Nauk SSSR*, **204**, 1180–1181.
12. Davtyan, S.P., Zhirkov, P.V., and Vol'fson, S.A. (1984) Problems of non-isothermal character in polymerisation processes. *Russ. Chem. Rev.*, **53**, 150–163.
13. Fiori, S., Mariani, A., Ricco, L., and Russo, S. (2003) First synthesis of a polyurethane by frontal polymerization. *Macromolecules*, **36**, 2674–2679.
14. Mariani, A., Bidali, S., Fiori, S., Malucelli, G., and Sanna, E. (2003) Synthesis and characterization of a polyurethane prepared by frontal polymerization. *e-Polymers*, **44**, 1–9.
15. Mariani, A., Fiori, S., Bidali, S., Alzari, V., and Malucelli, G. (2008) Frontal polymerization of diurethane diacrylates. *J. Polym. Sci. Part A Polym. Chem.*, **46**, 3344–3352.
16. Bidali, S., Fiori, S., Malucelli, G., and Mariani, A. (2003) Frontal atom transfer radical polymerization of tri-ethylene glycol) dimethacrylate. *e-Polymers*, **60**, 1–12.
17. Mariani, A., Fiori, S., Chekanov, Y., and Pojman, J.A. (2001) Frontal ring-opening metathesis polymerization of dicyclopentadiene. *Macromolecules*, **34**, 6539–6541.
18. Begishev, V.P., Volpert, V.A., Davtyan, S.P., and Malkin, A.Y. (1973) On some features of the process of anionic activated polymerization of E-caprolactans under conditions of wave propagation. *Dokl. Akad. Nauk SSSR*, **208**, 892.
19. Begishev, V.P., Volpert, V.A., Davtyan, S.P., and Malkin, A.Y. (1985) On some features of the anionic activated Ẽ caprolactam polymerization process under wave propagation conditions. *Dokl. Phys. Chem.*, **279**, 1075–1077.
20. Davtyan, S.P., Arutyunyan, K.A., Shkadinskii, K.G., Rozenberg, B.A., and Yenikolopyan, N.S. (1978) The mechanism of epoxide oligomer hardening by diamines under advancing reaction front conditions. *Polym. Sci. USSR*, **19**, 3149–3154.
21. Korotkov, V.N., Chekanov, Y.A., and Rozenberg, B.A. (1993) The simultaneous process of filament winding and curing for polymer composites. *Comput. Sci. Tech.*, **47**, 383–388.
22. White, S.R. and Kim, C. (1993) A simultaneous lay-up and in situ cure process for thick composites. *J. Reinf. Plast. Compos.*, **12**, 520–535.
23. Chekanov, Y., Arrington, D., Brust, G., and Pojman, J.A. (1997) Frontal curing of epoxy resin: comparison of mechanical and thermal properties to batch cured materials. *J. Appl. Polym. Sci.*, **66**, 1209–1216.
24. Fiori, S., Mariani, A., Ricco, L., and Russo, S. (2002) Interpenetrating polydicyclopentadiene/polyacrylate networks obtained by simultaneous non-interfering frontal polymerization. *e-Polymers*, **29**, 1–10.

25. Pojman, J.A., Griffith, J., and Nichols, H.A. (2004) Binary frontal polymerization: velocity dependence on initial composition. *e-Polymers*, **13**, 1–7.
26. Texter, J. and Ziemer, P. (2004) Polyurethanes via microemulsion polymerization. *Macromolecules*, **37**, 5841–5843.
27. Pojman, J.A., Varisli, B., Perryman, A., Edwards, C., and Hoyle, C. (2004) Frontal polymerization with thiol-ene systems. *Macromolecules*, **37**, 691–693.
28. Pujari, N.S., Vishwakarma, A.R., Kelkar, M.K., and Ponrathnam, S. (2004) Gel formation in frontal polymerization of 2-hydroxyethyl methacrylate. *e-Polymers*, **49**, 1–10.
29. Pujari, N.S., Vishwakarma, A.R., Pathak, T.S., Mule, S.A., and Ponrathnam, S. (2004) Frontal copolymerization of 2-hydroxyethyl methacrylate and ethylene glycol dimethacrylate without porogen: comparison with suspension polymerization. *Polym. Int.*, **53**, 2045–2050.
30. Pujari, N.S., Vishwakarma, A.R., Pathak, T.S., Kotha, A.M., and Ponrathnam, S. (2004) Functionalized polymer networks: synthesis of microporous polymers by frontal polymerization. *Bull. Mater. Sci.*, **27**, 529–536.
31. Chen, S., Sui, J., Chen, L., and Pojman, J.A. (2005) Polyurethane-nanosilica hybrid nanocomposites synthesized by frontal polymerization. *J. Polym. Sci., Part A: Polym. Chem.*, **43**, 1670–1680.
32. Chen, S.H., Sui, J., and Chen, L. (2005) Segmented polyurethane synthesized by frontal polymerization. *Colloid. Polym. Sci.*, **283**, 932–936.
33. Arutiunian, K.A., Davtyan, S.P., Rozenberg, B.A., and Enikolopyan, N.S. (1975) Curing of epoxy resins of bis-phenol a by amines under conditions of reaction front propagation. *Dokl. Akad. Nauk SSSR*, **223**, 657–660.
34. Surkov, N.F., Davtyan, S.P., Rozenberg, B.A., and Enikolopyan, N.S. (1976) Calculation of the steady velocity of the reaction front during hardening of epoxy oligomers by diamines. *Dokl. Phys. Chem.*, **228**, 435–438.
35. Kim, C., Teng, H., Tucker, C.L., and White, S.R. (1995) The continuous curing process for thermoset polymer composites. Part 1: modeling and demonstration. *J. Compos. Mater.*, **29**, 1222–1253.
36. Scognamillo, S., Bounds, C., Luger, M., Mariani, A., and Pojman, J.A. (2010) Frontal cationic curing of epoxy resins. *J. Polym. Sci., Part A: Polym. Chem.*, **48**, 2000–2005.
37. Crivello, J.V. (2005) Investigation of the photoactivated frontal polymerization of oxetanes using optical pyrometry. *Polymer*, **46**, 12109–12117.
38. Crivello, J.V. (2006) Design and synthesis of multifunctional glycidyl ethers that undergo frontal polymerization. *J. Poly. Sci., Part A: Polym. Chem.*, **44**, 6435–6448.
39. Crivello, J.V. (2007) Hybrid free radical/cationic frontal photopolymerizations. *J. Polym. Sci., Part A: Polym. Chem.*, **45**, 4331–4340.
40. Crivello, J.V. and Bulut, U. (2005) Photoactivated cationic ring-opening frontal polymerizations of oxetanes. *Des. Monomers Polym.*, **8**, 517–531.
41. Crivello, J.V., Falk, B., and Zonca, M.R. Jr. (2004) Photoinduced cationic ring-opening frontal polymerizations of oxetanes and oxiranes. *J. Poly. Sci., Part A: Polym. Chem.*, **42**, 1630–1646.
42. Falk, B., Zonca, M.R., and Crivello, J.V. (2005) Photoactivated cationic frontal polymerization. *Macromol. Symp.*, **226**, 97–108.
43. Pojman, J.A., Ilyashenko, V.M., and Khan, A.M. (1996) Free-radical frontal polymerization: self-propagating thermal reaction waves. *J. Chem. Soc., Faraday Trans.*, **92**, 2825–2837.
44. Barelko, V.V., Barkalov, I.M., Goldanskii, V.I., Kiryukhin, D.P., and Sanin, A.M. (1988) Autowave modes of conversion in low-temperature chemical reactions in solids. *Adv. Chem. Phys.*, **74**, 339–385.
45. Kiryukhin, D.P., Barelko, V.V., and Barkalov, I.M. (1999) Traveling waves of cryochemical reaction in radiolyzed systems (review). *High Energ. Chem.*, **33**, 133–144.

46. Kiryukhin, D.P., Kichigina, G.A., and Barelko, V.V. (2010) The self sustained wave regime of cryocopolymerization for filler containing systems. *Polym. Sci., Ser. B*, **52**, 221–226.
47. Nason, C., Roper, T., Hoyle, C., and Pojman, J.A. (2005) Uv-induced frontal polymerization of multifunctional (meth)acrylates. *Macromolecules*, **38**, 5506–5512.
48. Pojman, J.A., Nagy, I.P., and Salter, C. (1993) Traveling fronts of addition polymerization with a solid monomer. *J. Am. Chem. Soc.*, **115**, 11044–11045.
49. Fortenberry, D. and Pojman, J.A. (1997) Solvent-free processing of solid monomers utilizing frontal polymerization. *Polym. Prepr. (Am. Chem. Soc. Div. Polym. Chem.)*, **38** (2), 472–473.
50. Savostyanov, V.S., Kritskaya, D.A., Ponomarev, A.N., and Pomogailo, A.D. (1994) Thermally initiated frontal polymerization of transition metal nitrate acrylamide complexes. *J. Poly. Sci., Part A: Poly. Chem.*, **32**, 1201–1212.
51. Dzhardimalieva, G.I., Pomogailo, A.D., and Volpert, V.A. (2002) Frontal polymerization of metal-containing monomers: a topical review. *J. Inorg. Organomet. Polym.*, **12**, 1–21.
52. Barelko, V.V., Pomogailo, A.D., Dzhardimalieva, G.I., Evstratova, S.I., Rozenberg, A.S., and Uflyand, I.E. (1999) The autowave modes of solid phase polymerization of metal-containing monomers in two- and three-dimensional fiberglass-filled matrices. *Chaos*, **9**, 342–347.
53. Fortenberry, D.I. and Pojman, J.A. (2000) Solvent-free synthesis of acrylamide by frontal polymerization. *J. Polym. Sci., Part A: Polym Chem.*, **38**, 1129–1135.
54. Merzhanov, A.G. and Rumanov, E.N. (1999) Physics of reaction waves. *Rev. Mod. Phys.*, **71**, 1173–1211.
55. Lebrat, J.-P. and Varma, A. (1992) Self-propagating high-temperature synthesis of Ni_3Al. *Combust. Sci. Tech.*, **88**, 211–221.
56. Varma, A. and Lebrat, J.-P. (1992) Combustion synthesis of advanced materials. *Chem. Eng. Sci.*, **47**, 2179–2194.
57. Pojman, J.A., Curtis, G., and Ilyashenko, V.M. (1996) Frontal polymerization in solution. *J. Am. Chem. Soc.*, **118**, 3783–3784.
58. Pojman, J.A., Willis, J., Fortenberry, D., Ilyashenko, V., and Khan, A. (1995) Factors affecting propagating fronts of addition polymerization: velocity, front curvature, temperature profile, conversion and molecular weight distribution. *J. Polym. Sci., Part A: Polym Chem.*, **33**, 643–652.
59. Goldfeder, P.M., Volpert, V.A., Ilyashenko, V.M., Khan, A.M., Pojman, J.A., and Solovyov, S.E. (1997) Mathematical modeling of free-radical polymerization fronts. *J. Phys. Chem. B*, **101**, 3474–3482.
60. Masere, J., Chekanov, Y., Warren, J.R., Stewart, F., Al-Kaysi, R., Rasmussen, J.K., and Pojman, J.A. (2000) Gas-free initiators for high-temperature polymerization. *J. Poly. Sci., Part A: Polym. Chem.*, **38**, 3984–3990.
61. Washington, R.P. and Steinbock, O. (2003) Frontal free-radical polymerization: applications to materials synthesis. *Polym. News*, **28**, 303–310.
62. Tredici, A., Pecchini, R., and Morbidelli, M. (1998) Self-propagating frontal copolymerization. *J. Polym. Sci., Part A: Polym. Chem.*, **36**, 1117–1126.
63. Dixon, G.D. (1980) In depth curing of resins induced by Uv radiation. US Patent 4,222,835.
64. Gregory, S. (2001) Ultraviolet curable resin compositions having enhanced shadow cure properties. US Patent 6,245,827.
65. Pojman, J.A. and McCardle, T.W. (2000) Functionally gradient polymeric materials. US Patent 6,057,406.
66. Pojman, J.A. and McCardle, T.W. (2001) Functionally gradient polymeric materials. US Patent 6,313,237.
67. Pfeil, A., Burgel, T., Morbidelli, M., and Rosell, A. (2003) Mortar composition, curable by frontal polymerization, and a method for fastening tie bars. US Patent 6,533,503.
68. Bürgel, T. and Böck, M. (2004) Mortal composition of at least two components and curable by heat initiation and a method for fastening tie rods,

reinforcing steel for concrete or the like in solid substrates. US Patent 6,815,517.
69. Tu, J., Chen, L., Fang, Y., Wang, C., and Chen, S. (2010) Facile synthesis of amphiphilic gels by frontal free-radical polymerization. *J. Polym. Sci., Part A: Polym. Chem.*, **48**, 823–831.
70. Gavini, E., Mariani, A., Rassu, G., Bidali, S., Spada, G., Bonferoni, M.C., and Giunchedia, P. (2009) Frontal polymerization as a new method for developing drug controlled release systems (Dcrs) based on polyacrylamide. *Eur. Polym. J.*, **45**, 690–699.
71. Lu, G.D., Yan, Q.Z., and Ge, C.C. (2007) Preparation of porous polyacrylamide hydrogels by frontal polymerization. *Polym. Int.*, **56**, 1016–1020.
72. Chekanov, Y.A. and Pojman, J.A. (2000) Preparation of functionally gradient materials via frontal polymerization. *J. Appl. Polym. Sci.*, **78**, 2398–2404.
73. Yan, Q.Z., Lu, G.D., Zhang, W.F., Ma, X.H., and Ge, C.C. (2007) Frontal polymerization synthesis of monolithic macroporous polymers. *Adv. Funct. Mater.*, **17**, 3355–3362.
74. Vicini, S., Mariani, A., Princi, E., Bidali, S., Pincin, S., Fiori, S., Pedemonte, E., and Brunetti, A. (2005) Frontal polymerization of acrylic monomers for the consolidation of stone. *Polym. Adv. Technol.*, **16**, 293–298.
75. Mariani, A., Bidali, S., Cappelletti, P., Caria, G., and Colella, A. (2009) Frontal polymerization as a convenient technique for the consolidation of tuff. *e-Polymers*, **64**, 1–12.
76. Merzhanov, A.G. (1969) The theory of stable homogeneous combustion of condensed substances. *Combust. Flame*, **13**, 143–156.
77. Garbey, M., Taik, A., and Volpert, V. (1998) Influence of natural convection on stability of reaction fronts in liquids. *Quart. Appl. Math.*, **56**, 1–35.
78. Garbey, M., Taik, A., and Volpert, V. (1994) Influence of natural convection on stability of reaction fronts in liquids. *Prepr. CNRS*, **187**, 1–42.
79. Garbey, M., Taik, A., and Volpert, V. (1996) Linear stability analysis of reaction fronts in liquids. *Quart. Appl. Math.*, **54**, 225–247.
80. Bowden, G., Garbey, M., Ilyashenko, V.M., Pojman, J.A., Solovyov, S., Taik, A., and Volpert, V. (1997) The effect of convection on a propagating front with a solid product: comparison of theory and experiments. *J. Phys. Chem. B*, **101**, 678–686.
81. Bazile, M., Nichols, H.A., Pojman, J.A., and Volpert, V. Jr. (2002) The effect of orientation on thermoset frontal polymerization. *J. Polym. Sci., Part A: Polym. Chem.*, **40**, 3504–3508.
82. Pojman, J.A., Gunn, G., Owens, J., and Simmons, C. (1998) Frontal dispersion polymerization. *J. Phys. Chem. Part B*, **102**, 3927–3929.
83. McCaughey, B., Pojman, J.A., Simmons, C., and Volpert, V.A. (1998) The effect of convection on a propagating front with a liquid product: comparison of theory and experiments. *Chaos*, **8**, 520–529.
84. Texier-Picard, R., Pojman, J.A., and Volpert, V.A. (2000) Effect of interfacial tension on propagating polymerization fronts. *Chaos*, **10**, 224–230.
85. Ostrach, S. (1982) Low-gravity fluid flows. *Annu. Rev. Fluid Mech.*, **14**, 313–345.
86. Veretennikov, I., Indeikina, A., and Chang, H. (1998) Front dynamics and fingering of a driven contact line. *J. Fluid Mech.*, **373**, 81–110.
87. Eres, M.H., Schwartz, L.W., and Roy, R.V. (2000) Fingering phenomena for driven coating films. *Phys. Fluids*, **12**, 1278–1295.
88. Leizerson, I., Lipson, S.G., and Lyushnin, A.V. (2003) Finger instability in wetting-dewetting phenomena. *Langmuir*, **20**, 291–294.
89. Maksimov, Y.M., Pak, A.T., Lavrenchuk, G.V., Naiborodenko, Y.S., and Merzhanov, A.G. (1979) Spin combustion of gasless systems. *Combust. Expl. Shock Waves*, **15**, 415–418.
90. Pojman, J.A., Ilyashenko, V.M., and Khan, A.M. (1995) Spin mode instabilities in propagating fronts of polymerization. *Phys. D*, **84**, 260–268.

91. Ilyashenko, V.M. and Pojman, J.A. (1998) Single head spin modes in frontal polymerization. *Chaos*, **8**, 285–287.
92. Solovyov, S.E., Ilyashenko, V.M., and Pojman, J.A. (1997) Numerical modeling of self-propagating fronts of addition polymerization: the role of kinetics on front stability. *Chaos*, **7**, 331–340.
93. Schult, D.A. and Volpert, V.A. (1999) Linear stability analysis of thermal free radical polymerization waves. *Int. J. of SHS*, **8**, 417–440.
94. Spade, C.A. and Volpert, V.A. (2001) Linear stability analysis of nonadiabatic free radical polymerization waves. *Combust. Theory Modell.*, **5**, 21–39.
95. Gross, L.K. and Volpert, V.A. (2003) Weakly nonlinear stability analysis of frontal polymerization. *Stud. Appl. Math.*, **110**, 351–376.
96. Commissiong, D.M.G., Gross, L.K., and Volpert, V.A. (2003) in *Nonlinear Dynamics in Polymeric Systems*, Acs Symposium Series No. 869 (eds J.A. Pojman and Q. Tran-Cong-Miyata), American Chemical Society, Washington, DC, pp. 147–159.
97. Shkadinsky, K.G., Khaikin, B.I., and Merzhanov, A.G. (1971) Propagation of pulsating exothermic reaction front in the condensed phase. *Combust. Expl. Shock Waves*, **1**, 15–22.
98. Masere, J. and Pojman, J.A. (1998) Free radical-scavenging dyes as indicators of frontal polymerization dynamics. *J. Chem. Soc. Faraday Trans.*, **94**, 919–922.
99. Tryson, G.R. and Shultz, A.R. (1979) A calorimetric study of acrylate photopolymerization. *J. Polym. Sci. Polym. Phys. Ed.*, **17**, 2059–2075.
100. Gray, K.N. (1988) Photopolymerization kinetics of multifunctional acrylates. Master's Thesis, University of Southern Mississippi.
101. Masere, J., Stewart, F., Meehan, T., and Pojman, J.A. (1999) Period-doubling behavior in propagating polymerization fronts of multifunctional acrylates. *Chaos*, **9**, 315–322.
102. Manz, B., Masere, J., Pojman, J.A., and Volke, F. (2001) Magnetic resonance imaging of spiral patterns in crosslinked polymer gels produced via frontal polymerization. *J. Poly. Sci., Part A: Polym. Chem.*, **39**, 1075–1080.
103. Pojman, J.A., Masere, J., Petretto, E., Rustici, M., Huh, D.-S., Kim, M.S., and Volpert, V. (2002) The effect of reactor geometry on frontal polymerization spin modes. *Chaos*, **12**, 56–65.
104. Huh, D.S. and Kim, H.S. (2003) Bistability of propagating front with spin-mode in a frontal polymerization of trimethylolpropane triacrylate. *Polym. Int.*, **52**, 1900–1904.
105. Volpert, V.A., Volpert, V.A., and Pojman, J.A. (1994) Effect of thermal expansion on stability of reaction front propagation. *Chem. Eng. Sci.*, **14**, 2385–2388.
106. Pojman, J.A., Popwell, S., Fortenberry, D.I., Volpert, V.A., and Volpert, V.A. (2003) in *Nonlinear Dynamics in Polymeric Systems*, Acs Symposium Series No. 869 (eds J.A. Pojman and Q. Tran-Cong-Miyata), American Chemical Society, Washington, DC, pp. 106–120.
107. McFarland, B., Popwell, S., and Pojman, J.A. (2004) Free-radical frontal polymerization with a microencapsulated initiator. *Macromolecules*, **37**, 6670–6672.
108. McFarland, B., Popwell, S., and Pojman, J.A. (2006) Free-radical frontal polymerization with a microencapsulated initiator: characterization of microcapsules and their effect on pot life, front velocity and mechanical properties. *Macromolecules*, **39**, 53–63.
109. Pojman, J.A., Viner, V., Binici, B., Lavergne, S., Winsper, M., Golovaty, D., and Gross, L. (2007) Snell's law of refraction observed in thermal frontal polymerization. *Chaos*, **17**, 033125.
110. Binici, B., Fortenberry, D.I., Leard, K.C., Molden, M., Olten, N., Popwell, S., and Pojman, J.A. (2006) Spherically propagating thermal polymerization fronts. *J. Polym. Sci., Part A: Polym. Chem.*, **44**, 1387–1395.

5
Isothermal Frontal Polymerization
Lydia L. Lewis and Vladimir A. Volpert

5.1
Introduction

Frontal polymerization (FP) is a class of polymerization reactions that polymerize directionally, usually from one end of their container to another, instead of polymerizing in bulk (i.e., uniform polymerization throughout the container) as traditionally many polymerization reactions do. Because many more people are familiar with thermal frontal polymerization (TFP), we begin this chapter by comparing isothermal frontal polymerization (IFP) to TFP including differences in their mechanisms, reaction properties, and finished products. In addition, we review all previous work on IFP (experimental and mathematical) and present a brief summary of this information.

5.1.1
A Comparison between TFP and IFP: Their Mechanisms and Front Properties

Chapter 4 of this book discusses TFP. TFP is a polymerization reaction in which the solvent is the initiator and monomer (or monomer mixture, i.e., two or more monomers and initiator) and the polymerization reaction occurs directionally in a thermal wave [1–5]. The mechanism is described in the remainder of this paragraph [1–5]. An energy source (e.g., heat or UV light) is applied to one area of the reaction container, and this energy increases the initiator decomposition rate in that region, which in turn increases the polymerization rate in that region. Polymerization reactions are exothermic, and the energy from the area of increased polymerization rate diffuses into the surrounding area, increasing its polymerization rate. Under appropriate conditions, an autocatalytic thermal wave moves through the reaction vessel, leaving behind the polymerized product (Figure 5.1a).

A second category of FP exists, that is, IFP. IFP differs from TFP in that the entire process is contained in an isothermal environment and the mechanism requires a "polymer seed" (a piece of preformed polymer) [7, 8]. Koike discovered IFP and proposed its mechanism in 1988 with additional definitions in 1990 (Figure 5.1b). A reaction vessel contains a polymer seed at the bottom with a

Nonlinear Dynamics with Polymers: Fundamentals, Methods and Applications.
Edited by John A. Pojman and Qui Tran-Cong-Miyata
Copyright © 2010 WILEY-VCH Verlag GmbH & Co. KGaA, Weinheim
ISBN: 978-3-527-32529-0

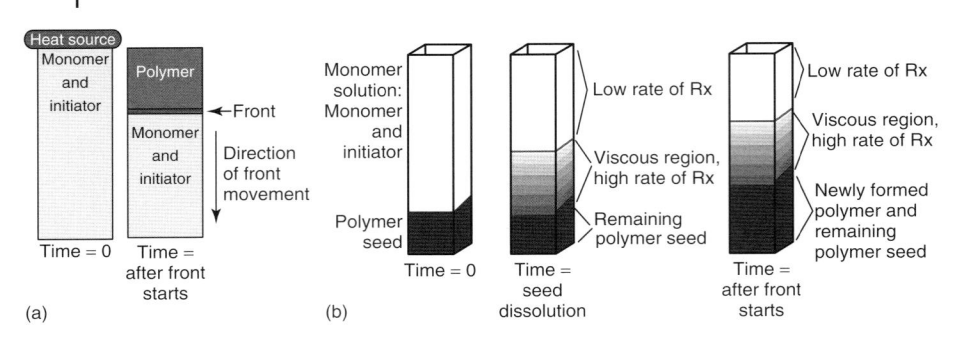

Figure 5.1 (a) A diagram of FP. (b) A diagram of IFP. (Reprinted with permission from Ref. [6]. Copyright 2005 Wiley.)

solution of monomer and thermal initiator (termed *"monomer solution"*) above the seed (Figure 5.1b, time = 0). Over time, the monomer solution diffuses into the polymer seed, swelling the uppermost layer of the seed to form a viscous (or gelatinous) region (Figure 5.1b, time = seed dissolution). Polymerization occurs in both the monomer solution and the viscous region but occurs faster in the viscous region because of the Trommsdorff, or gel, effect. As a new polymer is produced in the viscous region, the process of the monomer solution diffusing into this new polymer region, swelling it, and producing additional new polymer continues. The process goes on autocatalytically (Figure 5.1b, time = after front starts) until the monomer solution polymerizes sufficiently (i.e., high molecular weight chains) so that the monomer solution is no longer able to diffuse into the newly formed polymer [7, 8]. In 1988, Koike termed the process *interfacial-gel polymerization* (IGP) [7]. This term, as well as the terms IFP and nonthermal polymerization, are used throughout the literature.

FP systems must have certain conditions for the front to autocatalytically occur. The necessary and sufficient conditions for TFP include a monomer that will polymerize via free-radical polymerization and a thermal initiator (a photoinitiator may be used to start the reaction, but a thermal initiator is necessary to sustain the reaction) [9, 10]. The necessary and sufficient conditions for IFP include a monomer that will dissolve the polymer seed, polymerize via free-radical polymerization, and exhibit the gel effect; a thermal initiator; and a viscous region in which the gel effect can occur (i.e., the seed dissolving) [6, 11]. Ideally, another necessary IFP condition is a monomer–polymer system that produces an optically clear product because most IFP products are used in optical applications.

The mechanisms of TFP and IFP, as well as the properties of their finished products, define which chemical systems each uses, and, thus, the differences in their polymerization rates. Typically, TFP is faster than IFP [2–6], and, therefore, different free-radical polymerization chemical systems are used for each. A typical TFP system is methacrylic acid (MA) [2–5], whereas a typical IFP system is methyl methacrylate (MMA) [6, 7, 11, 12]. MA has less steric hindrance than MMA, allowing FP to proceed faster: Typical TFP times are on the order of minutes [2–5],

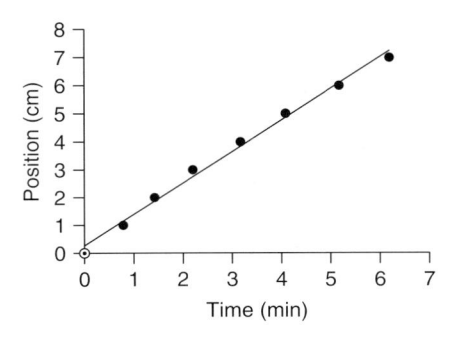

Figure 5.2 Representative plot of position versus time graphs of an FP system [21].

while typical IFP times are on the order of hours [6, 7, 12, 13]. One use for TFP products is coatings, which do not have to be optically clear, whereas IFP products are used for gradient refractive index materials (GRINs), which do require optical clarity [8, 14–19]. Therefore, IFP reactions need slower kinetics so that the newly forming polymer chains form optically clear polymers.

In addition, the necessary and sufficient conditions for both FPs depend on how the experiments are conducted. For TFP, scientists initiate the fronts at the top of the reaction vessel to minimize convection, which could destroy the front [2–5]. As is known so far, IFP takes one of three paths as reviewed by Koike: (i) radially inward (where the reaction vessel is a tube with the seed closest to the outside of the reaction vessel), (ii) vertically up the reaction vessel (where the reaction vessel is a test tube or cuvette with the seed at the bottom of the reaction vessel), or (iii) spherically outward from the center (where the reaction vessel is a sphere with a seed in the middle of the sphere) [20]. Because TFP systems reach a steady-state energy diffusion during their propagation, their velocities are constant (Figure 5.2) [21]. The velocities of IFP systems are not constant (Figure 5.3) as discussed by Lewis et al. While the product forms in the viscous region and propagates from this region, monomer solution of IFP polymerizes at a slower rate. As the monomer solution continues to polymerize, its viscosity increases, and this increased viscosity means that some of the material diffusing into the front has a higher molecular weight. As these higher molecular weight oligomers diffuse into the viscous region, the required time to reach the critical viscosity for the gel effect to occur diminishes. Thus, the front propagates faster [6].

5.1.2
Background

To fully discuss IFP, we discuss the three main IFP categories, the mathematical models, and the experimental systems, and highlight the scientific discoveries within these three main categories. In 1988, Koike first reported a study on IFP and its use to produce GRINs [7], which are materials that exhibit a spatial change in their refractive indices (Figure 5.4). Some uses of GRINs include lenses [8], attenuators [14, 15, 22], and switchable Bragg gratings [17]. In addition, much research has been reported on GRINs for fiber-optic cables for local area networks

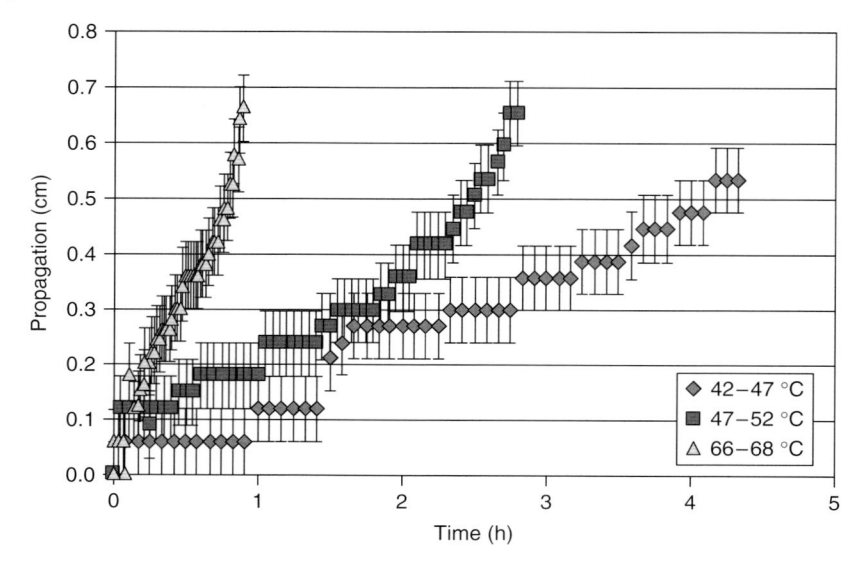

Figure 5.3 Position versus time graphs of the IFP of a system of 0.15% AIBN in MMA at three different temperatures. (Reprinted with permission from Ref. [6]. Copyright 2005 Wiley.)

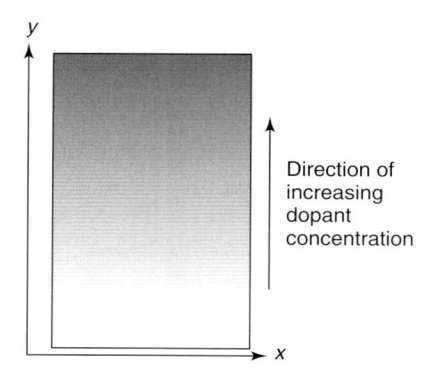

Figure 5.4 Diagram of axial GRIN.

(LANs) [23–26]. Before the use of IFP for the production of GRINs, they were being produced using a slab technique (polymer slabs of differing refractive indices were fused together to create the product) [20, 27, 28]. The advantage that IFP has over the slab technique is that IFP GRINs have a continuous change in the refractive index (RI), whereas the slab GRINs have a stepwise RI change [20, 27, 28]. Koike produced GRINs using the IFP technique by adding a dopant (i.e., a second monomer or dye whose RI is different from the initial monomer) to the monomer solution [7]. Schult and Volpert proposed a theory on how the dopant is incorporated into the propagating front to produce the GRIN. The dopant particles, which are uniformly distributed in the fresh mixture, are redistributed by the propagating polymerization front. The front can absorb only a fraction of the additives it encounters. The remaining fraction is propelled by the front, creating an additive concentration gradient in the bulk and hence in the polymer. The concentration gradient produces the RI gradient in the product [29].

Since Koike's discovery of IFP in 1988, many research groups have studied IFP in three main categories: (i) IFP containing no dopant (which produced a homogeneous RI product) [6, 7, 30–34], (ii) IFP containing a dopant, (which produced a GRIN) [7, 14, 17–20, 22, 23, 28, 29, 35–39], and (iii) IFP containing an inhibitor [12, 13, 34, 40, 41] (both a small-molecule inhibitor that can penetrate the viscous region and a large molecular weight inhibitor that cannot penetrate the viscous region). The development of these categories was driven by the desire to produce new GRIN products, and, thus, the timeline of discoveries does not follow the traditional pattern of least complex chemical system to more complex chemical systems. For discussion purposes, we will present the IFP categories in the order of least complex chemical system to a more complex chemical system.

An IFP system containing no dopant consists of the polymer seed, monomer, and thermal initiator. These systems produce a finished product containing no GRIN and ideally possessing a homogeneous RI. These reactions have primarily been studied to provide experimental evidence in favor of the IFP mechanism [6, 11, 41], to verify the predicted results of changing the experimental parameters within the system [6, 11], and to illustrate that the front is truly isothermal [41].

An IFP system containing a dopant consists of the polymer seed, monomer, thermal initiator, and dopant, at the minimum. These systems produce a finished product containing a GRIN, and many of their uses, as well as the theory of how the GRIN is produced by the incorporation of the dopant into the viscous region, have been stated in a previous paragraph. Koike reported in a review that these GRINs can have one of three shapes: axial (the GRIN is along the y-axis of the reaction vessel), radial (the GRIN is along the radius of the rod), and spherical (the GRIN extends from the outer to inner radius of a sphere [20]. The dopants for GRINs such as optical limiters (which are predominately axial GRINS) are typically dyes, whereas those for GRINS such as fiber-optic cables are typically a second monomer; fiber-optic cables are extruded from radial GRINs.

For any IFP system, the isothermal front stops when the viscosity of the polymerizing monomer solution reaches that of the propagating front [11, 30–32]. At this point, the monomer solution rapidly polymerizes (i.e., bulk polymerizes) because of the Trommsdorff (or gel) effect, and the propagating front is stopped. In 2002, Ivanov *et al.* termed these IFP reactions, where polymerization of the monomer solution halted the front, as *quasi-fronts* [13]. In order to affect the polymerization in the bulk, an inhibitor species can be added to the initial mixture. If the inhibitor is of low molecular weight (e.g., oxygen), it can diffuse into the front, react with the growing polymer chains in the viscous region, and slow (or stop) the front. If the inhibitor is of high molecular weight, it remains in the monomer solution and slows bulk polymerization, letting the front propagate longer. An IFP system containing an inhibitor, at the minimum, consists of the polymer seed, monomer, thermal initiator, and inhibitor. These systems produce a finished product containing a GRIN if a dopant is used and containing no GRIN in the absence of a dopant. These reactions have primarily been studied to determine how far the front can propagate under certain conditions as well as to study the kinetics of the said system [12, 13, 41, 42].

A goal of IFP research is to obtain a model that accurately describes the experimental results to be used as a predictive tool [6, 11, 30–32]. In the next sections, we review the most recent mathematical and experimental IFP research by discussing the mathematical models developed and their characteristics (e.g., properties calculated, trends within the data, and theories involved) and by discussing the experiments performed (e.g., data obtained, trends within the data, and techniques used to obtain the information).

5.2
Mathematical Models

In the simplest model of IFP [30], the basic free-radical polymerization mechanism

1. $I \xrightarrow{k_d} D_0$ (decomposition)
2. $D_n + M \xrightarrow{k_p} D_{n+1}$ (polymerization)
3. $D_n + D_m \xrightarrow{k_t} P$ (termination)

is used. Here I, M, and P denote initiator, monomer, and dead polymer species, respectively, and D_n, $n = 1, 2, \ldots$, denotes a radical containing n monomer molecules (with D_0 being a primary radical) [30].

The mathematical model [30] consists of the kinetic equations that represent mass balances for the reacting species and account for diffusion of the monomer:

$$\frac{\partial [I]}{\partial t} = -k_d [I] \tag{5.1a}$$

$$\frac{\partial [D]}{\partial t} = 2fk_d [I] - 2k_{tg}[D]^2 \tag{5.1b}$$

$$\frac{\partial [M]}{\partial t} = D_M \frac{\partial^2 [M]}{\partial y^2} - k_p [D][M] \tag{5.1c}$$

supplemented by the boundary conditions

$$\frac{\partial [M](0, t)}{\partial y} = 0, \quad \frac{\partial [M](L, t)}{\partial y} = 0 \tag{5.2}$$

and the initial conditions at $t = 0$

$$[M](y, 0) = [M]_0 H(y - y_0) \equiv [M_0] \quad \text{if } y > y_0,$$
$$[M](y, 0) = [M]_0 H(y - y_0) \equiv 0 \quad \text{if } y < y_0 \tag{5.3a}$$

$$[I](y, 0) = [I]_0, \quad [D](y, 0) = 0 \tag{5.3b}$$

Spade and Volpert formulated Eqs (5.1–5.3) for the unknown concentrations [I], [D], and [M] of the corresponding species, which are functions of the spatial variable y, $0 < y < L$, along the axis of the test tube (or cuvette, Figure 5.8). The model is one dimensional, that is, it has been assumed that there is no significant variation of the concentrations in each cross section of the test tube. t is the temporal variable, D_M is the monomer diffusion coefficient, which is assumed to be constant, and H denotes the Heaviside step function. The total radical concentration [D] contains an implied summation over n, that is, $[D] = \sum[D_n]$. The no-flux boundary conditions (5.2) state that the monomer cannot penetrate through the ends of the tube. The initial conditions (5.3) state that there are no radicals in the initial mixture, that the initial concentration of the initiator is a given constant I_0, and that monomer is present in the initial mixture in the region $y_0 < y < L$, with initial concentration M_0, while the remaining part of the mixture $0 < y < y_0$ is filled with a polymer substrate which is free of the monomer [30].

The authors model the quantity k_{tg}. This quantity is the termination reaction rate constant which accounts for the gel effect: as soon as the critical degree of conversion is reached, the gel effect occurs. The termination reaction rate constant in the gel region is much smaller than that in the region where the gel has not yet formed, and, thus, is a function of the degree of conversion $\eta = (M_0 - M)/M_0$. This dependence is modeled by an *ad hoc* function

$$k_{tg} = k_t k(\eta) \equiv k_t \left[\frac{1+\delta}{2} - \frac{1-\delta}{2} \tanh \frac{\alpha(\eta - \eta_{cr})}{2} \right] \tag{5.4}$$

which has a qualitatively correct behavior. Here k_t is the termination rate constant in the absence of the gel effect, η_{cr} is the critical degree of conversion at which the gel effect occurs, δ characterizes the drop in the termination reaction rate constant due to the gel effect, and α controls the width of the layer in η over which the gel effect ensues [30].

The authors explain that a typical numerical solution of the isothermal polymerization problem (5.1–5.4) (Figure 5.5) exhibits the spatial distribution of the degree of conversion η at different times. The growth of the region where $\eta = 1$, that is, the polymerization process came to completion, was observed as the time increased. The solution can be viewed as a slowly varying polymerization wave that propagates to the right. The wave ceases to propagate as the degree of conversion ahead of the front reaches the critical value η_{cr} at which the gel effect occurs. Starting at this instant, the polymerization process at the front is no longer faster than that in the bulk, ahead of the front, and the degree of conversion in the bulk increases rapidly. This event is referred to as the *breakdown of the front* [30]. Solutions in the form shown in Figure 5.5 can be referred to as *quasi-traveling-wave solutions* since their characteristics, such as the state ahead of the front and therefore the propagation velocity, vary slowly in time.

An approximate analytical solution of the problem yielded important characteristics of the process such as the propagation velocity of the front, the propagation time before the front broke down, and the total distance traveled by the front [30]. These characteristics are of interest, in particular, because they can be easily measured in

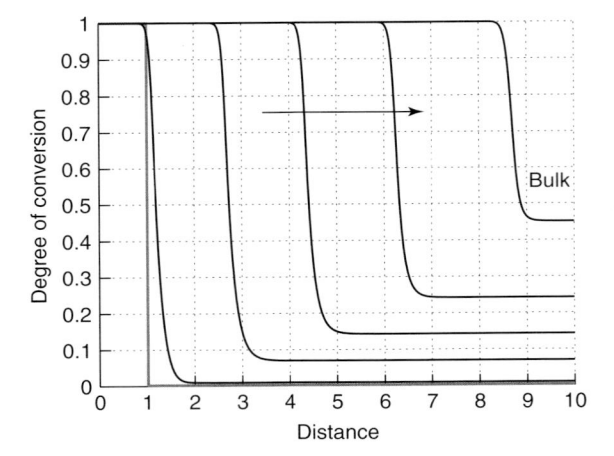

Figure 5.5 Numerical solution of the isothermal polymerization problem in Eqs (5.1–5.4) exhibiting the growth of the polymer region; see text. (Reprinted with permission from Ref. [34]. Copyright 2006 Springer.)

experiments. For example, the total distance d_{max} traveled by the front is

$$d_{max} \approx \frac{2}{\delta^{1/4}} \left[\sqrt{\eta_{cr} + B} - \sqrt{\frac{B}{1 - \eta_{cr}}} \right] y^* \tag{5.5}$$

$$B = \frac{1}{\alpha} \ln \frac{1}{3\delta}$$

where y^* is a spatial scale composed of the reaction rate constants, the initial concentration of the initiator, and the monomer diffusion coefficient, given by

$$y^* = \left(\frac{D_M}{k_p} \right)^{1/2} \left(\frac{2k_t}{k_d I_0} \right)^{1/4}$$

while the other factor on the right-hand side of Eq. (5.5) characterizes the gel effect [30]. The formula clearly demonstrates how the various parameters of the problem affect the total distance.

In order to obtain the analytical results, two simplifying assumptions were made in Ref. [30]. Specifically, a steady-state assumption (SSA) for the radical concentration reduced the mathematical model to a single differential equation for the degree of conversion. In addition, in deriving the expression (5.5) for the total propagation distance of the front, it was assumed that the initiator was far from being completely consumed in the polymerization process. Analysis of the simplified problem yielded results which were in good agreement with numerical results for the full problem for a limited range of parameters [30]. The deviation between the analytical and numerical results in Ref. [30] can be attributed mainly to the use of the SSA in the gel region and to the assumption that initiator decomposition was negligible. In Ref. [31] the limitations brought about by

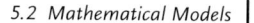

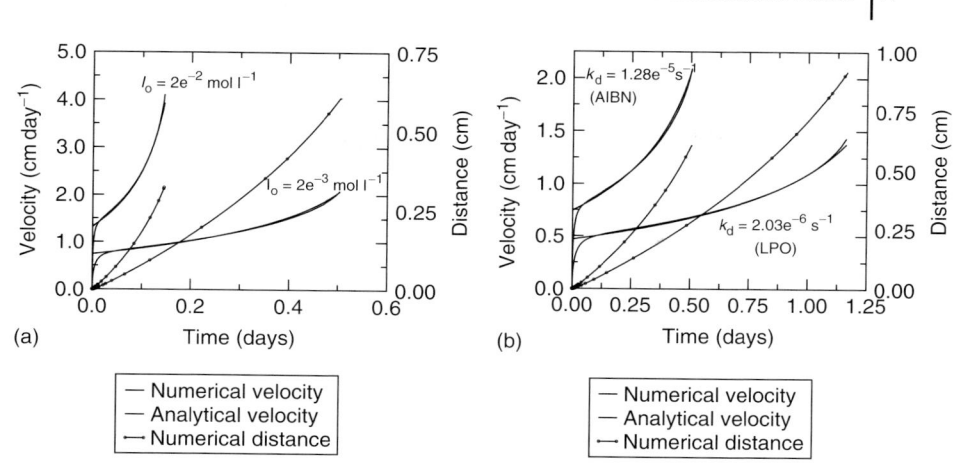

Figure 5.6 Propagation velocity and distance traveled by the front for various parameter values. Here the velocity curves are labeled with the parameter values for each data set, and the corresponding distance curves can be distinguished by matching the end times. (Reprinted with permission from Ref. [31]. Copyright 2000 Wiley.)

these two assumptions were resolved, and improved results were obtained. As an example, Figure 5.6 (see Ref. [31] for many more results) depicts the propagation velocity and distance traveled by the MMA polymerization front before it breaks down due to bulk polymerization.

Further modeling steps were made in Ref. [34]. First, the *ad hoc* function (5.4) that gives only a qualitative description of the gel effect [34] was replaced by a function that was derived from the expression [43] for the termination rate given in terms of self-diffusion rates of polymer radicals, which in turn are found by applying the free volume theory [44, 45]. Secondly, the effect of adding an inhibitor to the mixture was considered [34].

Additional works, both with mathematical models and experimental systems, have generated additions to the mathematical models in Refs. [30] and [31]. As discussed earlier, the breakdown of the propagating front occurs as a result of bulk polymerization [11, 30–32]. In order to affect the polymerization in the bulk, an inhibitor species can be added to the initial mixture. The inhibitor reacts with the growing polymer chains, terminating the polymerization. Thus, the inhibitor slows down bulk polymerization, letting the front propagate longer. On the other hand, if the inhibitor is able to penetrate the front, it affects the reactions at the front and slows down the front. In Ref. [34], the situation in which a small amount of a strong inhibitor that is able to penetrate the front is added to the initial mixture. The main conclusion of this part of the work is that adding a small amount of a strong inhibitor to the system simply delays the process of polymerization by a time interval τ and reduces the initiator concentration. The time τ is the time when the inhibitor has been completely consumed [34].

Though there are quite a few works devoted to either thermal or isothermal polymerization fronts, theoretical studies that address both types of FP occurring in the same system in the course of the same experiment are limited. However, both modes of polymerization wave propagation do occur in the same monomer–initiator system. In fact, attempts to generate an isothermal front sometimes resulted in a thermal front propagation. In Ref. [46], the authors placed an initiator and monomer mixture in a hot thermostatic bath intending to produce an isothermal front at the walls of the test tube that would propagate into the interior of the mixture. Instead, they observed the spontaneous formation of a thermal polymerization front at the axis of the tube and ensuing propagation of this front radially outward toward the walls of the tube [46]. This experiment was modeled in Ref. [47] and it was determined that initiation of a thermal polymerization front occurs (or fails to occur) depending on the test tube radius and the bath temperature [47]. In particular, for a fixed bath temperature, a reaction front is more likely to form in larger test tubes. Isothermal fronts were observed in a similar setting in Ref. [20]. The authors made test tubes composed of a polymer that played the role of the polymer seed [20]. The tube wall swelled and initiated an isothermal front that propagated toward the axis of the tube [20].

A model that encompassed both thermal and isothermal fronts was proposed in Ref. [33]. An experimental setup in which a polymer test tube containing its monomer and initiator mixture placed in a hot thermostatic bath was considered. The wall of the test tube acts as a polymer seed, which suggests that an isothermal front propagating from the wall of the tube toward the center may occur [33]. The front temperature will be that of the bath because the bath will quickly heat the mixture. However, since the reactions are exothermic, there may be a heat buildup that is sufficient to initiate a thermal front. One might expect that the heat buildup occurs at the axis of the tube and that the thermal front will propagate from the axis of the tube toward the wall. However, there are other scenarios as well, as illustrated in Figure 5.7. A parameter sensitivity study that reveals the qualitative trends of the front propagation is presented in Ref. [33]. In particular, the effects of varying the tube radius, the strength of the gel effect, the initiator concentration, and the monomer concentration are discussed [33].

As we have discussed, the practical interest in IFP is due to the use of the polymer product in optical applications [35]. A mathematical problem of the production of GRIN polymers can be regarded as consisting of two parts. The first part is a mathematical description of the polymerization process in which a reaction front propagates converting the monomer into the polymer. This propagation is what we have been discussing so far. The second part of the problem is to understand how the polymerization process affects the distribution of the dopant, which is a subject of investigation in Ref. [29], where a simple model that predicts the dopant distribution in the polymer was considered. The authors proposed a mechanism for propulsion of the dopant by the front. The phenomenon can be attributed to diffusion of the species. Indeed, diffusion of both the monomer and the dopant occurs during the polymerization process. However, the consequences of diffusion for the two species are quite different near the gel region. If the dopant diffuses

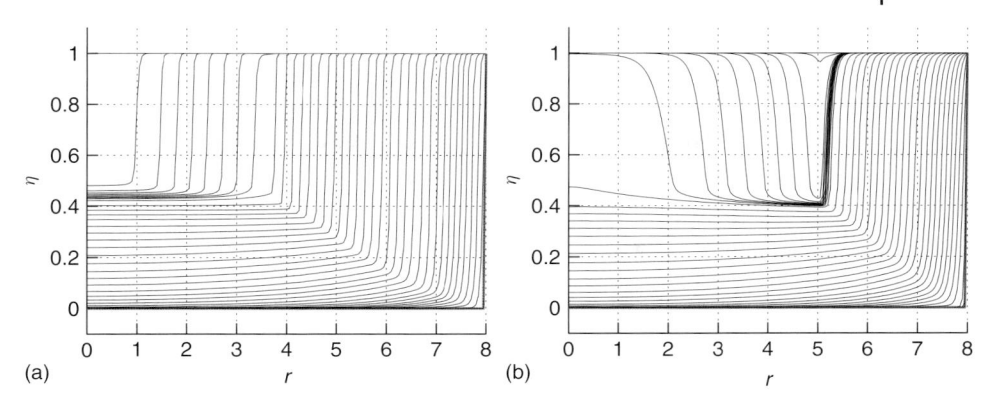

Figure 5.7 Spatiotemporal distribution of the degree of conversion. The front is initiated at the test tube wall ($r = 8$) and moves toward the tube axis ($r = 0$). The profiles are shown over equal time intervals. Densely drawn profiles propagating to the left from the test tube wall toward the axis correspond to the isothermal mechanism of propagation, which is characterized by a much smaller propagation speed than a thermal front. A thermal front is then formed that can propagate either to the left toward the tube axis (a) or to the right from the tube axis (b) depending on problem parameters. (Reprinted with permission from Ref. [33]. Copyright 2006 Elsevier.)

into a vacancy in the gel region, it may either remain there or diffuse back into the fresh mixture. If the monomer diffuses into the gel region, it will likely react there and permanently occupy the vacancy, preventing the dopant from entering the gel. The dopant can diffuse easily into the fresh mixture, but its diffusion into the gel is impeded because of the lack of vacancies. Thus there is a net diffusion flux of the dopant into the fresh mixture. The model leads to the result that the dopant distribution can be controlled by controlling the propagation speed of the polymerization process, for example, by adjusting the thermal conditions of the experiment. The distribution depends on the diffusion coefficient of the dopant in the initial mixture as well as on the segregation coefficient [29].

5.3
Experimental IFP

To our knowledge, the most comprehensive evidence in support of the IFP mechanism was in Lewis *et al.* [6] where the differences among three systems were shown: (i) an IFP system, (ii) an IFP system with a small-molecule inhibitor added so that only diffusion and no front occurred, and (iii) a system with only monomer solution and no seed so that bulk polymerization occurred. The differences in these systems were shown using Wiener's method [48, 49], a laser sheet deflection technique (LLD) that illuminated changes in RI. The experimental system of this study contained a polymer seed and its monomer solution in a cuvette (Figure 5.8a, side view) [6]. To explain the results of this experiment, the authors first explained how LLD illuminated an IFP sample: A low-power laser (about 8 mW) passed

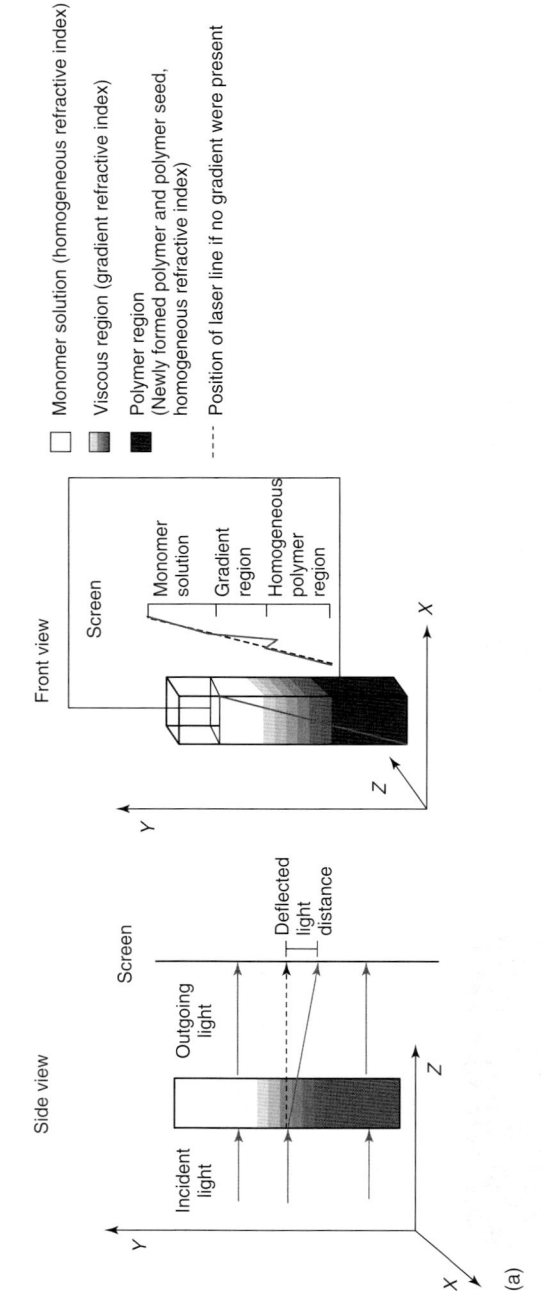

Figure 5.8 (a) Side and front views of Wiener's method (LLD). (b) Raw photo of an IFP system. (c) Plot of |exiting light − incoming light| vs y-axis of the cuvette. (Reprinted with permission from Ref. [6]. Copyright 2005 Wiley.)

5.3 Experimental IFP

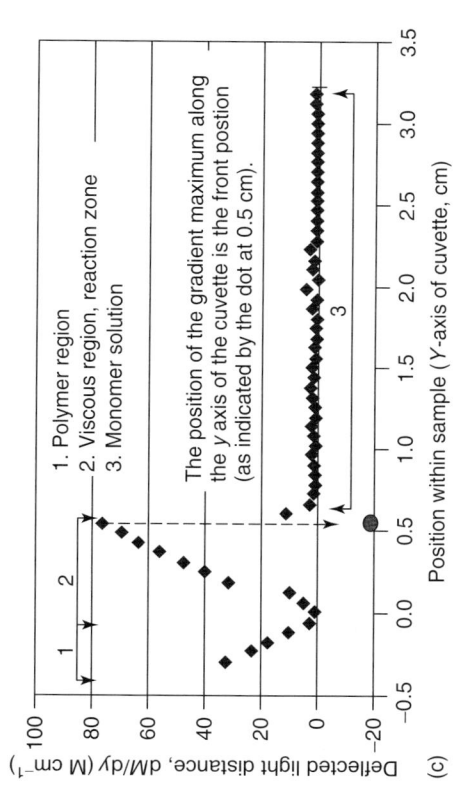

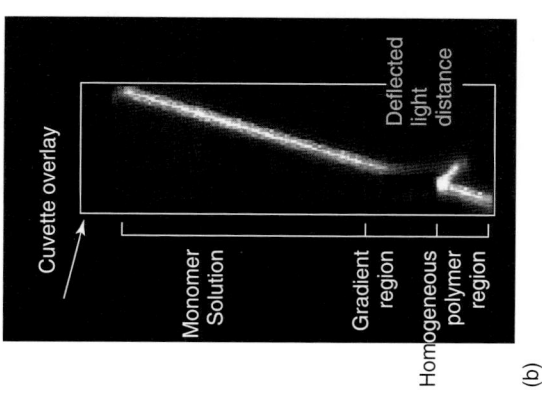

Figure 5.8 (*continued.*)

(b) Cuvette overlay — Monomer Solution, Gradient region, Homogeneous polymer region, Deflected light distance.

(c) Deflected light distance, dM/dy (M cm^{-1}) vs. Position within sample (Y-axis of cuvette, cm).

1. Polymer region
2. Viscous region, reaction zone
3. Monomer solution

The position of the gradient maximum along the y axis of the cuvette is the front position (as indicated by the dot at 0.5 cm).

through the system normal to the sample's front side. Because the monomer solution and polymer solution had homogeneous RIs, the incoming light exited at the same position. The viscous region had a continuous concentration change smaller than the wavelength of light, and, thus, a continuous RI change that caused the entering laser light to deflect in a downward arc and exit in a different position. Because LLD uses a laser sheet on the diagonal of the cuvette, the entire sample is illuminated at the same time, and the viscous region appears as deflected light on the sample screen (Figure 5.8a, front view). For an IFP system, this deflected light illuminates the viscous region as it propagates upward along the y-axis of the cuvette (Figure 5.8a, front view and Figure 5.8b), and the front position is defined as the position of maximum deflected light (Figure 5.8b,c (dot)). This position corresponds to the largest RI change, and front propagation is the movement of this position with respect to time [6].

The use of LLD allowed this reproducible definition of the front position, which enabled the authors to determine whether a system exhibited IFP or not. Figure 5.9 shows representative data for the three previously mentioned systems [6]. System 1: Figure 5.9a shows the IFP system and the light deflection move up the cuvette. This movement indicates the viscous region moving up the cuvette, which is the front propagation. System 2: For an IFP system with 0.03% of the small-molecule inhibitor 2, 2′, 6, 6′-tetramethyl-1-piperidinyloxy (TEMPO), the deflected light showed a viscous region that widened along the y-axis but never propagated (Figure 5.9b). This widening of the deflected light indicated a widening of the viscous region caused by diffusion of the monomer solution downward into the polymer and dissolution of the polymer upward into the monomer solution where the monomer diffused into the polymer seed. System 3: For a system with no seed, the deflected light showed no change (Figure 5.9c) throughout the polymerization. This indicated that the RI of the solution was homogeneous throughout the reaction. A hard polymer product was produced at the end of the experiment. These two facts indicate that the solution polymerized homogeneously [6].

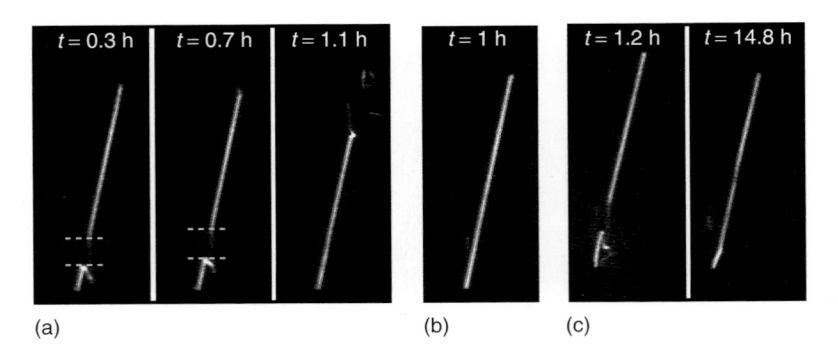

Figure 5.9 Wiener's method (LLD) illuminating an IFP sample (a), a sample with no seed that exhibits homogeneous polymerization (b), and an IFP sample with inhibitor added to prevent IFP (c). (Reprinted with permission from Ref. [6]. Copyright 2005 Wiley.)

The authors used the data from IFP systems illuminated by Wiener's method to generate two types of graphs, a gradient profile (Figure 5.8c, a plot of |amount of deflected light| vs the y-axis of the cuvette at a single point in time during the reaction) and propagation graphs (Figure 5.3, a plot of the maximum change in the |amount of deflected light| vs time to indicate the propagation of the front). From these graphs, the front's velocity, reaction zone width, and reaction zone appearance were determined. In addition, this study determined, or verified, how changes to the experimental variables (e.g., initiator concentration, cure temperature) changed the properties of the front. For example, an increase in the initiator concentration and/or temperature increased the polymerization rate but decreased the propagation time and length. In addition, an increase in initiator concentration and/or temperature caused a measurable decrease in the molecular weight of the newly formed polymer [6].

Evstratova *et al.* reported the verification of the fact that isothermal fronts were truly isothermal. A no-dopant system of MMA and its polymer, poly(methyl methacrylate (PMMA)), was used and monitored visibly with LLD and thermally with three thermocouples. The thermocouples were placed (i) directly above the polymer seed in the monomer solution, (ii) higher in the monomer solution, and (iii) in the thermostatic box in which the reactions were run. The heat produced from the exothermic polymerization diffused into the heated atmosphere of the thermostatic box (Figure 5.10). The constant temperature between the dashed lines indicates the time period of the front. The temperature spike at the later time was due to bulk polymerization of the monomer system [41]. This study also reported the results of the effect of the low molecular weight inhibitor of oxygen on IFP [41]. No-dopant IFP samples were purged with argon, and their propagation was compared to that of unpurged samples (control samples containing oxygen from the air dissolved in the monomer). The results qualitatively agreed with the

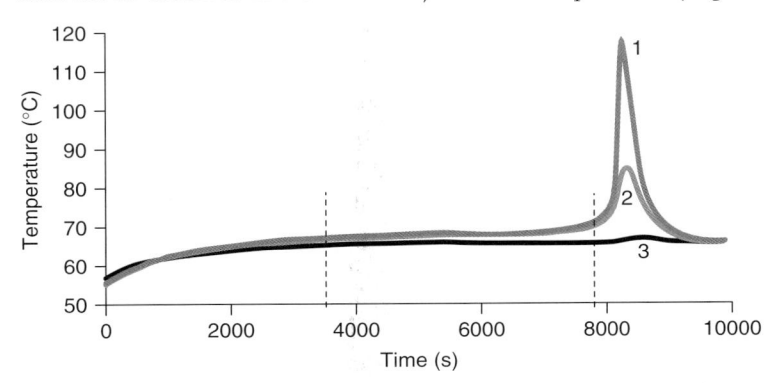

Figure 5.10 Temperature behavior of an IFP system. The dashed lines represent the time front propagation occurred, the peak numbers represent the position of the thermocouples, and the temperature spikes after 8000 s represent bulk polymerization of the monomer solution. (Reprinted with permission from Ref. [41]. Copyright 2006 Wiley.)

mathematical predictions in Ref. [33]. Unpurged samples showed a lag time in front propagation where there was no propagation until the inhibitor was depleted, and then these samples showed normal propagation afterwards [41].

In addition to experimental studies using low molecular weight inhibitors, there are several published studies using polymeric (high molecular weight) inhibitors: All reported work states that the polymeric inhibitor used is a PMMA backbone with TEMPO (or TEMPO-derivative) pendant groups replacing the methyl group at the ester [12, 13]. References [12] and [13] reported the use of polymeric inhibitors with and without dopant systems. Reference [12] used the polymeric inhibitor to extend the propagation length of the front up to 4 cm to study the effect of using cobalt–porphyrin complexes (a chemical that acts as a chain-transfer catalyst) to lower the molecular weight of the frontally produced polymer products. Their study indicated that, when only a hemotoporphyrin compound was used, the molecular weight was higher (on the order of 1×10^6 to 2.5×10^6) than when a cobalt–hematoporphyrin compound was used (on the order of 0.08×10^6 to 0.26×10^6) [12].

Ivanov *et al.* continued studies with no-dopant systems and polymeric inhibitors, and in 2002 they published a study on the effects of the seed composition and how it affected IFP. Various percentages of MMA were mixed with prepolymerized PMMA in a technique he called *priming*. (All studies also added 20 mM of the thermal initiator 2,2′-azobisisobutyronitrile (AIBN) to the seed.) The samples were run under two conditions: (i) an argon environment (where the system was not purged, and, thus, contained only the amount of the small-molecular inhibitor oxygen that was present in the air dissolved in the sample) and (ii) an oxygen-purged environment to introduce more of the small-molecule inhibitor oxygen into the monomer solution on top of the seed. Primed samples compared to unprimed samples (seeds with no MMA or AIBN initially present in them) had shorter times ("induction periods") before the front propagation began. In addition, the oxygen-purged samples indicated that the oxygen was depleted before the front propagation began, indicating longer induction periods than the argon-capped samples. (These longer induction times also provide experimental evidence in favor of a small-molecule inhibitor prolonging the induction period before front propagation began.) One of the goals of their study was to determine under what conditions only bulk polymerization would occur without any IFP when the necessary conditions for IFP were present (e.g., a seed). They discovered that the conditions for a primed seed were (i) "30–70% PMMA" in the seed, (ii) no initiator in the seed, and (iii) the addition of "5–30% PMMA" to the monomer solution on top of the seed. Another important discovery of this study was that the addition of other materials (e.g., "ground glass") to the seed would not hinder the formation of an isothermal front [13].

Other important hallmarks in the history of IFP include IFP systems with dopants. Ref. [15] reports the first use of deuterated and fluorinated MMA derivatives to produce GRINS of lower signal loss. Additionally, Refs. [24], [25] and [50] report the use of metal-doped plastic optical fibers (POFs) to increase attenuation. In addition, Ref. [26] reports the use of a two-step IFP process to produce POFs having a larger diameter than glass optical fibers, which decreases the cost of the LAN. The two-step process produces an "almost ideal RI profile" for GRIN POFs, which is an advantage over GRIN POFs produced by the one-step process [26]. An increase in the attenuation over a longer distance was achieved using this two-step process ("a gigabit data transmission over 300 m") [50]. In addition to using IFP systems with dopants to produce a specified GRIN product, researchers determined that IFP could exist outside traditional cure temperatures. Masere et al. produced IFP GRINs below room temperatures using the initiator tricaprylmethylammonium persulfate [51].

5.4
Comparison of Experimental and Mathematical IFP

One goal of IFP research is to obtain a model that produces the experimental results to use them as a predictive tool (i.e., one can know the initial conditions of the experimental parameters such as the initiator concentration and cure temperature) [11, 6, 30–32]. There have been several coordinated studies of experimental and mathematical IFP including those of no-dopant IFP, IFP with inhibitor, IFP systems that produced TFP under certain conditions, and IFP with dopant [11, 6, 38, 40, 42]. One of the most recent comparisons was a collaborative study of no-dopant IFP to compare propagation distances and gradient profiles for the parameter changes of initiator concentration and of cure temperature [6]. The experimental method used LLD to monitor the fronts and was discussed in previous paragraphs; the mathematical model was also discussed in the previous paragraphs. Here we present a short paraphrase of their results. Nine sets of reaction conditions were studied, and the mathematical results compared well with the experimental data. For an increase in the initiator concentration or cure temperature, the propagation time decreased, the overall behavior of the propagation rate increased, and the propagation distance shortened. Figure 5.11 shows a representative graph of the propagation distance with time. The front propagations obtained from the experimental and mathematical data were on the same order of magnitude for all reaction conditions, and their velocities (as indicated by the slopes of the graphs) increased with time. This increased velocity was explained in a previous paragraph by the monomer solution's increasing viscosity. In addition, gradient profiles were generated both experimentally and mathematically for various times throughout the sample. Figure 5.12 shows a typical comparison of various times throughout one front (e.g., 1.75, 4.50, and 6.00 h) for the reaction conditions of 0.03% AIBN in MMA at 47–52 °C cure temperature. From the point of reference of the gradient maximum (the front position), the right-hand sides of the gradient profiles indicate the untangling of polymer chains into the monomer solution, and the left-hand sides of the gradient profiles indicate monomer penetration into the seed. Both right- and left-hand sides indicate the change in RI for their respective regions. Both experimental and mathematical profiles show sharp changes in RI on the right side of the gradient maxima. The sharpness on the right side indicates

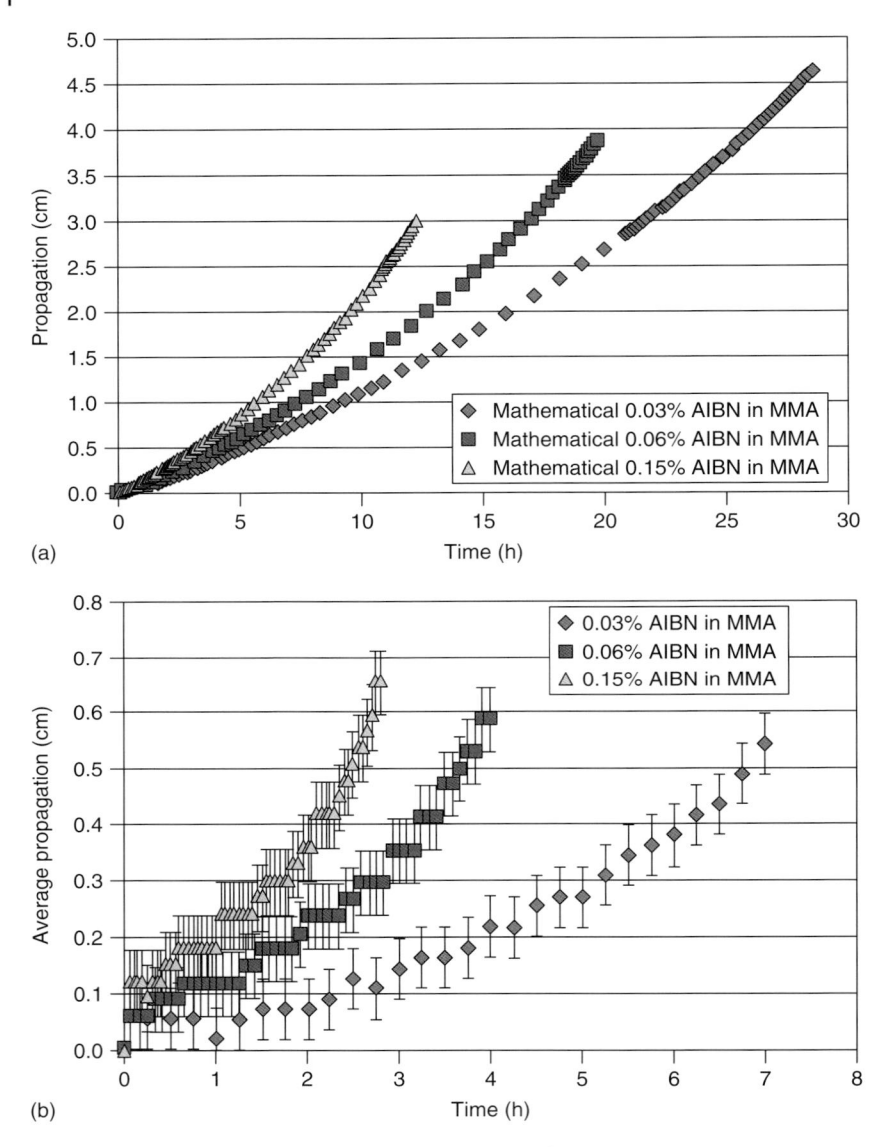

Figure 5.11 IFP front propagation with respect to time for the conditions of 0.06% AIBN in MMA for 47 – 52 °C cure temperature, theoretical (a) and experimental (b).

that the front propagation position is almost directly adjacent to the monomer solution and that there is little polymer untangling into the monomer solution. Both experimental and mathematical profiles show a more relaxed curve on the left side of the gradient maxima, indicating that the viscous region is behind the moving front and that the reaction zone is not infinitely thin. In addition, both

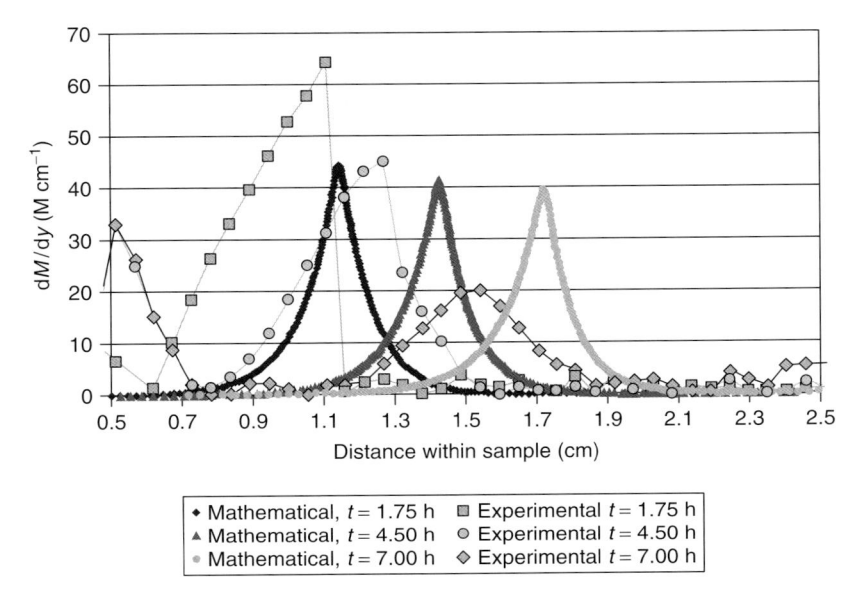

Figure 5.12 Experimental and modeled gradient profiles for 0.03% AIBN in MMA for various times (1.75, 4.50, and 6.00 h) for the cure temperature of 47 – 52 °C (The experimental and modeled gradient maxima were scaled (i.e., the gradient maxima of $t = 0$ was at the distance (x-axis) = 1.0 cm) for comparison purposes.) (Reprinted with permission from Ref. [6]. Copyright 2005 Wiley.)

experimental and mathematical profiles indicate a decrease in the overall change in RI with respect to time as indicated by the decreasing magnitude of the gradient profile. This decrease correlates with the polymerizing monomer solution, which has an increasing RI. Thus, the overall change in RI, as indicated by the height of the deflected light, decreases with time [6].

5.5 Conclusions

IFP has been studied both experimentally and mathematically in three categories of no-dopant IFP, dopant-IFP, and inhibitor-IFP. The two main emphases of IFP studies are to produce better GRINs and to produce a predictive tool (i.e., a mathematical model) to determine the ideal conditions to use to produce a desired material [11, 6, 30–32]. Experimental and mathematical work have been conducted from two aspects of the problem: (i) product-driven to produce better optical products and (ii) studies to better understand the underlying kinetics as well as the effects that the change in parameters (i.e., initiator concentration, cure temperature, seed composition, amount of small or large molecular weight inhibitor, and dopant type and concentration) have.

Acknowledgments

The authors would like to thank the Millsaps College Julian and Kathryn Wiener Medical-Mentoring Fund as well as their research students and colleagues.

References

1. Chechilo, N.M., Khvilivitskii, R.J., and Enikolopyan, N.S. (1972) On the Phenomenon of Polymerization Reaction Spreading. *Dokl. Akad. Nauk SSSR*, **204**, 1180–1181.
2. Davtyan, S.P., Zhirkov, P.V., and Vol'fson, S.A. (1984) Problems of non-isothermal character in polymerisation processes. *Russ. Chem. Rev. (Engl. Transl.)*, **53**, 150–163.
3. Pojman, J.A. (1991) Traveling fronts of methacrylic acid polymerization. *J. Am. Chem. Soc.*, **113**, 6284–6286.
4. Khan, A.M. and Pojman, J.A. (1996) The use of frontal polymerization in polymer synthesis. *Trends Polym. Sci.*, **4**, 253–257.
5. Pojman, J.A., Ilyashenko, V.M., and Khan, A.M. (1996) Free-radical frontal polymerization: self-propagating thermal reaction waves. *J. Chem. Soc., Faraday Trans.*, **92**, 2824–2836.
6. Lewis, L.L. (2003) The development and characterization of an optical monitoring technique and a mathematical algorithm for isothermal frontal polymerization. Dissertation, The University of Southern Mississippi.
7. Lewis, L.L., DeBisschop, C.S., Pojman, J.A., and Volpert, V.A. (2005) Isothermal frontal polymerization: confirmation of the mechanism and determination of factors affecting front velocity, front shape, and propagation distance with comparison to mathematical modeling. *J. Polym. Sci., Part A: Polym. Chem.*, **43**, 5774–5786.
8. Koike, Y., Takezawa, Y., and Ohtsuka, Y. (1988) New interfacial-Gel Copolymerization technique for steric grin polymer optical waveguides and lens arrays. *Appl. Opt.*, **27**, 486–491.
9. Koike, Y. and Ohtsuka, Y. (1990) Low-loss Gi plastic optical fiber and novel optical polymers. *Mat. Res. Soc. Symp. Proc.*, **172**, 247–252.
10. Pojman, J.A., Popwell, S., Fortenberry, D.I., and Volpert, V.A. (2004) in *Nonlinear Dynamics in Polymeric Systems*, ACS Symposium Series No. 869 (eds J.A. Pojman and Q. Tran-Con-Miyata), American Chemical Society, Washington, DC, pp. 106–120.
11. Nason, C., Roper, T., Hoyle, C., and Pojman, J.A. (2005) Uv-induced frontal polymerization of multifunctional (Meth)acrylates. *Macromolecules*, **38**, 5506–5512.
12. Ivanov, V.V., Stegno, E.V., and Pushchaeva, L.M. (1997) Non-thermal frontal polymerization of methyl methacrylate with a polymeric seed modified with cobalt-porphyrin complex additives. *Chem. Phys. Rep.*, **16**, 947–951.
13. Ivanov, V.V., Stegno, E.V., Mel'nikov, V.P., and Puschchaeva, L.M. (2002) Influence of reaction conditions on the frontal polymerization of methyl methacrylate. *Polym. Sci. Ser. A*, **44**, 1017–1022.
14. Tagaya, A., Koike, Y., Nihei, E., Teramoto, S., Fujii, K., Yamamoto, T., and Sasaki, K. (1995) Basic performance of an organic dye-doped polymer optical fiber amplifier. *Appl. Opt.*, **34**, 988–992.
15. Nihei, E., Ishigure, T., and Koike, Y. (1996) High-bandwidth, graded-index polymer optical fiber for near-infrared use. *Appl. Opt.*, **35**, 7085–7090.
16. Tagaya, A., Kobayashi, T., Nakatsuka, S., Nihei, E., Sasaki, K., and Koike, Y. (1997) High gain and high power organic dye-doped polymer optical fiber amplifiers: absorption and emission cross sections and gain characteristics. *Jap. Journal of Appl. Phys.*, **36**, 2705–2708.

17. Boxel, R.V., Verbiest, T., and Persoons, A. (2004) in *Optical Fibers and Passive Components* (ed. S. Shen), The International Society for Optical Engineering, Bellingham, Washington, DC, pp. 77–84.
18. Park, J.-H., Choi, W.S., Koo, H.Y., and Kim, D.-Y. (2005) Colloidal photonic crystal with graded refractive-index distribution. *Adv. Mater.*, **17**, 879–885.
19. Kailasnath, M., Sreejaya, T.S., Kumar, R., Vallabhan, C.P.G., Nampoori, V.P.N., and Radhakrishnan, P. (2008) Fluorescence characterization and gain studies on a dye-doped graded index polymer optical-fiber preform. *Opt. Laser Tech.*, **40**, 687–691.
20. Koike, Y. (1992) in *Polymers for Lightwave and Integrated Optics: Technology and Applications* (ed. L.A. Hornak), Marcel Dekker, Inc., Morgantown, WV, pp. 71–102.
21. Pojman, J.A., West, W.W., and Simmons, J. (1997) Propagating fronts of polymerization in the physical chemistry laboratory. *J. Chem. Educ.*, **74**, 727–730.
22. Tagaya, A., Teramoto, S., Nihei, E., Sasaki, K., and Koike, Y. (1997) High-power and high-gain organic dye-doped polymer optical fiber amplifiers: novel techniques for preparation and spectral investigation. *Appl. Opt.*, **36**, 572–578.
23. Koike, Y., Nihei, E., Tanio, N., and Ohtsuka, Y. (1990) Graded-index plastic opical fiber composed of methyl methacrylate and vinyl phenylacetate copolymers. *Appl. Opt.*, **29**, 2686–2691.
24. Zhang, Q., Wang, P., and Zhai, Y. (1998) Preparation of graded index plastic rods doped with Nd3+ by Interfacial-Gel polymerization. *J. Appl. Polym. Sci.*, **67**, 1431–1436.
25. Kuriki, K., Kobayashi, T., Imai, N., Tamura, T., Nishihara, S., Tagaya, A., Koike, Y., and Okamoto, Y. (2000) Fabrication and properties of polymer optical fibers containing Nd-chelate. *Photonics Tech. Lett.*, **12**, 989–991.
26. Ishigure, T., Tanaka, S., Kobayashi, E., and Koike, Y. (2002) Accurate refractive index profiling in a GRAded-Index plastic optical fiber exceeding gigabit transmission rates. *J. Lightwave Technol.*, **20**, 1449–1456.
27. Koike, Y. (1992) Graded-Index and single-mode polymer optical fibers. *Mat. Res. Soc. Symp. Proc.*, **247**, 817–828.
28. Ishigure, T., Nihei, E., and Koike, Y. (1994) Graded-Index polymer optical fiber for high-speed data communication. *Appl. Opt.*, **33**, 4261–4266.
29. Schult, D.A., Spade, C.A., and Volpert, V.A. (2002) Dopant distribution in isothermal frontal polymerization. *Appl. Math. Lett.*, **15**, 749–754.
30. Spade, C.A. and Volpert, V.A. (1999) Mathematical modeling of interfacial gel polymerization. *Math. Comput. Model.*, **30**, 67–73.
31. Spade, C.A. and Volpert, V.A. (2000) Mathematical modeling of interfacial gel polymerization for weak and strong gel effects. *Macromol. Theory Simul.*, **9**, 26–46.
32. Lewis, L.L., DeBisschop, C.A., Volpert, V.A., and Pojman, J.A. (2004) in *Nonlinear Dynamics in Polymeric Systems*, Acs Symposium Series No. 869 (eds J.A. Pojman and Q. Tran-Con-Miyata), American Chemical Society, Washington, DC, pp. 169–183.
33. Devadoss, D.E. and Volpert, V.A. (2006) Mathematical modeling of radially propagating polymerization waves with the gel effect. *Appl. Math. Comput.*, **172**, 1036–1053.
34. Devadoss, D.E. and Volpert, V.A. (2006) Modeling isothermal free-radical frontal polymerization with gel effect using free volume theory, with and without inhibition. *J. Math. Chem.*, **39**, 73–105.
35. Ishigure, T., Horibe, A., Nihei, E., and Koike, Y. (1994) High bandwidth and high numerical aperture graded-index polymer optical fibre. *Electron. Lett.*, **30**, 1169–1171.
36. Ishigure, T., Horibe, A., Nihei, E., and Koike, Y. (1994) Low-loss high-bandwidth Gi polymer optical fiber. Advanced Materials '93 Proceedings of the Symposia, Vol. 15A, pp. 181–184.
37. Yang, S.Y., Chang, Y.H., Ho, B.C., Chen, W.C., and Tseng, T.W. (1995) Studies on the preparation of gradient-index polymeric rods by

Interfacial-gel copolymerization. *J. Appl. Polym. Sci.*, **56**, 1179–1182.

38. Zhang, Q., Want, P., and Zhai, Y. (1997) Refractive index distribution of graded index poly(Methyl Methacrylate) preform made by Interfacial-gel polymerization. *Macromolecules*, **30**, 7874–7879.

39. Ishigure, T., Aruga, Y., and Koike, Y. (2007) High-bandwidth Pvdf-Clad Gi Pof with ultra-low bending loss. *J. Lightwave Technol.*, **25**, 335–345.

40. Smirnov, B.R., Min'ko, S.S., Lusinov, I.A., Sidorenko, A.A., Stegno, E.V., and Ivanov, V.V. (1993) Frontal radical polymerization in the presence of a polymeric inhibitor. *Polym. Sci.*, **35**, 423.

41. Evstratova, S.I., Antrim, D., Fillingane, C., and Pojman, J.A. (2006) Isothermal frontal polymerization: confirmation of the isothermal nature of the process and the effect of oxygen and polymer seed molecular weight on front propagation. *J. Polym. Sci. A*, **44**, 3601–3608.

42. Ivanov, V.V. and Stegno, E.V. (1995) A model of frontal radical polymerization in the presence of polymeric inhibitor. *Polym. Sci. B*, **37**, 50–52.

43. Chekal, B.P. and Torkelson, J.M. (2002) Relationship between chain length and the concentration dependence of polymer and oligomer self-diffusion in solution. *Macromolecules*, **35**, 8126–8138.

44. Vrentas, J.S. and Druda, J.L. (1977) Diffusion in polymer-solvent systems: 1) Re-examination of free-volume theory. *J. Polym. Sci. Polym. Phys.*, **15**, 403–416.

45. Vrentas, J.S. and Druda, J.L. (1977) Diffusion in polymer-solvent systems: 2) predicative theory for dependence of diffusion-coefficients on temperature, concentration, and molecular-weight. *J. Polym. Sci. Polym. Phys.*, **15**, 417–439.

46. Asakura, K., Nihei, E., Harasawa, H., Ikumo, A., and Osanai, S. (2004) Spontaneous frontal polymerization: propagating front spontaneously generated by locally autoaccelerated fee-radical polymerization. *Nonlinear Dyn. Polym. Syst.*, **869**, 135–146.

47. Ritter, L.R., Olmstead, W.E., and Volpert, V.A. (2003) Initiation of free-radical polymerization waves. *SIAM J. Appl. Math.*, **53**, 1831–1848.

48. Sommerfeld, A. (1954) *Optics: Lectures of Theoretical Physics*, Vol. **Iv**, Academic Press, New York.

49. Petitjeans, P. and Maxworthy, T. (1996) Miscible displacements in capillary tubes Part 1: experiments. *J. Fluid Mech.*, **326**, 37–56.

50. Kondo, A., Ishigure, T., and Koike, Y. (2005) Fabrication process and optical properties of perdeuterated graded-index polymer optical fiber. *J. Lightwave Technol.*, **23**, 2443–2448.

51. Masere, J., Lewis, L.L., and Pojman, J.A. (2001) Optical gradient materials produced via low-temperature isothermal frontal polymerization. *J. Appl. Polym. Sci.*, **80**, 686–691.

6
Reaction-Induced Phase Separation of Polymeric Systems under Stationary Nonequilibrium Conditions

Hideyuki Nakanishi, Daisuke Fujiki, Dan-Thuy Van-Pham, and Qui Tran-Cong-Miyata

6.1
Introduction

For most cases, materials processing is carried out under various conditions far from equilibrium where the temperature, concentration, or pressure is not constant, but is, in general, a function of time and space. Under such stationary nonequilibrium conditions, coupling among the different variables such as diffusion, concentration, temperature, density, or surface tension distribution can lead to a wide variety of instabilities such as double diffusion, Mullins–Sekerka, Rayleigh–Benard, Rayleigh–Taylor, and so on [1–3]. As a consequence, for a physicochemical system undergoing phase separation far from thermodynamic equilibrium, one can expect the emergence of various exotic morphologies whereby functional materials can be designed and utilized.

In this chapter, we will show that, by using ultraviolet (UV) irradiation to induce and maintain phase separation under stationary nonequilibrium conditions, morphologies that cannot be generated under thermal equilibrium conditions can be produced and controlled by taking advantage of the competition between phase separation and photochemical reactions. At first, theories and numerical calculation of phase separation kinetics in both nonreacting and reacting systems are briefly overviewed, together with the experimental results on reaction kinetics in polymeric mixtures. Here, the nonuniform kinetics observed for mixtures in both the bulk and liquid states containing monomers is summarized. Emphasis is particularly given for the autocatalytic behavior of the reaction and its consequence on the phase separation kinetics and the resulting morphology. By taking advantage of the gradient of light intensity in an irradiated sample, morphology with spatially graded structures is constructed and controlled. Emergence of an elastic strain field accompanying the cross-linking process in polymer mixtures is monitored and analyzed by using Mach–Zehnder interferometry. The development and relaxation of this reaction-induced strain field are described, together with its influence on the morphology of the reacting polymer mixtures. Finally, results on the temporal modulation experiments of phase separation are briefly summarized. The conclusions are provided at the end of this chapter, together with some

Nonlinear Dynamics with Polymers: Fundamentals, Methods and Applications.
Edited by John A. Pojman and Qui Tran-Cong-Miyata
Copyright © 2010 WILEY-VCH Verlag GmbH & Co. KGaA, Weinheim
ISBN: 978-3-527-32529-0

6.2
Overview of Theoretical Studies on Phase Separation Kinetics of Nonreactive and Reactive Binary Mixtures

6.2.1
Phase Separation of Nonreacting Mixtures

As in the case of metallic alloys, phase separation of polymer mixtures occurs via two different pathways depending on thermodynamic conditions: nucleation and growth and spinodal decomposition [4–6]. The former process occurs inside the metastable region where droplets with saturated composition are generated in the early stage and stabilized if their size is larger than a threshold value determined by the competition between their volume and surface energies. At a late stage, the growth of these droplets is governed by the $t^{1/3}$ power law derived by Liftshitz and Slyosov [7] and also, simultaneously, by Wagner [8]. In this process, the larger nuclei grow at the expense of the smaller ones via the so-called Ostwald ripening mechanism. On the other hand, in the spinodal decomposition process, concentration fluctuations with much smaller amplitudes first appear and start growing with time. Composition fluctuations take place with a finite range of wavelengths that become unstable and turn into periodic structures. These structures, known as the *spinodal structures*, coarsen with time and eventually approach the equilibrium composition. The relation between the thermodynamics of a binary mixture with a lower critical solution temperature (LCST) and the morphologies resulting from these two kinetic processes is schematically illustrated in Figure 6.1. Similar to other pattern-forming processes, theoretical analyses performed for the kinetics in the early stage of phase separation rely on linear stability analysis techniques [9]. At later stages, the wavelengths and the amplitudes of these periodic structures increase with time, and linearized theory is, therefore, no longer valid. Instead, it is replaced by other methods such as scaling analysis, except in some rare cases where analytical theory using nonlinear mode-coupling calculation has been applied and yielded qualitative agreements with the experimental data [10].

For polymer mixtures, the first linearized theory was formulated by de Gennes [12] and accomplished by Pincus [13]. In the later stage, the wavelengths of these structures increase with their amplitudes. As in the case of systems of small molecules, their time evolution is no longer expressible by the linearized theory. Instead, the scaling hypothesis has been proposed to describe the universal behavior of phase separation in the late stage, and has been successful in explaining the experimental data [4–6, 14]. Finally, the morphology becomes coarser and the mixture reaches phase equilibrium with random, two-phase structures. Most

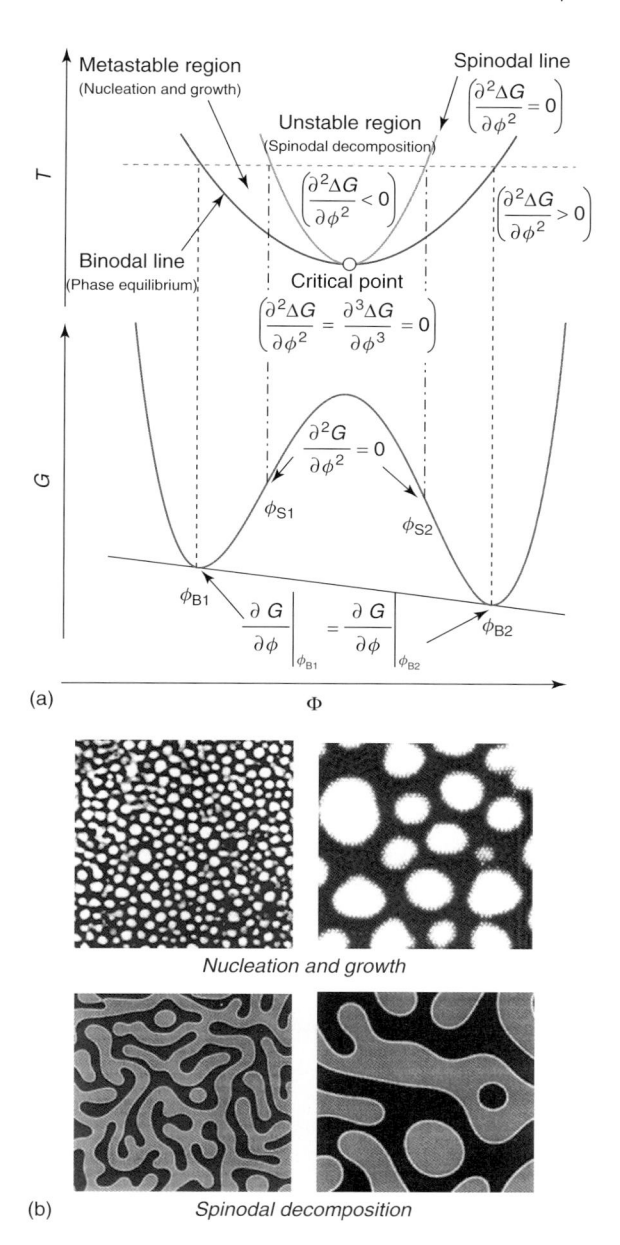

Figure 6.1 (a) Schematic presentation of the phase diagram and the corresponding Gibbs free energy for a binary mixture with a lower critical solution temperature (LCST). (b) Evolution of phase separation in a binary mixture as time progresses from left to right. (Upper) nucleation and growth (Lower) spinodal decomposition process [11].

theories developed for nonreactive mixtures have been carried out in the framework of systems with conserved order parameters using Model B in the classification of Hohenberg and Halperin [15].

From the application side, chemical reactions such as curing and/or transesterification [16] are frequently utilized for practical purposes such as hardening or compatibilization of multicomponent polymers. Among them, the reversible reaction is particularly interesting because of its potential for a new approach to polymer recycling. From the aspects of fundamental science, reaction-induced phase separation is strongly related to a mode-selection process where chemical reaction plays the role of a "mode-selector" for concentration fluctuations. This has been the strong motivation for a number of studies on reaction-induced phase separation with the attempt to regulate the symmetry or the characteristic length scales of morphology by which physical properties of the materials can be controlled [17].

6.2.2
Phase Separation of Reacting Mixtures

Theoretical studies on phase separation of chemically reacting polymer mixtures are currently still far from satisfactory because of the great complexity of the problem. First of all, chemical reactions do not proceed homogeneously in polymeric media; that is, they cannot be described by the mean-field kinetics, particularly in the bulk state. This feature of the reaction kinetics originates from the inhomogeneity persisting over the diffusion length scales required for the reaction to occur, the so-called "free volume distribution" or dynamic heterogeneity. Because relaxation of polymers is extremely slow, particularly in cross-linking systems, this inhomogeneity remains in the sample as a residual strain field that can feedback on the reaction and subsequently influence its kinetics. In general, this reaction-induced strain field participates in the phase-separation process as a long-range interaction, modifying the instability of the reacting mixture. Another important factor is molecular weight, that is, the size of the polymer molecule, which changes during the reaction, leading to a transient variation of the phase diagram and the mobility of the system. As a consequence, the mixture continues to move away from thermodynamic equilibrium during the reaction, whereas the mobility gradually decreases with time in the case of cross-linking reaction, until all the reactants are consumed, as schematically illustrated in Figure 6.2.

The first attempt to investigate the instability arising from a mixture of small (nonpolymeric) molecules undergoing phase separation induced by an autocatalytic reaction was made by Huberman in 1976 [18]. Two decades later, inspired by the regularity emerging from the competitions between antagonistic interactions in block copolymers, Glotzer and coworkers rediscovered the reaction-induced stabilization of long-wavelength fluctuations in an A/B mixture undergoing the reversible reaction $A \rightleftarrows B$. By linear stability analysis, these authors predicted that, in the presence of this reversible reaction, phase separation of an A/B mixture could proceed only to a certain extent and is eventually frozen because of the stabilization of long-wavelength fluctuations, as schematically illustrated in Figure 6.3a [11].

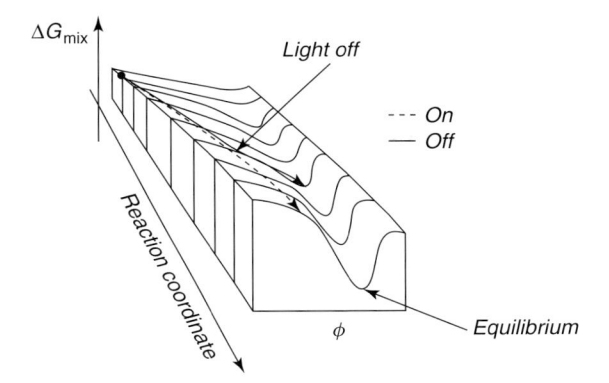

Figure 6.2 Variation of the free energy with chemical reaction for a reacting mixture.

This suppression also leads to an increase in the morphological regularity for a reacting mixture, as experimentally demonstrated by a sharp ring in Fourier space in Figure 6.3b. Here, phase separation of a polystyrene/poly(vinyl methyl ether) (PS/PVME) (20/80) mixture was driven by the trans–cis photoisomerization of stilbene moieties chemically labeled on the PS chains [19]. Since then, a number of similar studies have been carried out to elucidate the roles of reversible reactions in the phase separation of binary mixtures [20–25] and also for ternary mixtures with more complicated kinetic processes [26]. Recently, it was proposed that hierarchical and coherent morphologies could also be produced by dynamically coupling reversible reactions to phase separation [27]. Also, an alternative approach has been followed for the analysis of the couplings between reaction and phase separation at the surface of a catalyst [28]. However, from the macromolecular viewpoint, the reaction/diffusion formalism described above does not take into account several important polymer effects such as the feedback of the chemical reactions on the free energy of the mixture and vice versa [29–31], in particular, the reaction-induced elasticity, as shown later. Therefore, these theories are still far from the stage of quantitative comparison with the experimental data obtained for polymer systems [32].

From the macromolecular viewpoint, there are two important effects of chemical reactions on the critical phenomena of polymeric systems. The first is the feedback from chemical reactions on the thermodynamics such as entropy and enthalpy because of the change in the molecular weight as well as in the chemical structure of the polymer chains. This local effect can be confirmed by monitoring the change in the phase diagram during the reaction by means of an appropriate method such as light scattering. The second issue is the coupling of the reaction kinetics to the global properties of the system such as elasticity and/or viscoelasticity. This long-range effect can be experimentally verified by using Mach–Zehnder interferometry as described below. On the other hand, the reacting mixture is, in general, a nonconserved system because the composition of the reacting components keeps changing with reaction time, leading to a change in the free energy, as conceptually illustrated in Figure 6.2.

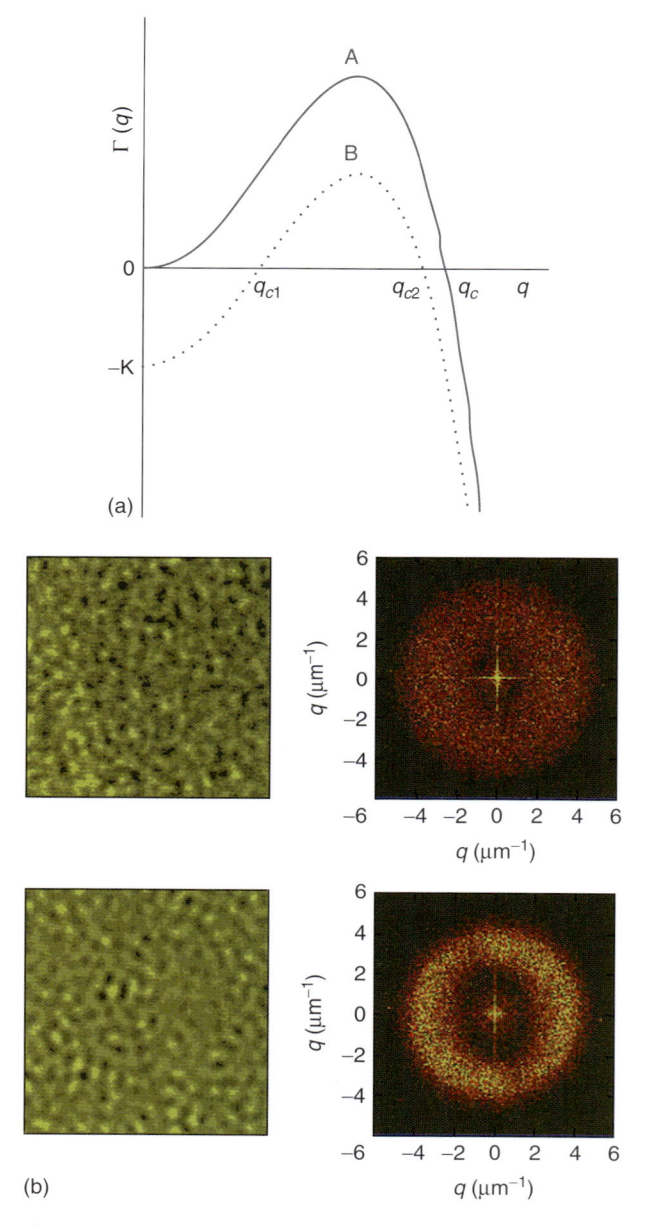

Figure 6.3 (a) Dispersion relation analytically obtained for phase separation with reversible reaction, indicating the suppression of long-wavelength modes in the region $0 \leqslant q \leqslant q_{c1}$. $\Gamma(q)$ is the growth rate of unstable modes (concentration fluctuations) [11]. (b) Morphology induced by photoisomerization of the trans-stilbene moieties labeled on polystyrene component in a polystyrene/poly(vinyl methyl ether) blend: (Upper left), phase-contrast optical micrograph obtained by heating; (Upper right), the corresponding 2D-FFT power spectra. (Lower left), morphology induced by the photoisomerization; (Lower right), the corresponding 2D-FFT power spectra [19].

The local strain (or deformation) arises from the coupling between the reaction kinetics and the inherent inhomogeneity of polymeric systems in the bulk state, known as *free-volume distribution* [33–35] or *dynamic inhomogeneity* [36, 37]. The effects of chemical reaction on the thermodynamics of a binary mixture in non-polymeric systems were rigorously tackled by Lefever and coworkers [38]. They found that, depending upon the competition between chemical reactions and phase separation, there exist four possible cases where one reproduces the prediction of Glotzer and other research groups on the reaction-induced stabilization of long-wavelength fluctuations. Interestingly, their calculation also leads to a special case where the reaction can destabilize the mixtures leading to a Turing-like instability [39] *inside* the thermodynamically stable region of the mixture. Nevertheless, this particular prediction has not yet been experimentally verified.

For nonreacting polymer mixtures, several efforts have been made to elucidate the effects of viscoelasticity on phase separation of polymer mixtures from a theoretical viewpoint [40, 41]. These particular effects on phase separation were shown to be important as a long-range interaction in the phase separation kinetics of polymer mixtures [42]. On the other hand, for reacting polymer systems, very recently, Ohta and coworkers took into account the elastic effects of the polymer on the phase separation kinetics of a mixture composed of liquid crystals dissolved in monomer undergoing polymerization [43].

6.3
Chemical Reactions in Polymeric Systems: the Non-Mean-Field Kinetics

Reaction kinetics was monitored for two cases: photodimerization of anthracene in the bulk state of binary polymer mixtures, and photopolymerization of methyl methacrylate (MMA) monomer containing a photoreactive polymer as a minor component in the liquid state.

6.3.1
Reaction Kinetics in the Bulk State of Polymer

Shown in Figure 6.4 is the cross-link kinetics of anthracene-labeled poly(ethyl acrylate) at 25 °C. The reaction kinetics was monitored via the decrease in the absorbance of anthracene with irradiation time by UV spectroscopy. For the irradiation intensity ranging from 1.0 to 5.0 mW cm^{-2}, the decay of absorbance of anthracene observed at 365 nm cannot be expressed by an exponential function of irradiation time, but is instead well fitted to the following modified Kohlrausch–Williams–Watts (KWW) function [44]:

$$OD(t) = (1 - B) \exp(-kt)^\beta + B \quad (6.1)$$

Here, B is a baseline expressing the limit of the photodimerization of anthracene, that is, the limiting cross-link density, in the mixture under a given irradiation condition; k is the average cross-link rate; and the exponent β is the inhomogeneity

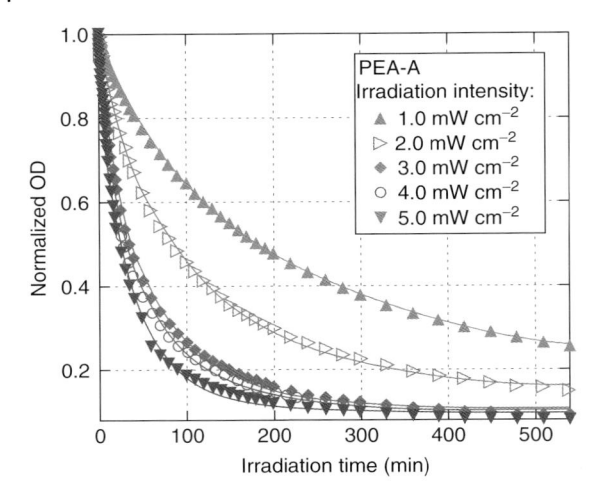

Figure 6.4 Photo-cross-link kinetics observed for an anthracene-labeled poly(ethyl acrylate) (PEA-A) under various light intensities observed at 25 °C [45].

index of the reaction. β is less than unity for an inhomogeneous reaction and becomes unity for the case of a homogeneous reaction. Under the irradiation conditions described above, β varies in the range 0.81–0.72 and decreases with increasing irradiation intensity. It is worth noting that this reaction inhomogeneity does not originate from the gradient of the light intensity inside the mixture because, even for a mixture with the same composition but with the thickness $\sim 1\,\mu m$ corresponding to the absorbance OD = 0.05, the reaction kinetics can be well fitted only to the KWW kinetics given in Eq. (6.1) with $\beta \sim 0.7$. These experimental results clearly indicate that photodimerization of anthracene, a diffusion-controlled reaction, proceeds inhomogeneously in the mixture. This inhomogeneity reflects the influence of segmental free-volume distribution or the dynamic heterogeneity in the blends described above. The information on the dynamic inhomogeneity of polymers has been given in great detail in a recent review [46].

6.3.2
Reaction Kinetics in the Liquid State of Polymer Mixtures

For polymerization in the liquid state, the reaction kinetics is even more complicated due to the heat and the increase in viscosity associated with the reaction, the so-called Tromsdorff effects [47]. The reaction accelerates itself as the polymerization proceeds as a result of a positive feedback loop generated by the increase in heat as well as in viscosity and the diffusion-controlled termination of the polymerization process [48, 49]. An example is illustrated in Figure 6.5 for a mixture of a polystyrene doubly labeled with anthracene and fluorescein (PSAF) dissolved in MMA monomer [50]. A PSAF/MMA (5/95) mixture containing 2 wt% of Lucirin TPO as a photoinitiator and 6 wt% of ethylene glycol dimethacrylate (EGDMA) as a

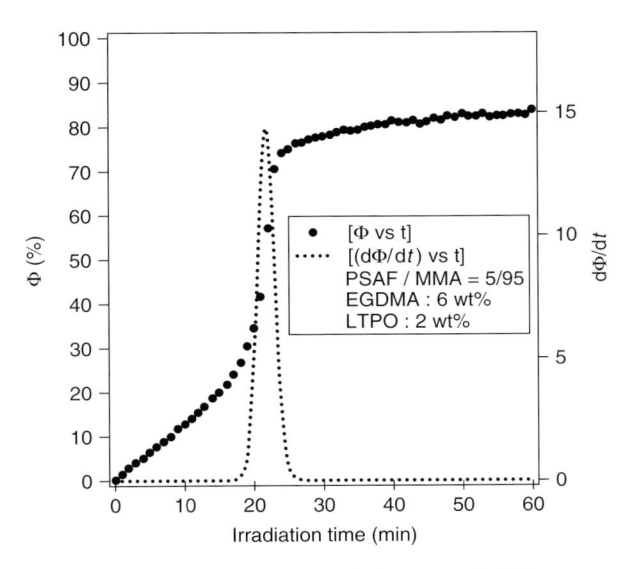

Figure 6.5 Photopolymerization and photo-cross-link kinetics of methyl methacrylate (MMA) in a PS/MMA (5/95) mixture *in situ* monitored by FT-IR at 25 °C. (Tasuku MURATA, Masters Dissertation, KIT, 2010.)

cross-linker for poly(methyl methacrylate) (PMMA) was irradiated with 365 nm UV light over 60 min at 25 °C. As clearly seen in Figure 6.5, the polymerization yield Φ obtained from the change in the absorbance of the C=C bond of MMA monomer monitored at 1640 cm^{-1} by Fourier transform infrared (FT-IR) under various irradiation times quickly rises at 22 min of irradiation, which corresponds to the peak of the derivative $(d\Phi/dt_{irr})$ with respect to irradiation time t_{irr}. This particular time dependence of the reaction yield indicates that the polymerization of MMA in the mixture proceeds autocatalytically with a maximum rate at 22 min after irradiation. As shown later, the phase separation induced by this reaction also exhibits autocatalytic behavior in response to this particular behavior of the polymerization.

Together with the reaction kinetics in the bulk state of polymer mixtures, we can conclude that, in general, chemical reaction proceeds inhomogeneously in polymeric media. As a consequence, a strain field associated with chemical reactions is generated in the polymer, and develops by this inhomogeneous kinetics. By using interferometry, the existence and relaxation of this strain field can be clarified, as described in the next section.

6.4
Reaction-Induced Elastic Strain and Its Relaxation Behavior

As seen in Figures 6.4 and 6.5, chemical reactions generally proceed inhomogeneously in polymeric systems. Particularly for the case of cross-linking reactions, the mixture gradually loses its mobility upon reacting because the whole mixture approaches the glassy state. Owing to this inhomogeneity, a transient local strain field can be generated and developed in the sample. By performing *in situ* experiments using a Mach–Zehnder interferometer [51], the relaxation of this strain field in both photo-cross-linked homopolymers [45] and in the bulk state of binary polymer mixtures was reported with its consequence on the morphology emerging in photo-cross-linked polymers [52, 53]. It should be noted that, in contrast to the bulk state, the effects of elastic strain are much more significant in liquid mixtures where the morphological regularity greatly decreases by periodic irradiation [54]. An example for this reaction-induced transient strain in polymeric systems is illustrated in Figure 6.6 for an anthracene-labeled poly(ethyl acrylate) cross-linked by photodimerization of anthracene. To monitor the development of the elastic strain generated by irradiation, the dependence of the interference patterns on the irradiation time was monitored under different irradiation intensities, that is, different cross-link densities. These results were obtained from the changes in the optical path length difference (OPLD) and in the refractive index between the irradiated and the nonirradiated parts in the same sample. Note that the strain or deformation ε induced by irradiation can be calculated from the following equation [45, 51]:

$$\varepsilon = \frac{\Delta d}{d_o} = \frac{OPLD}{(n_s - n_0)d_o} \tag{6.2}$$

Here, n_s and n_0 are, respectively, the refractive index of the sample and air; Δd and d_o are, respectively, the change in the sample thickness and the initial thickness of the sample. It is found that the change in the refractive index is almost negligible under the current experimental conditions as directly measured by the prism coupling method. As a result, the reaction-induced strain is finally obtained by Eq. (6.2) as an approximation [51]. Depending on the irradiation condition, the difference between the experimental temperature and the glass transition temperature of the cross-linked sample can be modified, leading to three distinct behaviors observed after stopping the irradiation: recovery with swelling, remaining unchanged, and further shrinkage with elapsed time upon increasing the irradiation intensity. From Figure 6.6, it can be concluded that the strain is unchanged, that is, there is no relaxation, after stopping irradiation at low cross-link density ($\gamma = 1$). However, the irradiated polymer continues to shrink under higher cross-link densities ($\gamma \geqslant 2$), a typical behavior of physical aging. For polymer mixtures undergoing phase separation, this reaction-induced strain can couple to fluctuations, leading to several interesting phenomena [52]. It is also found that the elastic deformation emerging under irradiation can be expressed by a universal function of the nondimensionalized aging rate, which is defined as the product between the aging rate k_a and the elapsed time t_e [45]. This aging rate is obtained by analyzing the decay of the elastic strain ε with the elapsed time after stopping irradiation.

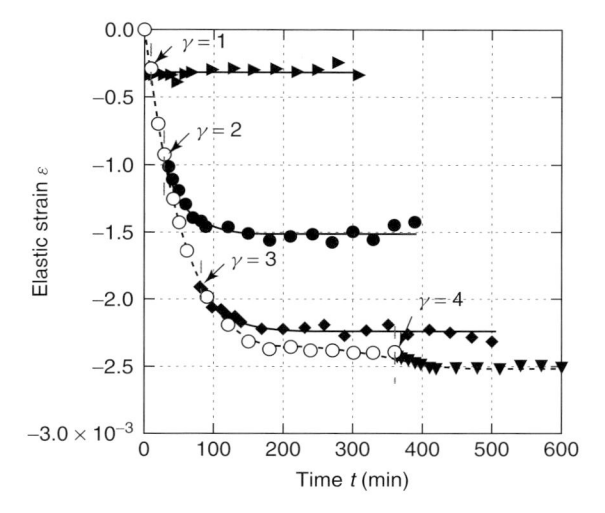

Figure 6.6 Relaxation of the elastic strain generated in a poly(ethyl acrylate) film observed by a Mach–Zehnder interferometer at 25 °C. γ: the mean cross-link density; dashed curve with open symbols: the elastic strain observed during irradiation; solid curves with closed symbols: elastic strain monitored after stopping irradiation [45].

6.5 Phase Separation under Nonuniform Conditions in Polymeric Systems

6.5.1 Polymers with Spatially Graded Continuous Structures

It is well known that, upon traveling through an absorbing medium, the light intensity exponentially decreases with the path length according to the Lambert–Beer law. On the other hand, for light-induced transition phenomena, such as light-induced phase separation, there exists a threshold of the control parameters for the transition to occur. Therefore, the transition occurs only after an induction time required for the control parameter to reach a critical threshold. From the viewpoint of critical phenomena, phase separation takes place as soon as this control parameter exceeds a critical value. The depth at which the control parameter exceeds a critical value, the so-called quench depth ΔT, determines the initial length scale ξ_o of the morphology. According to the linearized theory of phase separation [9, 12, 13], ξ_o is inversely proportional to this quench depth. For a given polymeric system, mass diffusion is always much slower than thermal diffusion. By combining the feature of photochemical reaction with the light-induced phase separation, we were able to demonstrate that it was possible to generate and control the morphologies with spatially graded structures by varying the light intensity.

It was found that, by irradiation using UV light with the intensity 0.01 mW/cm^{-2}, the resulting morphology observed in the front and back sides located at 4 µm from the two surfaces of a PSAF/MMA (10/90) mixture was exactly the same as revealed by the power spectra of the two-dimensional fast Fourier transform (2D-FFT) [55]. However, if the irradiation intensity was higher than 0.01 mW cm^{-2}, the characteristic length scales of the morphology observed in the front became smaller compared to that in the back, revealing the effects of the gradient of the irradiation intensity on the phase separation kinetics. The morphologies observed at different depths along the Z-coordinate are shown in Figure 6.7, together with the corresponding 3D images obtained for the two intensities ($I = 0.03$ and 0.08 mW cm^{-2}). These experimental results clearly indicate that the effects of light intensity gradient on the phase separation kinetics become significant under irradiation with high light intensity.

The phase separation kinetics observed *in situ* by laser-scanning confocal microscope (LSCM) at three different locations (lower, middle, and upper) in the mixture along the irradiation direction (Z) is illustrated in Figure 6.8 for a PSAF/MMA (10/90) mixture under $I = 0.03$ mW cm^{-2}. For all locations along the direction of propagation of the UV light, the co-continuous structures appear, then coarsen with

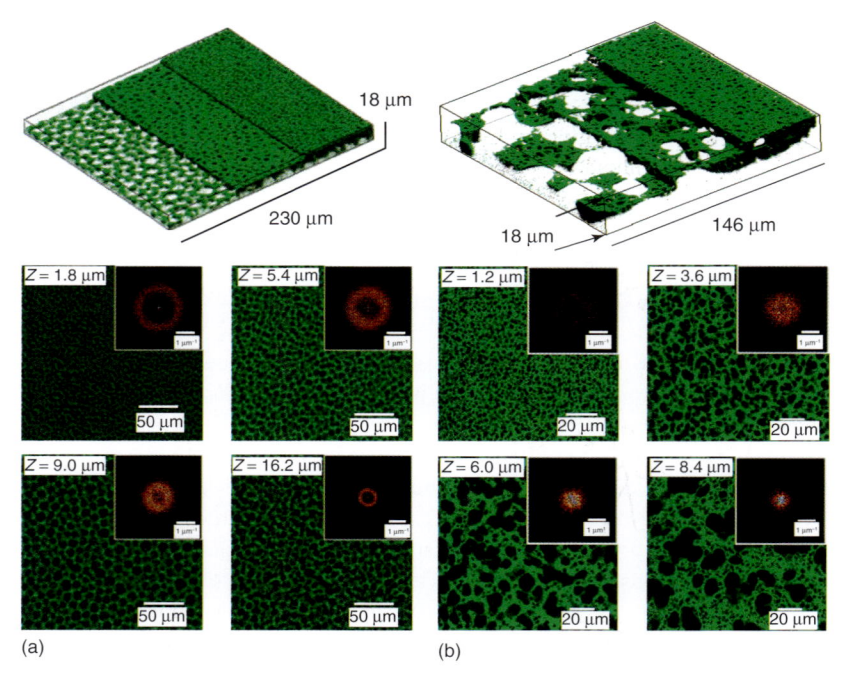

Figure 6.7 (a) (Upper) 3D morphology of the graded co-continuous morphology observed for a PSAF/MMA (10/90) mixture irradiated with 0.03 mW cm^{-2}. (Lower) graded co-continuous structures and the corresponding 2D-FFT power spectra (inset) observed at different depths along the irradiation direction. The number in each micrograph indicates the depth of observation [55]. (b) 3D morphology and its variation with the sample depth observed for the light intensity $I = 0.08$ mW cm^{-2}. PSAF-rich phase, bright, green; PMMA-rich phase, dark, gray [56].

irradiation time, and eventually reach a stationary state at a long irradiation time t_{st}. Structures observed at deeper locations attain this stationary state later compared to those in the shallower portions of the irradiated sample. This behavior can be explained by the response of the mixture to the intensity of the irradiating light. In general, the final morphologies (either co-continuous structures or droplets) are determined by the modification of the phase diagram, that is, the quench depth, resulting from the reaction and the decrease in mobility induced by the reaction. As a consequence, morphologies with controllable spatial gradients of length scales can be generated by adjusting this competition process [55].

An interesting aspect of the kinetics observed for this particular case is the direct coupling between the reaction autocatalysis and the phase-separation kinetics. Recently, it was found that this coupling also triggers and drives the phase separation in an autocatalytic fashion as revealed by the dependence of the derivative $(d\xi/dt_{irr})$ of the characteristic length scales ξ on irradiation time. As shown in Figure 6.8, the derivative $(d\xi/dt_{irr})$ exhibits a maximum at a particular irradiation time, which corresponds to the autocatalytic behavior of the polymerization [56]. This maximum is significant for phase separation at the position far from the light source where the mobility is less suppressed in comparison to other positions. The peak of the derivative of Figure 6.8 reveals the maximum rate of the phase-separation process. In other words, the autocatalysis of the polymerization of MMA triggers the autocatalytic phase separation of the mixture.

Variation of the spatial gradient of the characteristic length scales ξ with irradiation intensity is illustrated in Figure 6.9 for a PSAF/MMA (5/95) mixture irradiated with different UV intensities ranging from 0.06 to 0.10 mW cm^{-2}.

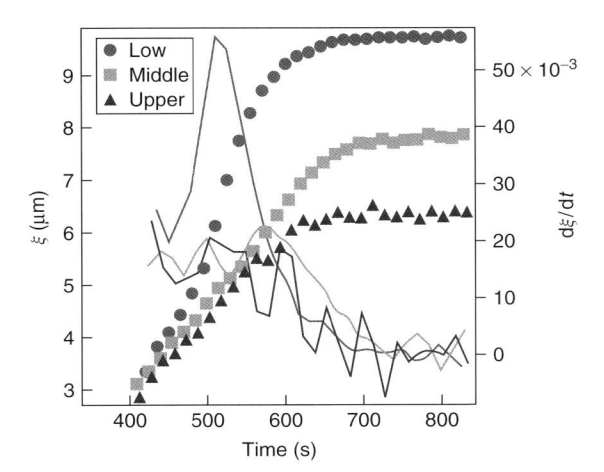

Figure 6.8 Symbols: irradiation-time dependence of the characteristic length scales of the co-continuous structures observed at three locations along the thickness of the mixture. Solid curves: the corresponding derivative of the characteristic length scales with respect to irradiation time [56].

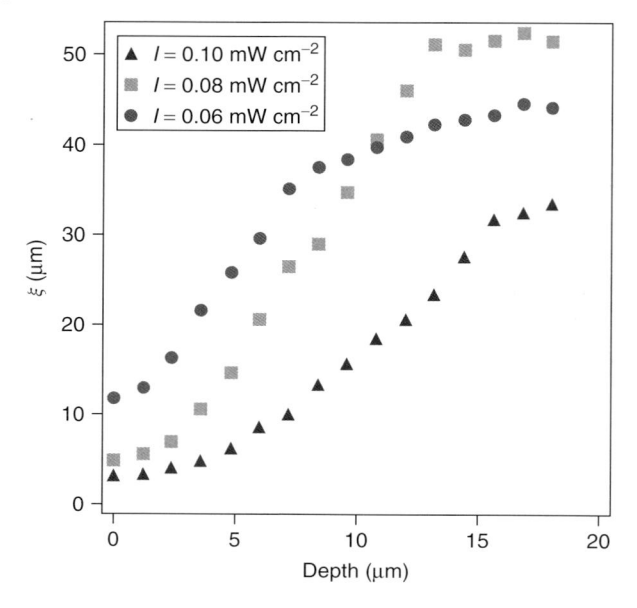

Figure 6.9 Variation of the morphological characteristic length scales along the sample depth as observed by laser-scanning confocal microscopy for a PSAF/MMA (10/90) mixture irradiated at 25 °C [55].

These spatial gradients share a common feature: in the vicinity of the two cover slips, ξ does not significantly change with irradiation intensity, whereas in the regions far from these two cover slips, ξ shows a strong dependence on the light intensity gradient. This behavior arises from the wetting effect induced by the affinity between the glass surface and the MMA component in the vicinity of the two covers, as revealed by 3D imaging using LSCM [57]. Another example for the construction of 3D spatially graded morphology using photo-cross-linking reactions is illustrated in Figure 6.10 for a poly(chlorinated isoprene)/poly(methyl methacrylate) (PI/PMMA) blend. Here, PI ($M_w = 4.5 \times 10^5$, $M_w/M_n = 1.8$) was dissolved in MMA monomer to form a uniform one-phase precursor for semi-interpenetrating polymer networks. The PMMA component was labeled with a trace of rhodamine B for fluorescence imaging using LSCM. Obviously, the morphology exhibits PI-rich domains (dark, black) dispersed in continuous PMMA-rich phases (bright, green) as imaged by fluorescence of rhodamine B. It should be noted that these spatial gradients of the morphology depend on both the variation of the phase diagram upon irradiation and the increase in viscosity of the mixture induced by the photo-cross-link and photopolymerization.

Except some recent studies [58, 59], preparation of polymers with spatially graded morphologies have so far heavily relied on polymerization of a guest monomer or on the spatial distribution of a guest polymer inside a host polymer in the bulk state generated through diffusion or sorption [60, 61]. The experimental data

Figure 6.10 3D morphology induced by photo-cross-link and photopolymerization. (a) Uniform co-continuous structure. (b) Graded co-continuous morphology obtained for a mixture of chlorinated polyisoprene and poly(methyl methacrylate) labeled with rhodamine (10/90) irradiated from above. The irradiation intensity at 365 nm was 0.06 mW cm^{-2}. The large domains of PMMA-rich phase due to wetting can be observed in the vicinity of the cover slip in the right-hand-side morphology. $L = 115$ µm; $d = 40$ µm. (Chuanming JING, Masters Dissertation, KIT, 2008.)

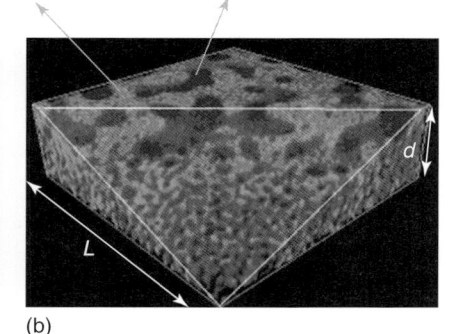

described above suggest that, besides the practical aspects such as generation of spatially graded materials, phase separation of photopolymerizing mixtures induced by a strong light intensity could provide a polymer system to study phase separation under stationary nonequilibrium conditions, particularly the couplings of instabilities developing differently at different locations along the direction of propagation of the exciting light.

6.5.2
Morphology with Arbitrary Symmetry and Distribution of Length Scales

In the previous section, it was shown that polymer materials with a unidirectional gradient in the characteristic length scales could be designed by coupling the Lambert–Beer law of the reactants to critical phenomena via spatially varying the quench depth. However, this method, together with irradiation using UV light intensity with a lateral gradient [62], can simply provide graded structures along the direction of propagation of light. In order to generate morphologies with arbitrary distribution of characteristic length scales in 3D, we have developed the so-called computer-assisted irradiation (CAI) method as described below.

6.5.2.1 The Computer-Assisted Irradiation Method
One of the features of reaction-induced phase separation is the existence of a critical threshold (or an induction time) for the reaction yield beyond which the mixture enters the two-phase region [63]. Since it is experimentally verified that this threshold strongly depends on the light intensity, the morphology resulting from reaction-induced phase separation can be spatially and temporally controlled by

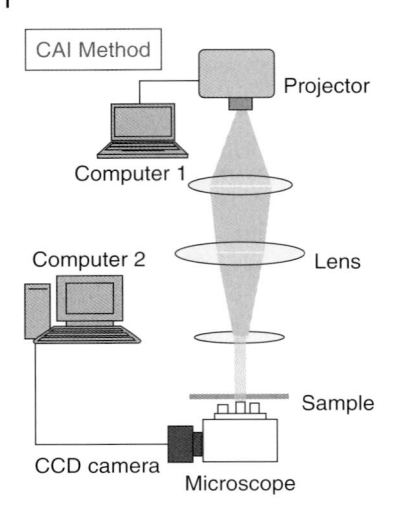

Figure 6.11 Block diagram of the computer-assisted irradiation (CAI) apparatus [65].

manipulating this threshold using UV irradiation. As a consequence, morphology with a distribution of characteristic length scales or a specific spatial symmetry can be generated and controlled by using static or dynamic patterns of UV light.

The above-mentioned purpose can be achieved by using the CAI setup illustrated in Figure 6.11 [64, 65], where a 2D light pattern $I(\lambda, \tau)$ with an arbitrary characteristic length scale λ and a lifetime τ was designed and generated on the first computer. Subsequently, this light pattern was transferred to a digital projector and was focused on a photoreactive blend situated under an optical microscope through an optical lens system, as schematically shown in Figure 6.11. The resulting morphologies were transferred and stored in a second computer prior to analysis. With the current optical system, this CAI apparatus can provide a light pattern with a minimum length scale of $\lambda = 20$ μm and a maximum frequency of 30 Hz.

6.5.2.2 Polymers with an Arbitrary Distribution of Characteristic Length Scales

As a demonstration of this CAI method, a concentric light pattern with different diameters and intensities, shown in Figure 6.12a, was generated by the computer 1 and impinged on a mixture of PSC (PolyStyrene labeled with *trans*-Cinnamic acid) and MMA containing the cross-linker reagent EGDMA. The response of the mixture to this light pattern after 60 min of irradiation is indicated in Figure 6.12b [63]. First of all, the mixture undergoes phase separation in response to the irradiation intensity, that is, phase separation proceeds faster in the region irradiated with a higher light intensity and becomes slower in the regions with lower intensities. Therefore, the morphology would vary with the irradiation intensity, providing morphological length scales with controllable distribution, as illustrated in Figure 6.12b. These experiments suggest two important points: the existence of different mode-selection processes arising from the competition between phase separation; and the reaction in different areas of the sample irradiated with different light intensities. As a consequence, it is possible to produce and control the distribution of physical quantities such as elastic modulus, conductivity, and so

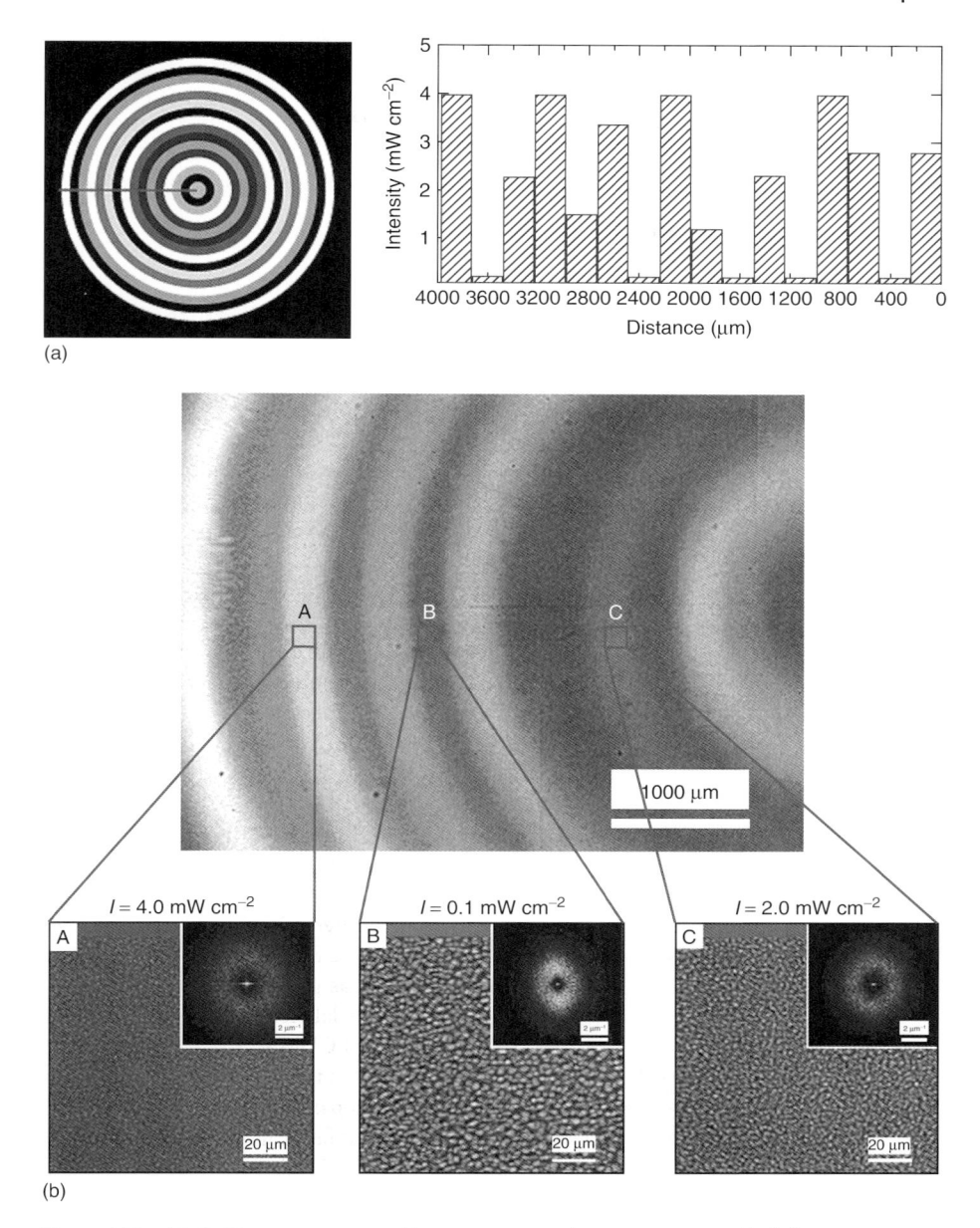

Figure 6.12 (a) (Left) a concentric visible light pattern generated by the computer 1. (Right) radial intensity distribution along the X-axis (indicated by the line). (b) (Upper part) Morphology with low magnification resulting from irradiating a PSC/MMA (20/80) mixture with a concentric visible light source shown in (a). (Lower part) morphology in restricted areas observed under high magnification with their corresponding 2D-FFT power spectra (inset) [63].

on, for given polymer mixtures or composites by manipulating these length scale distributions by using the CAI method.

The advantages of the CAI method for morphology control are convincingly revealed for the case of phase separation induced by temporal modulation [65]. Shown in Figure 6.13 is the morphology of a PSAF/MMA (5/95) mixture irradiated with two frequencies: 5 and (1/120) Hz. The modulation frequency was used to regulate the ON/OFF time of the irradiation process using the CAI apparatus. For irradiation with regular intervals, that is, (ON : OFF) = (1 : 1), the total number of photons received by the sample in the case of modulation is exactly one-half of those compared to the continuous irradiation, irrespective of the irradiation frequency. In order to compare the two experimental results obtained under different frequencies, the total number of photons input into the mixture for these two cases was fixed at a same value. It was found from Figure 6.13 that the regularity of the morphology is highest for the case of continuous irradiation, but exhibits the highest randomization when an ON : OFF ratio of 1 : 1 with a duration of 60 s (equivalent to 1/120 Hz) was applied. Unlike the case of polymer blends reacting in the bulk state, for a polymerizing mixture the resulting molecular

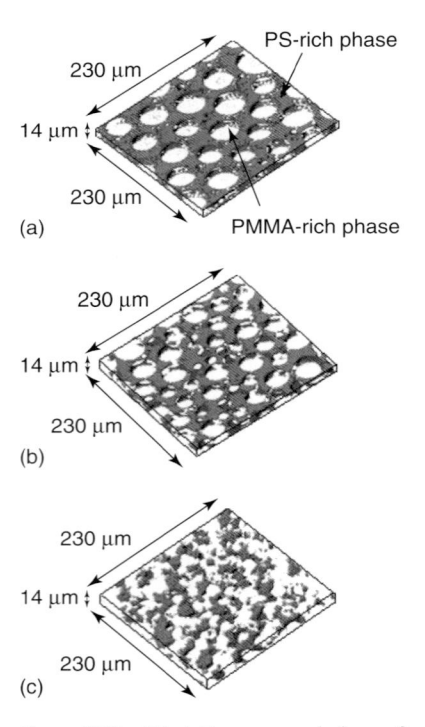

Figure 6.13 3D stationary morphology of a PSAF/MMA (5/95) blend obtained by irradiation with 365 nm under various conditions: (a) continuous irradiation; (b) periodic irradiation with a regular interval ON : OFF (1 : 1) and 0.1 s (5 Hz) duration; and (c) ON : OFF (1 : 1) and 60 s (1/120 Hz) duration. PSAF-rich phase, opaque; PMMA-rich phase, transparent [65].

weights and the structures of the PMMA networks are determined by the lifetime of MMA radicals, which in turn depends on the light intensity and the modulation frequency of the irradiation. The results illustrated in Figure 6.13 indicate that the irradiation frequency can be used as a tool to control the morphological regularity of phase-separating polymer mixtures [53]. This conclusion is supported by recent experiments using intermittent irradiation [54] and reversible photo-cross-linking reaction induced by two UV wavelengths [66] to drive the phase-separation process. In the former, the relaxation of the elastic strain associated with the polymerization of the monomer was relaxed step by step by varying the irradiation intermittency, whereas in the latter case the elastic strain generated by the photo-cross-linking reaction was positively relaxed by generating and annihilating (de-cross-linking) the forming polymer networks by using two different UV wavelengths. These experimental results show that the effects of periodic irradiation are significantly different for polymer mixtures in the liquid and bulk states. For reaction-induced phase separation phenomena, the reaction mechanism and its influences on the generation and relaxation processes of the elastic strain would play a key role in the resulting morphology.

6.6
Conclusions

Generally, chemical reactions do not follow the mean-field kinetics, and as a consequence proceed inhomogeneously in polymeric media. Upon coupling with the phase separation process, this inhomogeneity of the reaction plays an important role in the resulting morphology. By combining laser-confocal microscopy, light scattering, and Mach–Zehnder interferometry, the following results were obtained for this complex coupling and the competition between reaction and phase separation:

1) In the case of polymerization-induced phase separation, due to the positive feedback arising from the Tromsdorff effects, it was found that the phase separation induced by polymerization also becomes autocatalytic. The phase separation proceeds with a maximum rate in the vicinity of the irradiation time at which the rate of polymerization reaches its maximum. As a consequence, a local strain field is generated and develops as the reaction continues. For small-molecule systems, this reaction-induced strain quickly relaxes and its effects on the subsequent reaction and phase separation are not significant. For polymeric systems, this reaction-induced strain is not negligible and therefore affects the concentration fluctuations as a long-range interaction.
2) By using Mach–Zehnder interferometry, the development and relaxation of this reaction-induced strain field were detected and analyzed during and after the irradiation process. From these experimental results, we were able to correlate the resulting morphologies with the reaction-induced strain.
3) From the materials viewpoint, phase separation under stationary, nonequilibrium conditions may be used to design functional polymer materials having

spatially graded structures and morphologies with multiple length scales. From the nonlinear dynamic aspects, with the CAI method, the interference between the unstable modes evolving differently with irradiation time at different locations in the same sample can be elucidated to gain information on the mode-selection processes under stationary, nonequilibrium conditions.

Acknowledgments

We deeply appreciate the financial support from the Ministry of Education (MONKASHO), Japan, through grant-in-aid for the Priority Research Areas "Molecular Nano Dynamics" and "Soft Matter Physics." The grant-in-aid for Scientific Research Type B (to Q. T.-C.-M.) is also greatly acknowledged. The contents of this chapter are mainly taken from the masters dissertations of former graduate students in the Polymer Molecular Engineering Lab at KIT, in particular, the theses of Kyosuke Inoue (currently with Zeon Corp., Tokyo), Soh Ishino (currently with Sumitomo Rubber Industries, Hyogo), Nobuhiro Namikawa (currently with Shimadzu Corporation, Kyoto), Masahiro Sato (currently with Nippon Shokubai Co. Ltd., Osaka), and Chuanming Jing (Nippon Paint Co. Ltd., Osaka). The achievements in the past 5 years would not have been possible without the patience and constant enthusiasm of all the graduate students in the Polymer Molecular Engineering Lab. Finally, the technical assistance of Dr. Tomohisa Norisuye, Associate Professor in our group, is also greatly appreciated.

References

1. Drazin, P.G. and Reid, W.H. (1981) *Hydrodynamic Stability*, Cambridge University Press, Cambridge.
2. Vidal, C., Dewel, G., and Borckmans, P. (1990) *Au-dela de L'equilibre*, Masson, Paris.
3. Godreche, C. (ed.) (1992) *Solids Far from Equilibrium*, Cambridge University Press, Cambridge.
4. Hashimoto, T. (1988) Dynamics in spinodal decomposition of polymer mixtures. *Phase Transitions*, **12**, 47–119.
5. Binder, K. (1994) Phase-transitions in polymer blends and block – copolymer melts- some recent developments. *Adv. Polym. Sci.*, **112**, 181–299.
6. Onuki, A. (2002) *Phase Transition Dynamics*, Cambridge University Press, Cambridge.
7. Lifshitz, I.M. and Slyozov, V.V. (1961) The kinetics of precipitation from supersaturated solid solutions. *J. Phys. Chem. Solids*, **19**, 35–50.
8. Wagner, C.Z. (1961) Theorie der Alterung von Niederschlagen durch Umlosen. *Z. Elektrochem.*, **65**, 581–594.
9. (a) Cahn, J.W. (1961) On spinodal decomposition. *Acta Metall.*, **9**, 795–801; (b) Cahn, J.W. (1965) Phase separation by spinodal decomposition in isotropic systems. *J. Chem. Phys.*, **42**, 93–99.
10. Ackasu, A.Z., Bahar, I., Erman, B., Feng, Y., and Han, C.C. (1992) Theoretical and experimental study of dissolution of inhomogeneities formed during spinodal decomposition in polymer mixtures. *J. Chem. Phys.*, **97**, 5782–5793.
11. Glotzer, S.C., DiMarzio, E.A., and Muthukumar, M. (1995) Reaction-controlled morphology of phase-separating mixtures. *Phys. Rev. Lett.*, **74**, 2034–2037.
12. de Gennes, P.-G. (1980) Dynamics of fluctuations and spinodal decomposition

in polymer blends. *J. Chem. Phys.*, **72**, 4756–4763.
13. Pincus, P. (1981) Dynamics of fluctuations and spinodal decomposition in polymer blends II. *J. Chem. Phys.*, **75**, 1996–2000.
14. For example, see Furukawa, H. (1998) in *Structure and Properties of Multiphase Polymeric Materials*, Chapter 2 (eds T. Araki., Q. Tran-Cong, and M. Shibayama), Marcel Dekker, New York, pp. 35–66.
15. Hohenberg, P.C. and Halparin, B.I. (1977) Theory of dynamic critical phenomena. *Rev. Mod. Phys.*, **49**, 35–479.
16. Kimura, M., Porter, R.S., and Salee, G. (1983) Blends of poly(butylene terephthalate) and a polyacrylate before and after transesterification. *J. Polym. Sci. Polym. Phys. Ed.*, **21**, 367–378.
17. Balazs, A.C. (2007) Modeling self-assembly and phase behavior in complex mixtures. *Annu. Rev. Phys. Chem.*, **58**, 211–233.
18. Huberman, B.A. (1976) Striations in chemical reactions. *J. Chem. Phys.*, **65**, 2013–2019.
19. (a) Ohta, T., Urakawa, O., and Tran-Cong, Q. (1998) Phase separation of binary polymer blends driven by photoisomerization: an example for a wavelength-selection process in polymers. *Macromolecules*, **31**, 6845–6854; (b) Ohta, T. (1998) *Wavelength-Selection Process Driven by a Reversible Photoreaction in Polymer Blends and Its Implication for Morphology Control*, Master Dissertation, Department of Polymer Science and Engineering, Kyoto Institute of Technology, Kyoto, Japan, March 1998.
20. Verdasca, J., Borckmans, P., and Dewel, G. (1995) Chemically frozen phase separation in an adsorbed layer. *Phys. Rev. E*, **52**, R4616–R4619.
21. Christensen, J.J., Elder, K., and Fogedby, H.C. (1996) Phase segregation dynamics of a chemically reactive binary mixture. *Phys. Rev. E*, **54**, R2212–R2215.
22. Motoyama, M. and Ohta, T. (1997) Morphology of phase separating binary mixtures with chemical reaction. *J. Phys. Soc. Jpn.*, **66**, 2715–2725.
23. Huo, Y.L., Zhang, H.D., and Yang, Y.L. (2004) Effects of reversible chemical reaction on morphology and domain growth of phase separating binary mixtures with viscosity difference. *Macromol. Theor. Simul.*, **13**, 280–289.
24. Luo, K.F. (2006) The morphology and dynamics of polymerization-induced phase separation. *Eur. Polym. J.*, **42**, 1499–1505.
25. Travasso, R.D.M., Kuksenok, O., and Balazs, A.C. (2005) Harnessing light to create defect-free, hierarchically structured polymeric materials. *Langmuir*, **21**, 10912–10915.
26. Dayal, P., Kuksenok, O., and Balazs, A.C. (2008) Using a mask to create multiple patterns in three-component, photoreactive blends. *Langmuir*, **24**, 1621–1624.
27. Travasso, R.D.M., Kuksenok, O., and Balazs, A.C. (2006) Exploiting photoinduced reaction in polymer blends to create hierarchically ordered, defect-free materials. *Langmuir*, **22**, 2620–2628.
28. Mikhailov, A.S., Hildebrand, M., and Ertl, G. (2001) in *Coherent Structures in Classical Systems*, Lecture Notes in Physics, Vol. 567 (eds D. Reguera, L.L. Bonilla, and J.M. Rubi), Springer, New York, pp. 252–269.
29. Tran-Cong, Q., Meisyo, K., Ishida, Y., Yano, O., Shibayama, M., and Soen, T. (1992) Effects of critical concentration fluctuations on the photocyclization of a bichromophoric molecules in the one-phase region of polystyrene/poly(vinyl methyl ether) blends. *Macromolecules*, **25**, 2330–2335.
30. Tran-Cong, Q., Ishida, Y., Tanaka, A., and Soen, T. (1992) Further evidence for the effects of critical concentration fluctuations on the diffusion-controlled reaction-kinetics in binary polymer mixtures. *Polym. Bull.*, **29**, 89–96.
31. Tran-Cong, Q., Harada, A., Kataoka, K., Ohta, T., and Urakawa, O. (1997) Positive feedback driven by concentration fluctuations in asymmetrically photo-cross-linked polymer mixtures. *Phys. Rev. E*, **55**, R6340–R6343.
32. See, for review, Tran-Cong-Miyata, Q. and Nakanishi, H. (2009) in *Polymers, Liquids and Colloids in Electric Field*, Chapter 6 (eds Y. Tsori and

33. Paik, C.S. and Morawetz, H. (1972) Photochemical and thermal isomerization of aromatic residues in the side chains and the backbone of polymers in bulk. *Macromolecules*, **5**, 171–177.
34. Yu, W.-C., Sung, C.S.P., and Robertson, R.E. (1988) Site-specific labeling and the distribution of free volume in glassy polystyrene. *Macromolecules*, **21**, 355–364.
35. Yoshizawa, H., Ashikaga, K., Yamamoto, M., and Tran-Cong, Q. (1989) Photocyclomerization of bis (9-anthrylmethyl) ether in solid polymers. *Polymer*, **30**, 534–539.
36. Urakawa, O., Fuse, Y., Hori, H., Tran-Cong, Q., and Yano, O. (2001) A dielectric study on the local dynamics of miscible blends: poly(2-chlorostyrene)/poly(vinyl methyl ether). *Polymer*, **42**, 765–773.
37. Urakawa, O. (2004) Studies on dynamic heterogeneity in miscible polymer blends and dynamics of flexible polymers. *J. Soc. Rheol. Jpn.*, **32**, 265–270.
38. (a) Lefever, R., Carati, D., and Hassani, N. (1995) Monte-carlo simulations of phase separation in chemically reactive binary mixtures-comment. *Phys. Rev. Lett.*, **75**, 1674; (b) see, also Carati, D. and Lefever, R. (1997) Chemical freezing of phase separation in immiscible binary mixtures. *Phys. Rev. E*, **56**, 3127–3136.
39. Turing, A.M. (1952) The chemical basis of morphogenesis. *Phil. Trans. R. Soc. London B.*, **237**, 37–72.
40. Doi, M. and Onuki, A. (1992) Dynamic coupling between stress and composition in polymer solutions and blends. *J. Phys. II (France)*, **2**, 1631–1656.
41. For review, see Tanaka, H. (2000) Viscoelastic phase separation. *J. Phys.: Condens. Matter*, **12**, R207–R264.
42. Onuki, A. and Taniguchi, T. (1997) Viscoelastic effects in early stage phase separation in polymeric systems. *J. Chem. Phys.*, **106**, 5761–5770.
43. Nakazawa, H., Fujinami, S., Motoyama, M., Ohta, T., Araki, T., Tanaka, H., Fujisawa, T., Nakada, H., Hayashi, M., and Aizawa, M. (2001) Phase separation and gelation of polymer-dispersed liquid crystals. *Comp. Theor. Polym. Sci.*, **11**, 445–458.
44. (a) Kataoka, K., Harada, A., Tamai, T., and Tran-Cong, Q. (1998) Phase behavior and photo-cross-linking kinetics of *semi*-interpenetrating polymer networks prepared from miscible polymer blends. *J. Polym. Sci. Polym. Phys.*, **36**, 455–462; (b) see, also: Williams, G. and Watts, D.C. (1970) Non-symmetrical dielectric relaxation behaviour arising from a simple empirical decay function. *Trans. Faraday Soc.*, **66**, 80–85.
45. Van-Pham, D.-T., Sorioka, K., Norisuye, T., and Tran-Cong-Miyata, Q. (2009) Physical aging of photo-crosslinked poly (ethyl acrylate) observed in the nanometer scales by Mach-Zehnder interferometry. *Polym. J. (Tokyo)*, **41**, 260–265.
46. Colmenero, J. and Arbe, A. (2007) Segmental dynamics in miscible polymer blends: recent results and open questions. *Soft Matter*, **3**, 1474–1485.
47. Billmeyer, F.W. Jr. (1984) *Textbook of Polymer Science*, 3rd edn, Chapter 3, John Wiley & Sons, Inc., New York.
48. Pojman, J.A., Ilyashenko, V.M., and Khan, A.M. (1996) Free radical frontal polymerization: self-propagating thermal reaction waves. *J. Chem. Soc. Faraday Trans.*, **92**, 2824–2836.
49. Pojman, J.A., Popwell, S., Fortenberry, D.I., Volpert, V.I., and Volpert, V.L. (2003) in *Nonlinear Dynamics in Polymeric Systems*, ACS Symposium Series No. 869, Chapter 9 (eds J.A. Pojman and Q. Tran-Cong-Miyata), American Chemical Society, Washington, DC, pp. 106–120.
50. Nakanishi, H., Satoh, M., Norisuye, T., and Tran-Cong-Miyata, Q. (2006) Phase separation of interpenetrating polymer networks synthesized by using an autocatalytic reaction. *Macromolecules*, **39**, 9456–9466.
51. Inoue, K., Komatsu, S., Trinh, X.-A., Norisuye, T., and Tran-Cong-Miyata, Q. (2005) Local deformation in photo-crosslinked polymer blends monitored by Mach-Zehnder interferometry. *J. Polym. Sci. Polym. Phys.*, **43**, 2898–2913.

52. Tran-Cong-Miyata, Q., Nishigami, S., Ito, T., Komatsu, S., and Norisuye, T. (2004) Controlling the morphology of polymer blends using periodic irradiation. *Nat. Mater.*, **3**, 448–451.
53. Noma, K. (2009) Controlling phase separation of interpenetrating polymer networks using photopolymerization induced by periodic irradiation. Master Dissertation, Department of Macromolecular Science and Engineering, Graduate School of Science and Technology, Kyoto Institute of Technology, Kyoto, Japan, March 2009, (To be published).
54. Murata, K., Murata, T., Nakanishi, H., Norisuye, T., and Tran-Cong-Miyata, Q. (2009) Effects of light-induced regularity on the physical properties of multiphase polymers. *Macromol. Mat. Eng.*, **294**, 163–164.
55. Nakanishi, H., Namikawa, N., Norisuye, T., and Tran-Cong-Miyata, Q. (2006) Autocatalytic phase separation and graded co-continuous morphology generated by photocuring. *Soft Matter*, **2**, 149–156.
56. Nakanishi, H., Namikawa, N., Norisuye, T., and Tran-Cong-Miyata, Q. (2006) Interpenetrating polymer networks with spatially graded morphology controlled by UV-radiation curing. *Macromol. Symp.*, **242**, 157–164.
57. Nakanishi, H. (2007) Generation and manipulation of ordered structures in interpenetrating polymer networks by using photochemical reactions. PhD Dissertation, Department of Polymer Science and Engineering, Kyoto Institute of Technology, Kyoto, July 2007.
58. Okinaka, J. and Tran-Cong, Q. (1995) Directional phase separation of a binary mixture driven by a temperature gradient. *Phys. D*, **84**, 23–30.
59. Noblet, G., Desilles, N., Lecamp, L., Lebaudy, P., and Bunel, C. (2006) Gradient structure polymer obtained from a homogeneous mixture: synthesis and mechanical properties. *Macromol. Chem. Phys.*, **207**, 426–433.
60. Jasso, C.F., Martizez, J.J., Mendizabal, E., and Laguna, O. (1995) Mechanical and rheological properties of styrene/acrylic gradient polymers. *J. Appl. Polym. Sci.*, **58**, 2207–2212.
61. Akovali, G. (1999) Studies with gradient polymers of polystyrene and poly(methyl acrylate). *J. Appl. Polym. Sci.*, **73**, 1721–1725.
62. Nishioka, H., Kida, K., Yano, O., and Tran-Cong, Q. (2000) Phase separation of a polymer mixture driven by a gradient of light intensity. *Macromolecules*, **33**, 4301–4303.
63. Ishino, S. (2006) A novel method of morphology control for polymer alloys using computer-assisted irradiation. Master Dissertation, Department of Polymer Science and Engineering, Kyoto Institute of Technology, Kyoto, Japan, March 2006.
64. Ishino, S., Nakanishi, H., Norisuye, T., and Tran-Cong-Miyata, Q. (2006) Designing a polymer blend with phase separation tunable by visible light for computer-assisted irradiation experiments. *Macromol. Rapid Commun.*, **27**, 758–762.
65. Tran-Cong-Miyata, Q., Van-Pham, D.-T., Noma, K., Norisuye, T., and Nakanishi, H. (2009) The roles of reaction inhomogeneity in phase separation kinetics and morphology of reactive polymer blends. *Chinese J. Polym. Sci.*, **27**, 23–36. (Feature Article).
66. Trinh, X.A., Fukuda, J., Adachi, Y., Nakanishi, H., Norisuye, T., and Tran-Cong-Miyata, Q. (2007) Effects of elastic deformation on phase separation of a polymer blend driven by a reversible photo-cross-linking reaction. *Macromolecules*, **40**, 5566–5574.

7
Gels Coupled to Oscillatory Reactions
Ryo Yoshida

7.1
Introduction

Polymer gels is a research field of polymer science that has seen rapid progress during the past 20–30 years. The term *gel* can be widely defined as a cross-linked polymer network that swells by absorbing large amounts of a solvent such as water. Theoretical study of the characteristics of gel had already proceeded in the 1940s, and the principle of swelling by water absorption based on thermodynamics had been clarified by Flory. As an application of the research on gels, soft contact lenses were developed in the 1960s; subsequently, gels have been widely used in the medical and pharmaceutical fields. Since a polymer that can absorb about 1000 times as much water as its own weight was developed in the US in the 1970s, gels have been applied as superabsorbent polymers in several industrial fields, mainly in application to sanitary items, disposable diapers, and so on. Further, in 1978, it was discovered by Tanaka [1] that gels change volume reversibly and discontinuously in response to environmental changes such as solvent composition, temperature, pH change, and so on, (called *volume phase transition* phenomena). With this discovery as a turning point, research to use gels as functional materials for artificial muscle, robot hands (actuator), stimuli-responsive drug delivery systems (DDSs), separation or purification, cell culture, biosensors, shape memory materials, and so on, was activated [2–8].

Until now, fundamental and applied research, which includes many different fields such as elucidation of gelation mechanisms, analysis of physical properties and structure, functional control by molecular design, and so on, have been carried out. Especially, from the early 1990s, new functional gels that include the following three functions in themselves, namely, sensing an external signal (sensor function), judging it (processor function), and taking action (actuator function), have been developed by many researchers as "intelligent gels" or "smart gels."

Further, in recent years, the usefulness of gels has also been shown in the field of micromachines and nanotechnology. In addition to new synthetic methods to give unique functions by molecular design in the nanoscale including supramolecular design, the design and construction of micro- or nanomaterial systems with the

Nonlinear Dynamics with Polymers: Fundamentals, Methods and Applications.
Edited by John A. Pojman and Qui Tran-Cong-Miyata
Copyright © 2010 WILEY-VCH Verlag GmbH & Co. KGaA, Weinheim
ISBN: 978-3-527-32529-0

biomimetic functions of motion, mass transport, transformation and transmission of information, molecular recognition, and so on, have been attempted.

So far, many researchers have developed stimuli-responsive polymer gels that change volume abruptly in response to a change in their surroundings such as solvent composition, temperature, pH, and supply of electric field, and so on. Their ability to swell and deswell according to conditions makes them an interesting proposition for use in new intelligent materials. In particular, their applications in biomedical fields have been studied extensively. One of the strategies of these applications is to develop biomimetic material systems with stimuli-responding function; that is, systems in which the materials sense environmental changes by themselves and go into action. For these systems, the on–off switching of external stimuli is essential to instigate the action of the gel. Upon switching, the gels provide only one unique action, either swelling or deswelling.

This stimuli-responding behavior is a temporary action toward an equilibrium state. In contrast, there are many physiological phenomena in our body that continue their own native cyclic changes. These phenomena exist over a wide range from the cell to the body level, as represented by the cell cycle, cyclic reaction in glycolysis, pulsatile secretion of hormones, pulsatile potential of nerve cells, brain waves, heartbeat, peristaltic motion in the digestive tract, and human biorhythms, and so on. If such self-oscillation could be achieved for gels, possibilities would emerge for new biomimetic intelligent materials that exhibit autonomous rhythmical motion.

In this chapter, a new design concept for polymer gels that exhibit spontaneous and autonomous periodic swelling–deswelling changes under constant conditions without on–off switching of external stimuli will be introduced. In the materials design, nonlinear dynamics of chemical reactions and characteristics of gels as open systems play an important role.

7.2
Design of Self-Oscillating Gel

In order to realize the autonomous polymer system by tailormade molecular design, we focused on the Belousov–Zhabotinsky (BZ) reaction [9, 10], which is well known for exhibiting temporal and spatiotemporal oscillating phenomena. We attempted to convert the chemical oscillation of the BZ reaction to the mechanical changes of gels and generate an autonomic swelling–deswelling oscillation under nonoscillatory outer conditions [11–14]. A copolymer gel that consists of N-isopropylacrylamide (NIPAAm) and ruthenium tris(2, 2′-bipyridine) ($Ru(bpy)_3^{2+}$) was prepared. $Ru(bpy)_3^{2+}$, acting as a catalyst for the BZ reaction, is pendent to the polymer chains of NIPAAm (Figure 7.1). The poly(NIPAAm-co-$Ru(bpy)_3^{2+}$) gel has a phase transition temperature because of thermosensitive constituent NIPAAm. The oxidation of the $Ru(bpy)_3^{2+}$ moiety causes not only an increase in the swelling degree of the gel, but also a rise in the transition temperature. These characteristics may be interpreted by considering an increase in hydrophilicity of the polymer chains due to the oxidation of Ru(II)

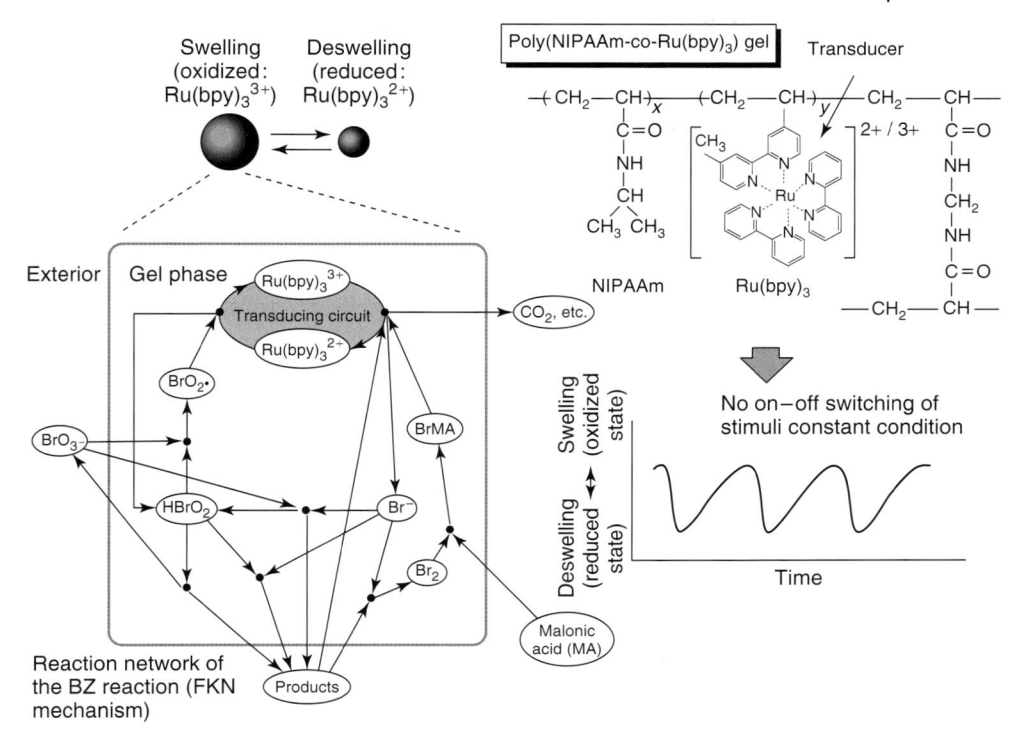

Figure 7.1 Mechanism of self-oscillation for poly(NIPAAm-co-Ru(bpy)$_3^{2+}$) gel coupled with the Belousov–Zhabotinsky reaction.

to Ru(III) in the Ru(bpy)$_3$ moiety. As a result, it is expected that the gel will undergo a cyclic swelling–deswelling alteration when the Ru(bpy)$_3$ moiety is periodically oxidized and reduced under constant temperature. When the gel is immersed in an aqueous solution containing the substrates of the BZ reaction (molonic acid and oxidant) except for the catalyst, the substrates penetrate into the polymer network and the BZ reaction occurs in the gel. Consequently, periodic redox changes induced by the BZ reaction produce periodic swelling–deswelling changes of the gel (Figure 7.1).

7.3 Self-Oscillating Behaviors of the Gel

7.3.1 Self-Oscillation of the Miniature Bulk Gel

Figure 7.2a shows the observed oscillating behavior under a microscope for the miniature cubic poly(NIPAAm-co-Ru(bpy)$_3^{2+}$) gel (each of length of about 0.5 mm).

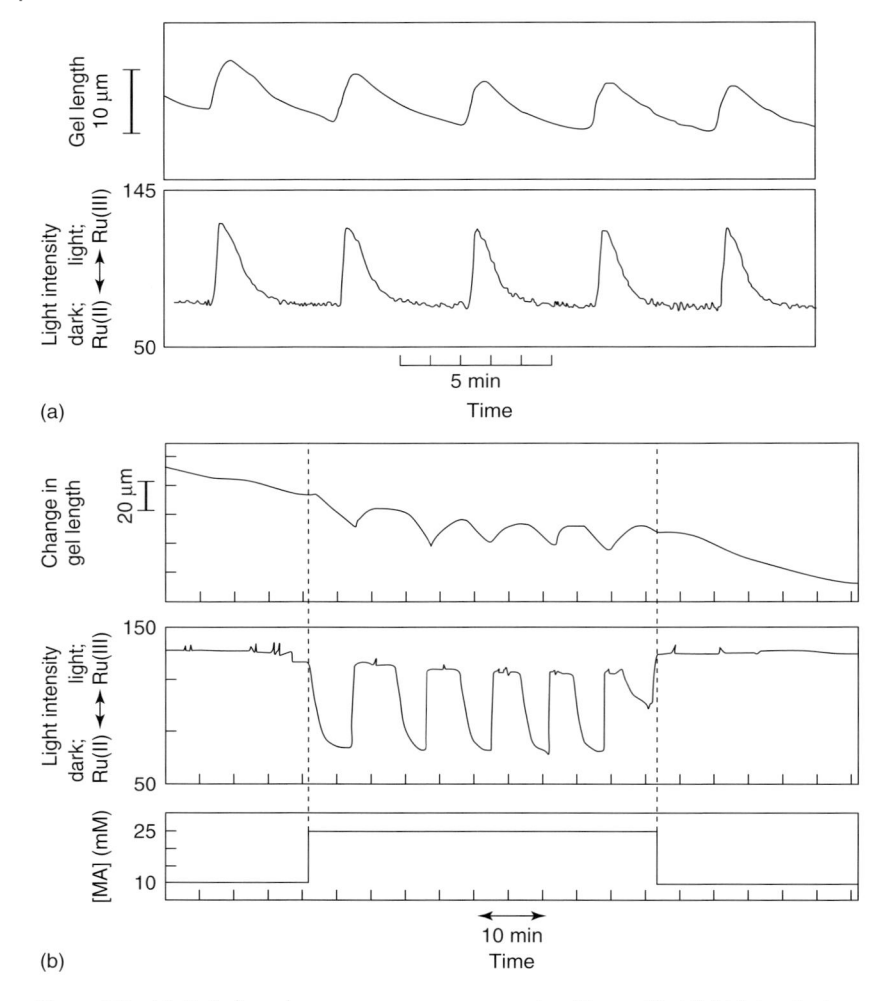

Figure 7.2 (a) Periodic redox changes of the miniature cubic poly(NIPAAm-co-Ru(bpy)$_3$$^{2+}$) gel (lower) and the swelling–deswelling oscillation (upper) at 20 °C. Color changes of the gel accompanied by redox oscillations (orange: reduced state, light green: the oxidized state) were converted to 8-bit grayscale changes (dark: reduced, light: oxidized) by image processing. Transmitted light intensity is expressed as an 8-bit grayscale value. Outer solution: [MA] = 62.5 mM; [NaBrO$_3$] = 84 mM; [HNO$_3$] = 0.6 M. (b) Change in oscillating behavior of the gel in response to the stepwise concentration changes of MA between 10 and 25 mM (others: [NaBrO$_3$] = 84 mM, [HNO$_3$] = 0.3 M, 20 °C).

In miniature gels sufficiently smaller than the wavelength of the chemical wave (typically several millimeters), the redox change of ruthenium catalyst can be regarded to occur homogeneously without pattern formation [15]. Because of the redox oscillation of the immobilized Ru(bpy)$_3$$^{2+}$, mechanical swelling–deswelling oscillation of the gel autonomously occurs with the same period as for the redox oscillation. The volume change is isotropic and the gel beats as a whole, like a heart muscle cell. The chemical and mechanical oscillations are synchronized without a phase difference (i.e., the gel exhibits swelling during the oxidized state and deswelling during the reduced state).

7.3.2
Control of Oscillating Behaviors

Typically, the oscillation period increases with a decrease in the initial concentration of the substrates. Further, in general, the oscillation frequency (the reciprocal of the period) of the BZ reaction tends to increase as the temperature increases, in accordance with the Arrhenius equation. The swelling–deswelling amplitude of the gel increases with an increase in the oscillation period and amplitude of the redox changes. Therefore, the swelling–deswelling amplitude of the gel is controllable by changing the initial concentration of the substrates as well as the temperature.

As an inherent behavior of the BZ reaction, the abrupt transition from a steady state (nonoscillating state) to an oscillating state occurs with a change in controlling parameter such as chemical composition, light, and so on. By utilizing these characteristics, reversible on–off regulation of self-beating triggered by addition and removal of MA was successfully achieved [16]. Figure 7.2b shows the oscillating behavior of the gel when the stepwise change in MA concentration was repeated between a lower concentration (10 mM) in steady state and a higher concentration (25 mM) in the oscillating state. At [MA] = 10 mM, the redox oscillation does not occur and, consequently, the gel exhibited no swelling–deswellng changes. Then the concentration was quickly increased to 25 mM. Immediately after the concentration was increased, the gel started self-beating. The beating stopped again as soon as the concentration was decreased back to the initial value. In this way, reversible on–off regulation of self-beating triggered by MA was successfully achieved. Since there are some organic acids that can act as substrate for the BZ reaction (e.g., citric acid), the same regulation of beating is possible by using those organic acids instead of MA. Furthermore, since the gel has thermosensitivity due to the NIPAAm component, the beating rhythm can be also controlled by temperature [17].

7.3.3
Peristaltic Motion of Gels with Propagation of Chemical Wave

When the gel size is larger than chemical wavelength, the chemical wave propagates in the gel by coupling with diffusion of intermediates. Then, a peristaltic motion of the gel is created. Figure 7.3 shows a cylindrical gel that is immersed in an aqueous solution containing the three reactants of the BZ reaction. The chemical waves propagate in the gel at a constant speed in the direction of the gel length [18]. Considering the orange (Ru(II)) and green (Ru(III)) zones as representing simply the shrunken and swollen parts, respectively, the locally swollen and shrunken

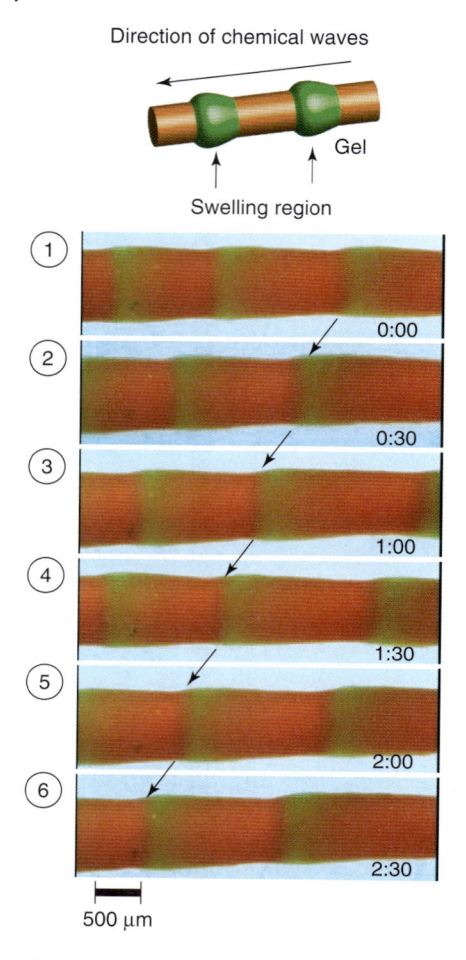

Figure 7.3 Time course of peristaltic motion of poly(NIPAAm-co-Ru(bpy)$_3^{2+}$-co-AMPS) gel in a solution of the BZ substrates (MA, sodium bromate, and nitric acid, 18 °C). The green and orange colors correspond to the oxidized and reduced states of the Ru moiety in the gel, respectively.

parts move with the chemical wave, like the peristaltic motion of living worms. The tensile force of the cylindrical gel with oscillation was also measured [19].

It is well known that the period of oscillation is affected by light illumination for the Ru(bpy)$_3^{2+}$-catalyzed BZ reaction [20]. Therefore, we can intentionally make a pacemaker with a desired period (or wavelength) by local illumination of the gel by a laser beam; or we can change the period (or wavelength) by local illumination of a pacemaker that already exists in the gel [21].

7.3.4
Self-Oscillation with Structural Color Changes

We prepared a periodically ordered mesoporous gel that reveals "structural color" depending on its swelling ratio. To obtain the gel, we used as a template the closest packing colloidal crystal composed of spherical silica particles (Figure 7.4). The porous gel obtained exhibits a bright color under white light and undergoes fast and drastic changes in color in response to temperature. The color is caused by the Bragg diffraction of visible light from the ordered voids regarded as crystallites. Because this coloring is due primarily to the structures formed in the crystal-like structure, we call it *structural color*. The peak values of the reflection spectra, λ_{max}, for the porous gel are obtained by

$$\lambda_{max} = 1.633(d/m)(D/D_0)n_a$$

where d is the diameter of the colloidal particles used, m is the order of the Bragg reflection, D and D_0 are the characteristic sizes of the gel in the equilibrium state at a certain condition and in the preparative state, respectively, and n_a is the average refractive index of the porous gel. D/D_0 is defined as the degree of equilibrium swelling of the gel. The value of n_a can be treated as a constant under varying degrees of D/D_0. It follows that the variable thickness L of the gel membrane can be calculated by using the spatiotemporally determined value of λ_{max}, according to

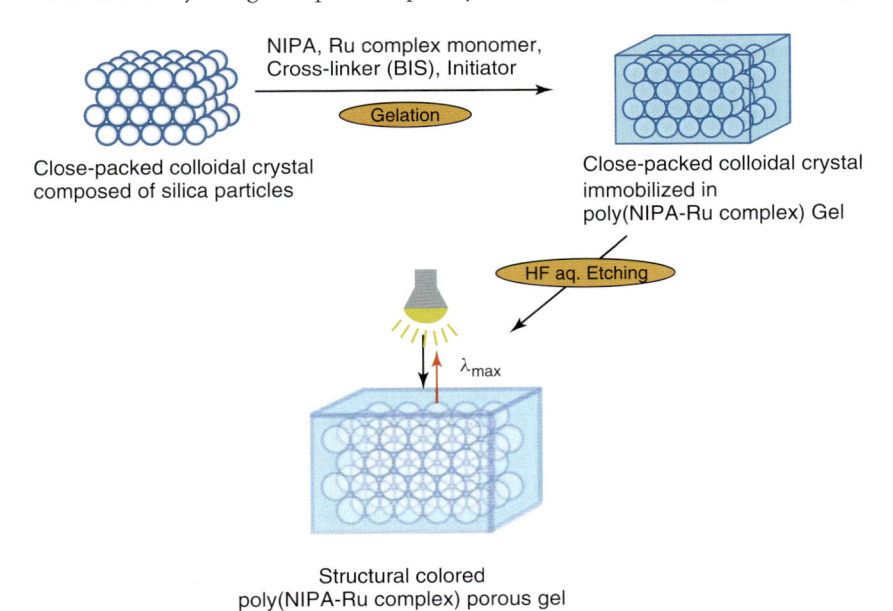

Figure 7.4 Preparation of a periodically ordered interconnecting porous poly(NIPAAm-*co*-Ru(bpy)$_3$) gel using a close-packed colloidal silica crystal as a template.

the equation

$$L = L_0(D/D_0) = L_0 m\lambda_{max}/1.633 dn_a$$

Here, L_0 is the preparative thickness of the gel membrane.

The templated periodical porous structure in the gel can reversibly change its spacing in response to temperature changes impacting the volume of the gels and will signal these changes through iridescence. The structural color is visible even when the structural color overlaps with the pigmented color of the Ru complex.

Considering these results, we expected that the self-sustained peristaltic motion on the surface of a gel could be observed through the change in the structural color during the BZ reaction. To demonstrate this oscillating change in the structural color, the porous disk-shaped gel was immersed in a reaction solution containing nitric acid, malonic acid, and sodium bromate at certain temperatures (Figure 7.5). At 19 °C, the structural color of the gel shifted to the ultraviolet region. Concentric brilliant blue rings, caused by the swelling of the gel due to the oxidation of the Ru complex spread out from the center with the BZ reaction, could be observed on the surface of the gel. Faint red structural colored concentric rings developed and spread out on the porous gel at 4 °C, whereas the green concentric waves which were also indistinct from the surrounding color were observed at 12 °C. The change in the reflection spectra was monitored by fixing a reflection probe in the appropriate position above the porous gel. A periodic swinging of the reflection spectra was observed during the BZ reaction. From the equation mentioned above, the temporal change in the thickness of the gel can be estimated. By this evaluation technique, chemical and optical control of the self-sustaining peristaltic motion of a structural colored porous gel was demonstrated (Figure 7.6) [22–24].

7.4
Design of Biomimetic Micro-/Nanoactuator Using Self-Oscillating Polymer and Gel

7.4.1
Self-Walking Gel

Further, we successfully developed a novel biomimetic walking-gel actuator made of the self-oscillating gel [25]. To produce directional movement of a gel, asymmetrical swelling–deswelling is desired. For these purposes, as a third component, hydrophilic 2-acrylamido-2-methylpropanesulfonic acid (AMPS) was copolymerized into the polymer to lubricate the gel and to cause anisotropic contraction. During polymerization, the monomer solution faces two different surfaces of plates: a hydrophilic glass surface and a hydrophobic Teflon surface. Since $Ru(bpy)_3^{2+}$ monomer is hydrophobic, it easily migrates to the Teflon surface side. As a result, a nonuniform distribution along the height is formed by the components, and the resulting gel has a gradient distribution for the content of each component in the polymer network.

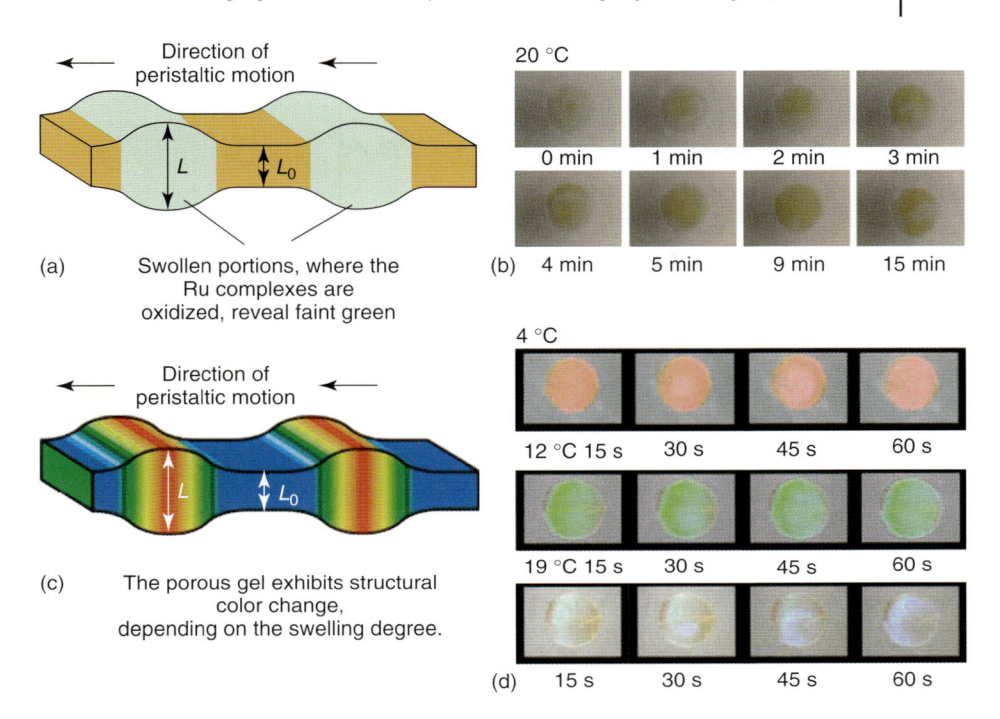

Figure 7.5 Models and pictures of peristaltic motion of the surface of poly(NIPAAm-co-Ru(bpy)$_3$) gel membranes. (a) and (b) Spatiotemporal color patterns of oscillating behavior for a rectangular (a: model) and a disk (b: pictures) bulk poly(NIPAAm-co-Ru(bpy)$_3$) gel can be observed because of the difference in color between the reduced state and the oxidized state of the Ru complex. (c) and (d) Spatiotemporal structural color changes of a rectangular-shaped (c: model) and a disk (d: pictures) porous poly(NIPAAm-co-Ru(bpy)$_3$) gel during the BZ reaction shows that the gel moves like an earthworm. The pictures show the time change in the pigment color of the bulk gel and the structural color of the porous gel during the BZ reaction at several temperatures. Both disk-shaped gels (4 mm diameter and 0.5 mm thickness) were immersed in an aqueous solution (20 ml) containing malonic acid (0.0625 M), sodium bromate (0.084 M), and nitric acid (0.890 M) in a temperature-controlled glass cell.

In order to convert the bending and stretching changes to one-directional motion, we employed a ratchet mechanism. A ratchet base with an asymmetrical surface structure was fabricated. On the ratchet base, the gel repeatedly bends and stretches autonomously resulting in the forward motion of the gel, while the teeth of the ratchet prevent the slide backwards. Figure 7.7 shows successive profiles of the "self-walking" motion of the gel like a looper in the BZ substrate solution under constant temperature. The walking velocity of the gel actuator was approximately 170 µm min^{-1}. Since the oscillating period and the propagating velocity of chemical wave change with concentration of substrates in the outer solution, the walking velocity of the gel can be controlled. By using a gel with a gradient structure, another type of actuator that generates a pendulum motion is also realized [26].

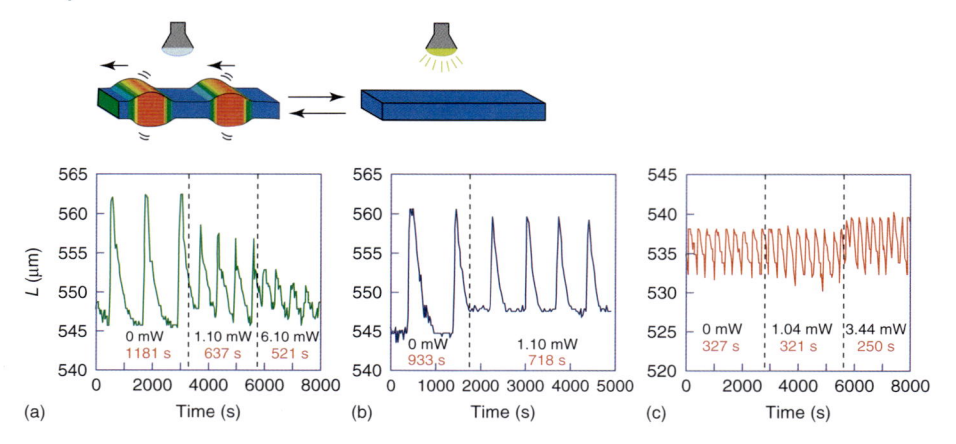

Figure 7.6 Irradiated light intensity and sodium bromate concentration tuning of self-sustaining peristaltic motion of the porous gel at 13 °C. Outer solution: [MA] = 62.5 mM; [HNO$_3$] = 0.890 M; [NaBrO$_3$] = 42 mM (a), 60 mM (b), and 84 mM (c). The periods of the self-sustaining peristaltic motions for each condition are shown.

7.4.2
Mass Transport Surface Utilizing Peristaltic Motion of Gel

Further, we attempted to transport an object by utilizing the peristaltic motion of poly(NIPAAm-co-Ru(bpy)$_3$-co-AMPS) gels. As a model object, a cylindrical poly(acrylamide) (PAAm) gel was put on the gel surface. It was observed that the PAAm gel was transported on the gel surface with the propagation of the chemical wave as it rolled [27] (Figure 7.8). We have proposed a model to describe the mass transport phenomena based on the Hertz contact theory, and investigated the relation between the transportability and the peristaltic motion. The functional gel surface generating autonomous and periodic peristaltic motion has the potential for several new applications such as a conveyor to transport soft materials, a formation process for ordered structures of micro- and/or nanomaterials, a self-cleaning surface, and so on.

7.4.3
Microfabrication of Self-Oscillating Gel for Microdevices

Recently, microfabrication techniques, such as photolithography, are also attempted for preparation of microgels. Since any shape of gel can be created by these methods, application as a new manufacturing method for soft microactuator, microgel valve, gel display, and so on, is expected. Microfabrication of self-oscillating gel has also been attempted by photolithography for application to such microdevices [28, 29].

In these devices, possible application to DDSs is a self-oscillatory drug release microchip. If the microfabricated self-oscillating gel can be used as beating

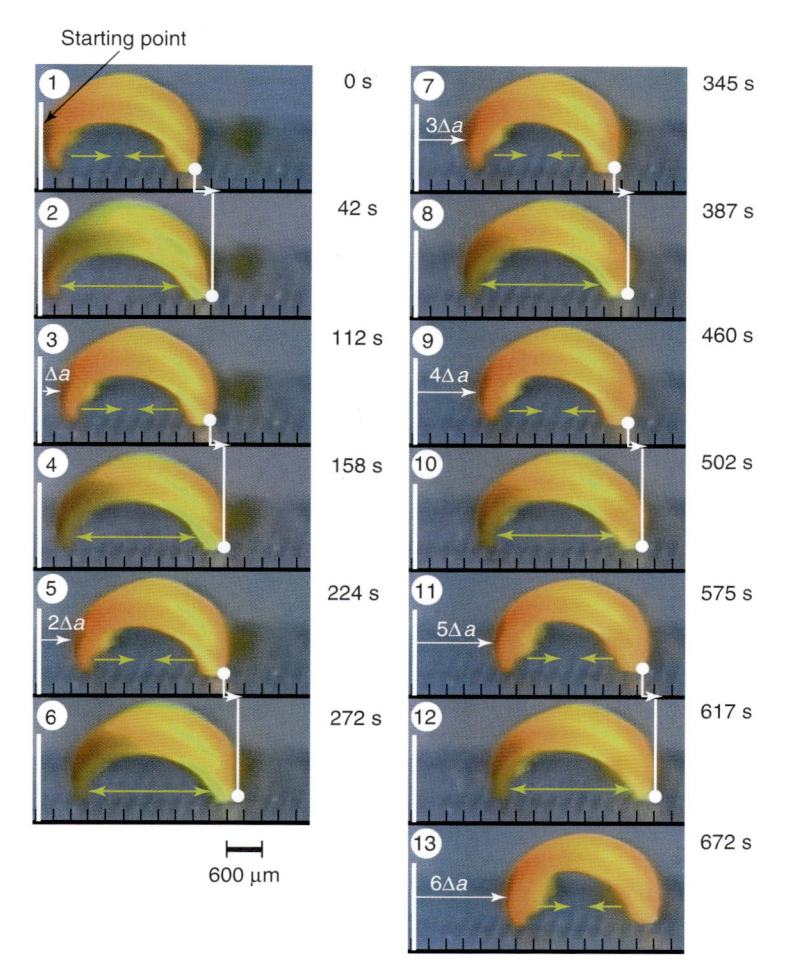

Figure 7.7 Time course of self-walking motion of the gel actuator. During stretching, the front edge can slide forward on the base, but the rear edge is prevented from sliding backward. Oppositely, during bending, the front edge is prevented from sliding backward while the rear edge can slide forward. This action is repeated, and as a result, the gel walks forward. Outer solution: [MA] = 62.5 mM, [NaBrO$_3$] = 84 mM, [HNO$_3$] = 0.894 M, 18 °C.

micropump to push and pull a diaphragm that separates drug reservoir in microchip, the microchip to release a drug periodically would be realized. Since the self-beating of the gel occurs in the closed solution containing the BZ reactants as energy source, complete stand-alone microchip without electric wiring and external apparatus is possible. Periodic release of drugs with preprogrammed periods under constant condition can be achieved, and it would lead to several effects: application to chronopharmacotherapy to release hormones synchronized with biorhythms, decreasing drug tolerance, and so on.

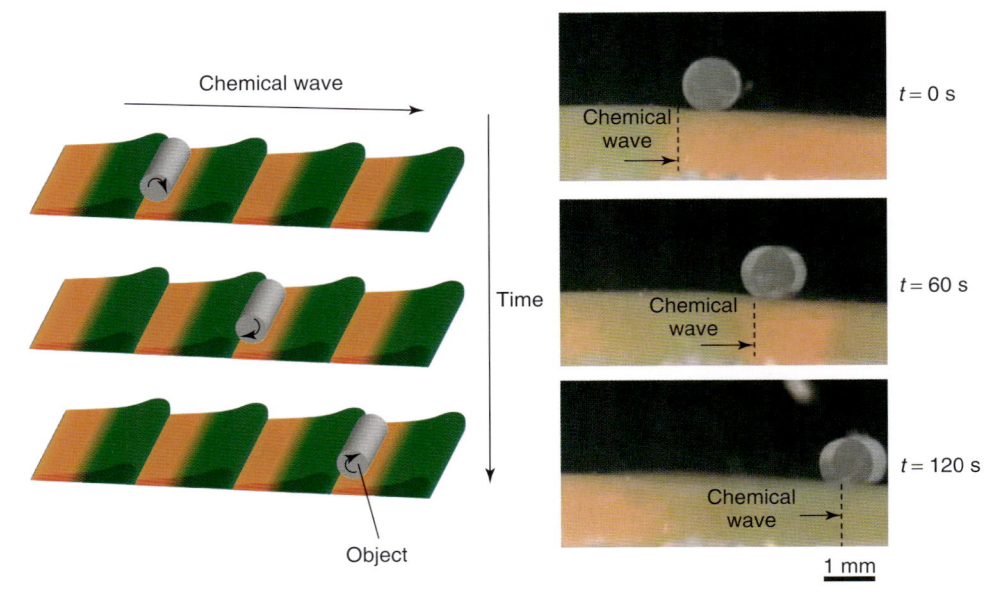

Figure 7.8 Schematic illustration of mass transport on the peristaltic surface (a) and observed transport of cylindrical PAAm gel on the poly(NIPAAm-co-Ru(bpy)$_3^{2+}$-co-AMPS) gel sheet (b).

Microfabrication of self-oscillating gel has also been attempted by lithography for application to a ciliary motion actuator (artificial cilia) [30]. The gel membrane with microprojection array on the surface was fabricated by utilizing X-lay lithography (LIGA). With the propagation of chemical wave, the microprojection array exhibits dynamic rhythmic motion like cilia. The actuator may also serve as a microconveyor.

7.4.4
Control of Chemical Wave Propagation in Self-Oscillating Gel Array

A chemomechanical actuator utilizing a reaction/diffusion wave across a gap was constructed in a novel mirconveyor by a micropatterned self-oscillating gel array [31]. Unidirectional propagation of the chemical wave the BZ reaction was induced on gel arrays. In the case of using a triangle-shaped gel as an element of the array, the chemical wave propagated from the corner of the triangle gel to the plane side of the other gel (C-to-P) across the gap junction, whereas it propagated from the plane side to the corner (P-to-C) in the case of the pentagonal gel array. Numerical analysis based on theoretical model was carried out for understanding the mechanism of unidirectional propagation in triangular and pentagonal gel arrays. By fabricating different shapes of gel arrays, one can possibly control of the direction. The swelling and deswelling changes of the gels followed the unidirectional propagation of the chemical wave. Application to novel microconveyors is expected.

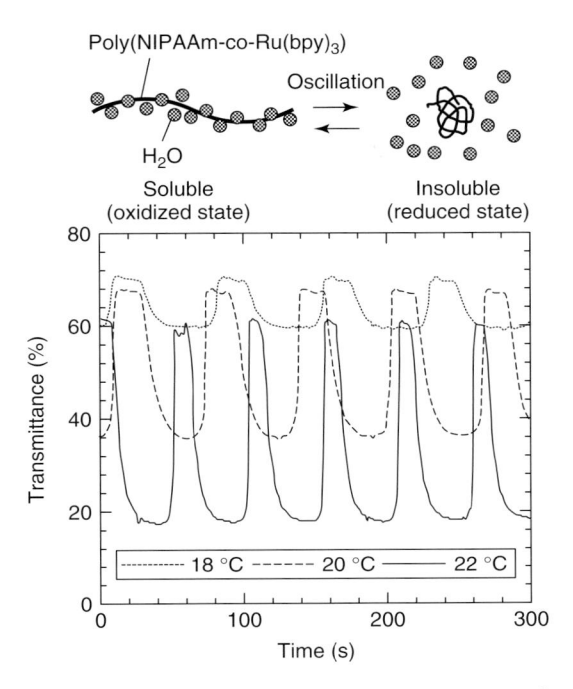

Figure 7.9 Oscillating profiles of optical transmittance for poly(NIPAAm-co-Ru(bpy)$_3$$^{2+}$)(Ru(bpy)$_3$$^{2+}$ = 5 wt%) solution at constant temperatures.

7.4.5
Self-Oscillating Polymer Chains as "Nano-Oscillators"

The periodic changes of linear and un-cross-linked polymer chains can be easily observed as cyclic transparent and opaque changes for the polymer solution with color changes due to the redox oscillation of the catalyst [32]. Synchronized with the periodical changes between Ru(II) and Ru(III) states of the Ru(bpy)$_3$$^{2+}$ site, the polymer becomes hydrophobic and hydrophilic, and exhibits cyclic soluble–insoluble changes (Figure 7.9).

By grafting the polymers or arraying the gel beads on the surface of substrates, we have attempted to design self-oscillating surfaces as nanoconvoyers (Figure 7.10). The self-oscillating polymer was covalently immobilized on a glass surface, and self-oscillation was directly observed at a molecular level by atomic force microscopy (AFM) [33]. The self-oscillating polymer with N-succinimidyl group was immobilized on an aminosilane-coupled glass plate. While no oscillation was observed in pure water, nanoscale oscillation was observed in an aqueous solution containing the BZ substrates (Figure 7.11). The amplitude was about 10–15 nm and the period was about 70 s, although some irregular behavior was observed because of no stirring. The amplitude was less than that in solution, as observed by dynamic light scattering (DLS)

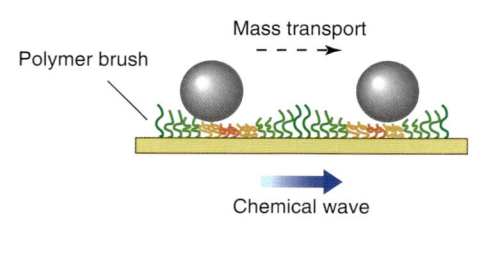

Figure 7.10 Concept of functional surface (nanoconveyor) using self-oscillating polymer and gel beads.

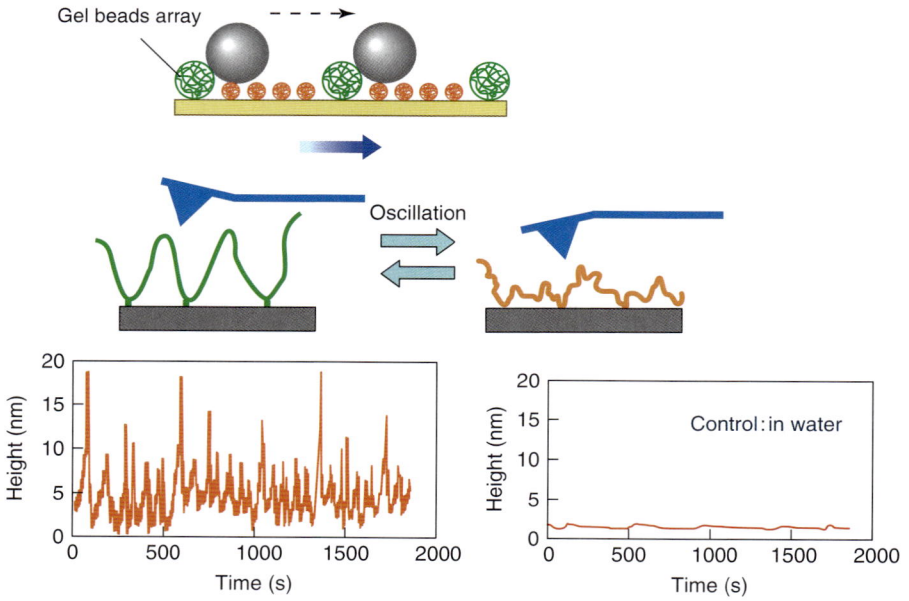

Figure 7.11 Self-oscillating behavior of immobilized polymer in the BZ substrate solution ([MA] = 0.1 M, [NaBrO$_3$] = 0.3 M, [HNO$_3$] = 0.3 M) measured by AFM.

(23.9 and 59.6 nm). This smaller amplitude may be because the structure of the immobilized polymer was a loop-train-tail: the moving regions were shorter than that of the soluble polymer, as illustrated in Figure 7.6. The amplitude and frequency were controlled by the concentration of reactant, as observed in the solution. Here, nanoscale molecular self-oscillation was observed for the first time. The oscillating polymer chain may be used as a component of a nanoclock or a nanomachine.

7.4.6
Self-Flocculating/Dispersing Oscillation of Microgels

We then prepared submicron-sized poly(NIPAAm-co-Ru(bpy)$_3^{2+}$) gel beads by surfactant-free aqueous precipitation polymerization, and analyzed their oscillating behaviors [34–37]. Figure 7.12 shows the oscillation profiles of transmittance

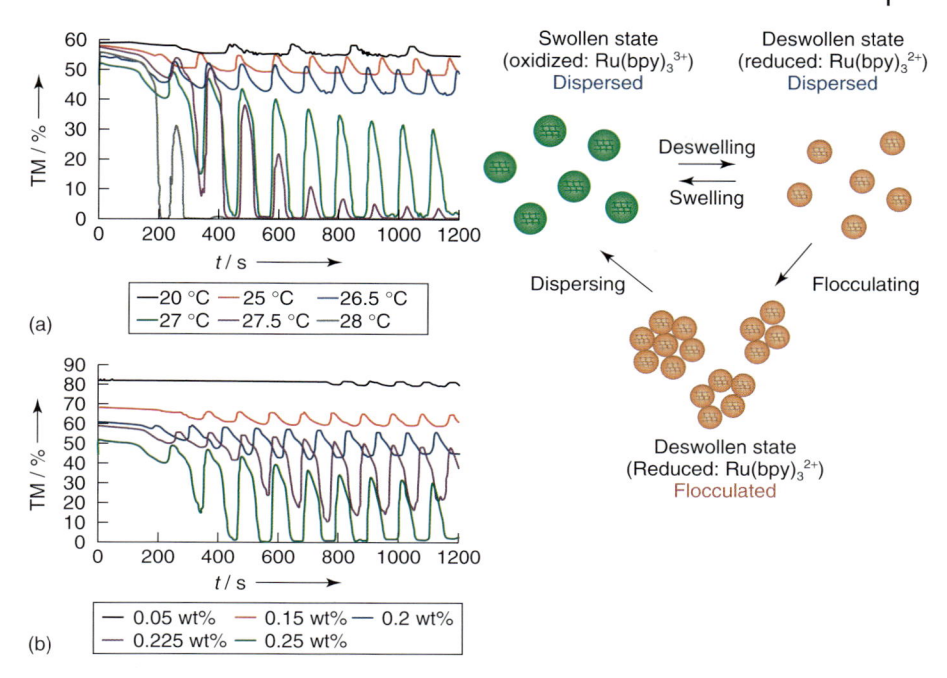

Figure 7.12 Self-oscillating profiles of optical transmittance for microgel dispersions. The microgels were dispersed in aqueous solutions containing MA (62.5 mM), NaBrO$_3$ (84 mM), and HNO$_3$ (0.3 M). Microgel concentration was 0.25 wt%. (a) Profiles measured at different temperatures. (b) Profiles measured at different microgel dispersion concentrations at 27 °C.

for the microgel dispersions. At low temperatures (20–26.5 °C), on raising the temperature, the amplitude of the oscillation became larger. The increase in amplitude is due to increased deviation of the hydrodynamic diameter between the Ru(II) and Ru(III) states. Furthermore, a remarkable change in waveform was observed between 26.5 and 27 °C. Then the amplitude of the oscillations dramatically decreased at 27.5 °C, and finally the periodic transmittance changes could no longer be observed at 28 °C. The sudden change in oscillation waveform should be related to the difference in colloidal stability between the Ru(II) and Ru(III) states. Here, the microgels should be flocculated as a result of the lack of electrostatic repulsion when the microgels were deswollen. The remarkable change in waveform was only observed at higher dispersion concentrations (greater than 0.225 wt%). The self-oscillating property makes microgels attractive for future developments such as microgel assembly, optical and rheological applications, and so on.

7.4.7
Fabrication of Microgel Beads Monolayer

As discussed in the previous section, we have been interested in the construction of micro-/nanoconveyors by grafting or arraying self-oscillating polymer or gel beads.

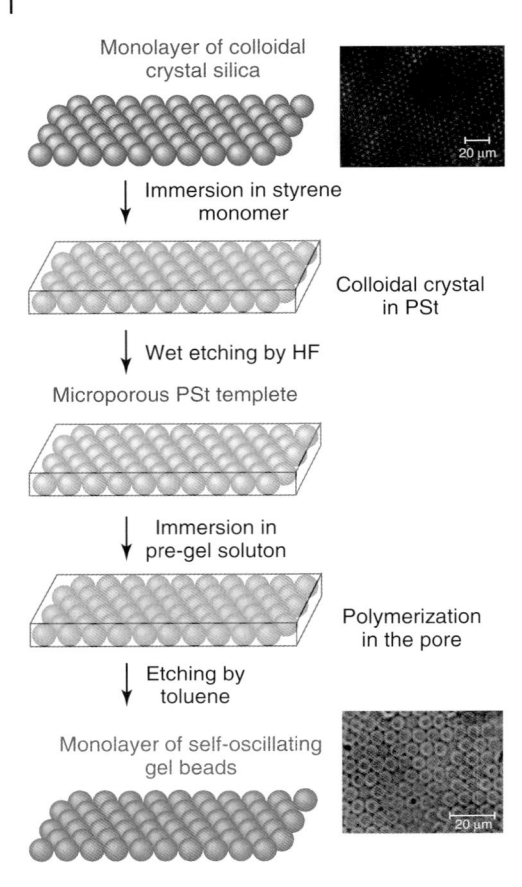

Figure 7.13 Preparation of self-oscillating gel bead monolayer by two-step template polymerization.

For this purpose, a fabrication method for organized monolayers of microgel beads was investigated [38]. A 2D close-packed array of thermosensitive microgel beads was prepared by double-template polymerization (Figure 7.13). First, a 2D colloidal crystal of silica beads with 10 μm diameter was obtained by solvent evaporation. This monolayer of colloidal crystal can serve as the first template for the preparation of macroporous polystyrene. The macroporous polystyrene trapping the crystalline order can be used as a negative template for fabricating a gel bead array. By this double-template polymerization method, functional surfaces using thermosensitive PNIPAAm gel beads were fabricated. It was observed that the topography of the surface changed with temperature. The fabrication method demonstrated here was so versatile that any kind of gel beads could be obtained. This method may be a key technology to create new functional surfaces. Actually, a monolayer of self-oscillating microgel beads was fabricated by this method and the chemical wave propagation on the monolayer was observed.

7.4.8
Attempts of Self-Oscillation under Physiological Conditions

However, in this self-oscillating polymer system the operating conditions are limited to the nonphysiological environment where the strong acid and the oxidant coexist. For extending the application field to biomaterials, more sophisticated molecular design to cause self-oscillation under physiological condition is needed. For this purpose, we constructed an integrated polymer system where all of the BZ substrates other than the biorelated organic substrate were incorporated into the polymer chain [39]. We synthesized the quaternary copolymer, which includes both of the pH-control and oxidant-supplying sites in the poly(NIPAAm-co-Ru(bpy)$_3$) chain at the same time. In the polymer, AMPS was incorporated as a pH-control site, and methacrylamidopropyltrimethylammonium chloride (MAPTAC) with a

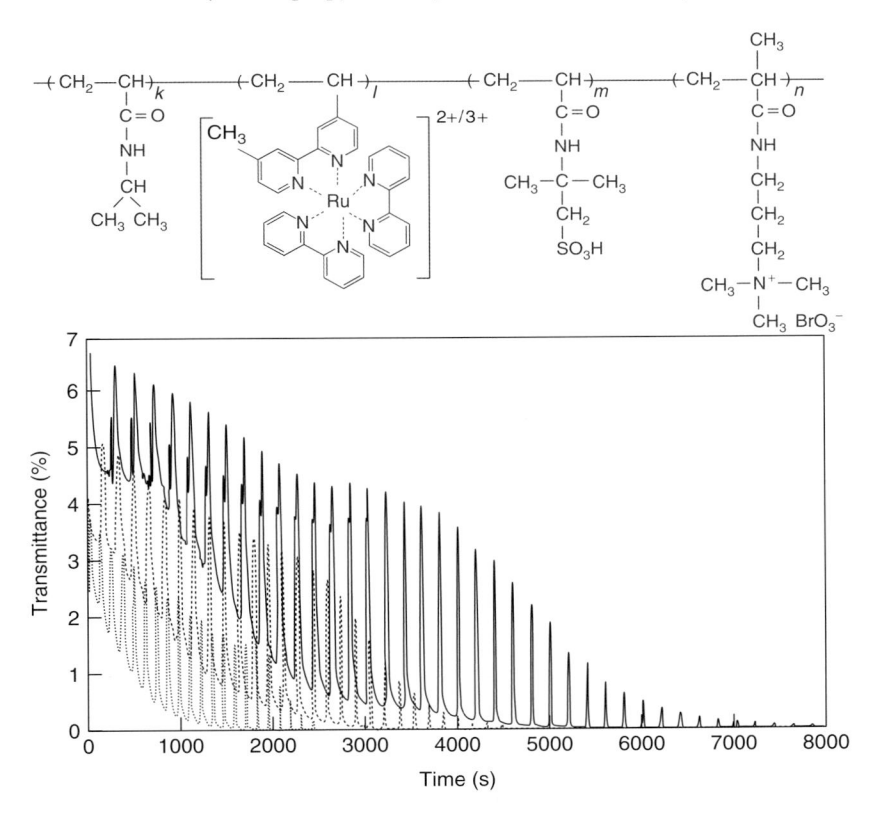

Figure 7.14 Chemical structure of poly(NIPAAm-co-Ru(bpy)$_3$$^{2+}$-co-AMPS-co-MAPTAC) (a) and the oscillating profiles of the optical transmittance for the polymer solution at 12 °C when only MA (0.7 M: fine dotted line, 0.5 M: rough dotted line, 0.3 M: solid line) is added to the solution (b).

positively charged group was incorporated as a capture site for an anionic oxidizing agent (bromate ion). By using the polymer, self-oscillation under biological conditions where only the organic acid (malonic acid) exists was actually achieved (Figure 7.14).

7.5
Conclusion

As pointed out in this chapter, novel biomimetic gels with self-oscillating function have been developed. The self-oscillating gel may be useful in a number of important applications, such as pulse generators or chemical pacemakers, self-walking (auto-mobile) actuators or micropumps with autonomous beating or peristaltic motion, device for signal transmission utilizing propagation of chemical waves, self-oscillating viscous fluid [40–42], functional microgels [43], oscillatory drug release synchronized with cell cycles or human biorhythms, and so on. Further studies on the control of oscillating behavior as well as practical applications are expected.

References

1. (a) Tanaka, T. (1978) Collapse of gels and the critical endpoint. *Phys. Rev. Lett.*, **40**, 820–823; (b) Tanaka, T. (1981) Gels. *Sci. Am.*, **244**, 124–136.
2. Yoshida, R. (2005) Design of functional polymer gels and their application to biomimetic materials. *Curr. Org. Chem.*, **9**, 1617–1641.
3. Yoshida, R., Sakai, K., Okano, T., and Sakurai, Y. (1993) Modern hydrogel delivery systems. *Adv. Drug Deliv. Rev.*, **11**, 85–108.
4. Okano, T. (ed.) (1998) *Biorelated Polymers and Gels: Controlled Release and Applications in Biomedical Engineering*, Academic Press.
5. Miyata, T. (2002) in *Supramolecular Design for Biological Applications* (ed. N. Yui), CRC Press, Boca Raton, pp. 191–225.
6. Osada, Y. and Khokhlov, A.R. (eds) (2002) *Polymer Gels and Networks*, Marcel Dekker, New York.
7. Yui, N., Mrsny, R.J., and Park, K. (eds) (2004) *Reflexive Polymers and Hydrogels – Understanding and Designing Fast Responsive Polymeric Systems*, CRC Press, Boca Raton.
8. Peppas, N.A., Hilt, J.Z., Khademhosseini, A., and Langer, R. (2006) Hydrogels in biology and medicine: from molecular principles to bionanotechnology. *Adv. Mater.*, **18**, 1345–1360.
9. Field, R.J. and Burger, M. (eds) (1985) *Oscillations and Traveling Waves in Chemical Systems*, John Wiley & Sons, Inc., New York.
10. Epstein, I.R. and Pojman, J.A. (1998) *An Introduction to Nonlinear Chemical Dynamics: Oscillations, Waves, Patterns, and Chaos*, Oxford University Press, New York.
11. Yoshida, R., Takahashi, T., Yamaguchi, T., and Ichijo, H. (1996) Self-oscillating gel. *J. Am. Chem. Soc.*, **118**, 5134–5135.
12. Yoshida, R., Takahashi, T., Yamaguchi, T., and Ichijo, H. (1997) Self-oscillating gels. *Adv. Mater.*, **9**, 175–178.
13. Yoshida, R. (2008) Self-oscillating polymer and gels as novel biomimetic materials. *Bull. Chem. Soc. Jpn.*, **81**, 676–688.
14. Yoshida, R., Sakai, T., Hara, Y., Maeda, S., Hashimoto, S., Suzuki, D., and Murase, Y. (2009) Self-oscillating gel as

novel biomimetic materials. *J. Controlled Release*, **140**, 186–193.

15. Yoshida, R., Tanaka, M., Onodera, S., Yamaguchi, T., and Kokufuta, E. (2000) In-phase synchronization of chemical and mechanical oscillations in self-oscillating gels. *J. Phys. Chem. A*, **104**, 7549–7555.

16. Yoshida, R., Takei, K., and Yamaguchi, T. (2003) Self-beating motion of gels and modulation of oscillation rhythm synchronized with organic acid. *Macromolecules*, **36**, 1759–1761.

17. Ito, Y., Nogawa, N., and Yoshida, R. (2003) Temperature control of the Belousov-Zhabotinsky reaction using a thermo-responsive polymer. *Langmuir*, **19**, 9577–9579.

18. Maeda, S., Hara, Y., Yoshida, R., and Hashimoto, S. (2008) Peristaltic motion of polymer gels. *Angew. Chem. Int. Ed.*, **47**, 6690–6693.

19. Sasaki, S., Koga, S., Yoshida, R., and Yamaguchi, T. (2003) Mechanical oscillation coupled with the Belousov-Zhabotinsky reaction in gel. *Langmuir*, **19**, 5595–5600.

20. Amemiya, T., Ohmori, T., and Yamaguchi, T. (2000) An Oregonator-class model for photoinduced behavior in the $Ru(bpy)_3^{2+}$-catalyzed Belousov-Zhabotinsky reaction. *J. Phys. Chem. A*, **104**, 336–344.

21. Yoshida, R., Sakai, T., Tabata, O., and Yamaguchi, T. (2002) Design of novel biomimetic polymer gels with self-oscillating function. *Sci. Tech. Adv. Mater.*, **3**, 95–102.

22. Takeoka, Y., Watanabe, M., and Yoshida, R. (2003) Self-sustaining peristaltic motion on the surface of a porous gel. *J. Am. Chem. Soc.*, **125**, 13320–13321.

23. Shinohara, S., Seki, T., Sakai, T., Yoshida, R., and Takeoka, Y. (2008) Chemical and optical control of peristaltic actuator based on self-oscillating porous gel. *Chem. Commun.*, 4735–4737.

24. Shinohara, S., Seki, T., Sakai, T., Yoshida, R., and Takeoka, Y. (2008) Photoregulated wormlike motion of a gel. *Angew. Chem. Int. Ed.*, **47**, 9039–9043.

25. Maeda, S., Hara, Y., Sakai, T., Yoshida, R., and Hashimoto, S. (2007) Self-walking gel. *Adv. Mater.*, **19**, 3480–3484.

26. Maeda, S., Hara, Y., Yoshida, R., and Hashimoto, S. (2008) Control of dynamic motion of a gel actuator driven by the Belousov-Zhabotinsky reaction. *Macromol. Rapid Commun.*, **29**, 401–405.

27. Murase, Y., Maeda, S., Hashimoto, S., and Yoshida, R. (2009) Design of a mass transport surface utilizing peristaltic motion of a self-oscillating gel. *Langmuir*, **25**, 483–489.

28. Yoshida, R., Omata, K., Yamaura, K., Ebata, M., Tanaka, M., and Takai, M. (2006) Maskless microfabrication of thermosensitive gels using a microscope and application to a controlled release microchip. *Lab Chip*, **6**, 1384–1386.

29. Yoshida, R., Omata, K., Yamaura, K., Sakai, T., Hara, Y., Maeda, S., and Hashimoto, S. (2006) Microfabrication of functional polymer gels and their application to novel biomimetic materials. *J. Photopolym. Sci. Technol.*, **19**, 441–444.

30. Tabata, O., Hirasawa, H., Aoki, S., Yoshida, R., and Kokufuta, E. (2002) Ciliary motion actuator using self-oscillating gel. *Sens. Actuators A*, **95**, 234–238.

31. Tateyama, S., Shibuta, Y., and Yoshida, R. (2008) Direction control of chemical wave propagation in self-oscillating gel array. *J. Phys. Chem. B*, **112**, 1777–1782.

32. Yoshida, R., Sakai, T., Ito, S., and Yamaguchi, T. (2002) Self-oscillation of polymer chains with rhythmical soluble-insoluble changes. *J. Am. Chem. Soc.*, **124**, 8095–8098.

33. Ito, Y., Hara, Y., Uetsuka, H., Hasuda, H., Onishi, H., Arakawa, H., Ikai, A., and Yoshida, R. (2006) AFM observation of immobilized self-oscillating polymer. *J. Phys. Chem. B*, **110**, 5170–5173.

34. Suzuki, D., Sakai, T., and Yoshida, R. (2008) Self-flocculating/self-dispersing oscillation of microgels. *Angew. Chem. Int. Ed.*, **47**, 917–920.

35. Suzuki, D. and Yoshida, R. (2008) Temporal control of self-oscillation

for microgels by cross-linking network structure. *Macromolecules*, **41**, 5830–5838.
36. Suzuki, D. and Yoshida, R. (2008) Effect of initial substrate concentration of the Belousov-Zhabotinsky reaction on self-oscillation for microgel system. *J. Phys. Chem. B*, **112**, 12618–12624.
37. Sakai, T. and Yoshida, R. (2004) Self-oscillating nanogel particles. *Langmuir*, **20**, 1036–1038.
38. Sakai, T., Takeoka, Y., Seki, T., and Yoshida, R. (2007) Organized monolayer of thermosensitive microgel beads prepared by double-template polymerization. *Langmuir*, **23**, 8651–8654.
39. Hara, Y. and Yoshida, R. (2008) Self-oscillating polymer fueled by organic acid. *J. Phys. Chem. B*, **112**, 8427–8429.
40. Hara, Y. and Yoshida, R. (2008) A viscosity self-oscillation of polymer solution induced by the BZ reaction under acid-free condition. *J. Chem. Phys.*, **128**, 224904.
41. Suzuki, D., Taniguchi, H. and Yoshida, R. (2009) Autonomously oscillating viscosity in microgel dispersions. *J. Am. Chem. Soc.*, **131**, 12058–12059.
42. Taniguchi, H., Suzuki, D. and Yoshida, R. (2010) Characterization of autonomously oscillating viscosity induced by swelling/deswelling oscillation of the microgels. *J. Phys. Chem. B*, **114**, 2405–2410.
43. Suzuki, D. and Yoshida, R. (2010) Self-oscillating core/shell microgels: effect of a crosslinked nanoshell on autonomous oscillation of the core. *Polym. J.*, **42**, 501–508.

8
Self-Oscillating Gels as Biomimetic Soft Materials
Olga Kuksenok, Victor V. Yashin, Pratyush Dayal, and Anna C. Balazs

8.1
Introduction

One of the hallmarks of living systems is irritability, the ability to sense, and respond to a potentially harmful stimulus. The most characteristic response in biological systems to such a threat is to simply move away. For example, the leaves of a mimosa plant fold away from the touch of a hand, and the same hand pulls away from a flame. One of the challenges in designing synthetic biomimetic systems that exhibit analogous behavior is creating macroscopic objects that not only sense an "adverse" condition but also undergo autonomous, directed motion in the presence of this condition. Recently, we have developed theoretical and computational models [1–3] for chemoresponsive polymer gels and, through these models, have been attempting to design such adaptive systems. Our efforts have focused on a particular class of responsive gels, namely, those undergoing the Belousov–Zhabotinsky (BZ) reaction [4]. As we show below, by harnessing the inherent properties of these polymer networks, we can design materials that emit a chemical "alarm signal" and directed motion in response to a mechanical deformation or impact [5]. We can also design a polymeric "worm" that moves away from light of a certain wavelength, which is an adverse stimulus for the BZ reaction [6].

The unique attributes of BZ gels make them ideal candidates for displaying biomimetic behavior, that is, these polymer networks can expand and contract periodically without external stimuli. This autonomous, self-oscillatory behavior is due to a ruthenium catalyst that is covalently bonded to the polymers [7–24]. The BZ reaction generates a periodic oxidation and reduction of the anchored metal ion, which changes the hydrophilicity of the polymer chains, and in this manner the chemical oscillations induce the rhythmic swelling and deswelling in the gel [7–24]. In other words, the chemical energy from BZ reaction fuels the mechanical oscillations of these gels. These self-oscillating gels can perform sustained work until the reagents in the host solution are consumed and can be simply "refueled" by replenishing these solutes. Millimeter-sized pieces of the BZ gels can actually undergo self-oscillations on the order of hours without replenishment of reagents [22, 23].

While this chemomechanical transduction has been demonstrated for BZ gels [7, 8, 24], mechanochemical transduction, where a mechanical deformation excites chemical oscillations within these polymers, has to date not been experimentally observed. Yet, it is this form of energy transduction that would be highly useful in creating a synthetic "skinlike" coating, which could send a signal throughout the system in response to a mechanical impact, and thereby exhibit a novel form of biomimetic irritability. There are, however, two prior studies that provide a tantalizing hint that BZ gels could display the desired functionality. First, experiments on a Nafion membrane loaded with a BZ catalyst showed that pressure from a glass stick could induce transient circular waves that emanated from the pressed area [25]. Second, our two-dimensional (2D) simulations revealed that confined films of BZ gels exhibit a complex dynamical response to a uniform compression [26].

The 2D simulations, however, are not sufficient for capturing the impact of a spatially localized force; this requires a full three-dimensional (3D) model. Such extensive calculations were only recently made possible by our development of the 3D gel lattice spring model (gLSM) [3], which now allows us to pinpoint the effects of nonuniform mechanical deformations. Using our 3D gLSM model, we show below that BZ gels can convert the impact from a spatially localized force into a global signal, which encompasses both chemical waves and surface "ripples" that propagate across the entire sample [5]. We also demonstrate that this response depends on the magnitude, duration, and location of the impact [5]. Thus, BZ gel coatings could be harnessed to execute the concerted, biomimetic functions of "sensing" a local mechanical impact, transmitting information about the deformation, and transporting species along the sample's surface via the impact-generated ripples.

In the above studies, we assume that the catalyst in the gel is homogeneously distributed within the sample. We have, however, also modeled the behavior of heterogeneous BZ gels that contain a well-tailored distribution of the BZ catalyst. Our interest in these materials is inspired by the structural heterogeneity and hierarchy found in responsive or adaptive biological materials. One example of this vital heterogeneity is provided by the structure of skeletal muscle, where the characteristic striations contribute to the functionality of this tissue. Another example is apparent in the morphology of bone, where the structural heterogeneity and hierarchy again contribute to the functioning of the material. By considering active gels that exhibit spatial heterogeneity, we can establish guidelines for expanding the utility of these gels or integrating a number of functions (e.g., sensing, communication, shape changing, and actuation) into one sample.

These heterogeneous systems are also of interest because the patterning of the gel provides a route for designing systems with well-defined dynamical behavior. For example, our 2D simulations of heterogeneous gels revealed that the lateral separation Δx between two BZ patches controls the synchronization dynamics of these patches. By varying Δx, we produced a switching between the in-phase and out-of-phase modes of synchronization in heterogeneous gels containing a linear array of BZ patches [27].

The lateral separation between patches in a sample of a heterogeneous BZ gel would be fixed during the fabrication of the gel. After fabrication, however, the distance could be changed by an applied mechanical deformation of the gel or a swelling or shrinking of the polymer network, which is induced by an external stimulus (e.g., changes in pH or temperature). The resulting mechanical strain could potentially drive a system into or out of the oscillatory state or alter the regime of synchronization between multiple oscillating patches. If this form of mechanochemistry were shown to occur, the heterogeneous BZ gels could be utilized as a novel material that alters its functionality in a response to an external mechanical action.

In order to design the responsive material described above, we investigate the effects of deforming heterogeneous BZ gels; to facilitate the theoretical analysis, we consider a one-dimensional (1D) model of these gels [28]. This 1D model describes BZ gels that are confined within a narrow tube, or capillary. Using this model, we show that the oscillations can be switched "on" and "off" by an applied mechanical strain [28]. For samples that contain two catalyst-containing patches, we also show that the synchronization between the two oscillating patches can be controlled by the imposed mechanical strain.

In the final section, we describe our findings from the first computational study on the influence of light on BZ gels. Using these results, we design a polymeric "worm" that performs not only self-sustained motion but also marked reorientation in response to spatial variations in illumination [6]. As we show below, this interplay between the chemoresponsive gels and the photosensitive reaction [29] can be exploited to initiate the movement of millimeter-sized gels toward the dark.

Before describing our specific findings, we briefly describe the governing equations and numerical approaches that make up our 3D gLSM approach. We then describe our findings on the sensitivity of these adaptive BZ gels to both mechanical deformation and light (of a specific wavelength).

8.2 Methodology

8.2.1 Continuum Equations

The dynamics of this complex BZ gel system can be described [1, 2] by a modified version of the original Oregonator model for the BZ reaction in solution [30] and equations for the elastodynamics of the polymer network. The Oregonator model describes the kinetics of the BZ reaction in terms of the dimensionless concentrations of the oxidized catalyst v and the key reaction intermediate (the activator) u. In the modified version, the equations explicitly depend on the volume fraction of polymer ϕ, which acts as a neutral diluent. The resulting equations for the gel dynamics are [2, 3]:

$$\frac{d_p \phi}{dt} = -\phi \nabla \cdot \mathbf{v}^{(p)} \tag{8.1}$$

$$\frac{d_p v}{dt} = -v \nabla \cdot \mathbf{v}^{(p)} + \varepsilon G(u, v, \phi) \tag{8.2}$$

$$\frac{d_p u}{dt} = -u \nabla \cdot \mathbf{v}^{(p)} + \nabla \cdot \left[\mathbf{v}^{(p)} \frac{u}{1-\phi} \right] + \nabla \cdot \left[(1-\phi) \nabla \frac{u}{1-\phi} \right] + F(u, v, \phi) \tag{8.3}$$

Here, $d_p/dt \equiv \partial/\partial t + \mathbf{v}^{(p)} \cdot \nabla$ denotes the material time derivative, where $\mathbf{v}^{(p)}$ is the velocity of the polymer network. The functions $G(u, v, \phi)$ and $F(u, v, \phi)$ describe the BZ reaction occurring within the gel [31]:

$$G(u, v, \phi) = (1-\phi)^2 u - (1-\phi)v \tag{8.4}$$

$$F(u, v, \phi) = (1-\phi)^2 u - u^2 - (1-\phi) f v \frac{u - q(1-\phi)^2}{u + q(1-\phi)^2} \tag{8.5}$$

The parameters f, q, and ε in the above equations have the same meaning as in the original Oregonator model. The stoichiometric parameter f effectively controls the concentration of oxidized catalyst v in the steady state and affects the amplitude of the oscillations in the oscillatory state [32].

Equation (8.1) is the continuity equation for the polymer; Eq. (8.2) describes the evolution of v due to the BZ reaction and the transport of the catalyst with the movement of the polymer network. Recall that the catalyst is tethered to the chains and, thus, does not diffuse through the solution. Equation (8.3) characterizes the changes in u due to the BZ reaction, the transport of this activator with the solvent, and the diffusion of the activator within the solvent. We assumed for simplicity that it is solely the polymer–solvent interdiffusion that contributes to the gel dynamics; hence, in Eq. (8.3), we took into account that $\phi \mathbf{v}^{(p)} + (1-\phi) \mathbf{v}^{(s)} = 0$, where $\mathbf{v}^{(s)}$ is the solvent velocity [2].

The dynamics of the polymer network is assumed to be purely relaxational, so that the forces acting on the deformed gel are balanced by the frictional drag due to the motion of the solvent [33]. Thus, we can write [2]:

$$\mathbf{v}^{(p)} = \Lambda_0 (1-\phi)(\phi/\phi_0)^{-3/2} \nabla \cdot \hat{\sigma} \tag{8.6}$$

Here, Λ_0 is the dimensionless kinetic coefficient and $\hat{\sigma}$ is the dimensionless stress tensor measured in units of $v_0^{-1} T$, where v_0 is the volume of a monomeric unit, T is temperature measured in energy units, and ϕ_0 is the volume fraction of polymer in the undeformed state. The factor $(\phi/\phi_0)^{-3/2}$ in Eq. (8.6) takes into account the ϕ dependence of the polymer–solvent friction in swollen polymer gels [33]. The stress tensor can be derived [2] from the free energy density U of the deformed gel, which consists of the elastic energy density associated with the deformations U_{el} and the polymer-solvent interaction energy density U_{FH}, that is, $U = U_{el}(I_1, I_3) + U_{FH}(I_3)$ where $I_1 = \text{tr}\hat{B}$ and $I_3 = \det \hat{B}$ are the invariants of the left Cauchy–Green (Finger) strain tensor $\hat{B}$ [34]. The invariant I_3 characterizes the volumetric changes in the deformed gel [34]. The local volume fractions of polymer in the deformed and undeformed states are related in the following way: $\phi = \phi_0 I_3^{-1/2}$.

The elastic energy contribution U_{el} describes the rubber elasticity of the cross-linked polymer chains, and is proportional to the cross-link density c_0, which is the number density of elastic strands in the undeformed polymer network. We use the Flory model [35] to specify U_{el}:

$$U_{el} = \frac{c_0 v_0}{2}(I_1 - 3 - \ln I_3^{1/2}) \tag{8.7}$$

The energy of the polymer–solvent interaction is taken to be of the following Flory–Huggins form [2]:

$$U_{FH} = \sqrt{I_3}\,[(1-\phi)\ln(1-\phi) + \chi_{FH}(\phi)\phi(1-\phi) - \chi^* v(1-\phi)] \tag{8.8}$$

where $\chi^* > 0$ describes the hydrating effect of the metal-ion catalyst and captures the coupling between the gel dynamics and the BZ reaction, and $\chi_{FH}(\phi)$ is the polymer–solvent interaction parameter. In Eq. (8.8), the coefficient $I_3^{1/2}$ appears in front of the conventional Flory–Huggins energy because the energy density is defined per unit volume of gel in the undeformed state. Finally, using Eqs (8.7) and (8.8), one can derive the following constitutive equation for the chemoresponsive polymer gels [2]:

$$\hat{\sigma} = -P(\phi, v)\hat{\mathbf{I}} + c_0 v_0 \frac{\phi}{\phi_0}\hat{\mathbf{B}} \tag{8.9}$$

where $\hat{\mathbf{I}}$ is the unit tensor, and the isotropic pressure $P(\phi, v)$ is defined as

$$P(\phi, v) = -[\phi + \ln(1-\phi) + \chi(\phi)\phi^2] + c_0 v_0 \phi(2\phi_0)^{-1} + \chi^* v\phi \tag{8.10}$$

where $\chi(\phi) = \chi_0 + \chi_1 \phi$ is derived from the Flory–Huggins parameter $\chi_{FH}(\phi)$ [36].

We note that the gel attains a steady state if the elastic stresses are balanced by the osmotic pressure and, simultaneously, the reaction exhibits a stationary regime. Such stationary solutions ($\phi_{st}, u_{st}, v_{st}$) are found by solving the following equations:

$$c_0 v_0 \left[\left(\frac{\phi_{st}}{\phi_0}\right)^{1/3} - \frac{\phi_{st}}{2\phi_0}\right] = \pi_{osm}(\phi_{st}, v_{st});$$
$$F(u_{st}, v_{st}, \phi_{st}) = 0;\ G(u_{st}, v_{st}, \phi_{st}) = 0 \tag{8.11}$$

The left-hand side of the first equation in Eq. (8.11) represents an elastic stress [2].

The values of most of the parameters in Eqs (8.4–8.11) were based on available experimental data [2]. Thus, the Oregonator parameters q and ε were estimated to have the values of $q = 9.52 \times 10^{-5}$ and $\varepsilon = 0.354$, respectively [2]. The polymer gel can be characterized by $\phi_0 = 0.139$, $c_0 = 1.3 \times 10^{-3}$, and $\Lambda_0 \sim 10^2$, whereas the function $\chi(\phi) = 0.338 + 0.518\phi$ can be used to describe the polymer–solvent interactions at the temperature of 20 °C [36]. The stoichiometric parameter f and the chemomechanical coupling constant χ^* are the adjustable parameters. With this choice of parameters, the characteristic time and length scales in our simulations are $T_0 \sim 1$ s and $L_0 \sim 40$ μm, respectively [1, 2].

8.2.2
Formulation of the Gel Lattice Spring Model (gLSM)

To study the dynamic behavior of the BZ gels, we numerically integrate Eqs (8.1–8.3) in two [1, 2] or three [3] dimensions using our recently developed "gLSM." This method combines a finite-element approach for the spatial discretization of the elastodynamic equations and a finite-difference approximation for the reaction and diffusion terms. We used the gLSM approach to examine 2D confined films and 3D bulk samples; here, we briefly discuss the more general 3D formulation [3].

We represent a 3D reactive, deformable gel by a set of general linear hexahedral elements [37, 38] (see Figure 8.1a). Initially, the sample is undeformed and consists of $(L_x - 1) \times (L_y - 1) \times (L_z - 1)$ identical cubic elements; here L_i is the number of nodes in the i-direction, $i = x, y, z$. The linear size of the elements in the undistorted state is given by Δ. (In the simulations presented below, we set $\Delta = 1$.) For a homogeneous BZ gel in the undeformed state, the polymer and cross-links are uniformly distributed over the gel sample: that is, each undeformed element is characterized by the same volume fraction ϕ_0 and cross-link density c_0. Upon deformation, the elements move together with the polymer network so that the amount of polymer and number of cross-links within each hexahedral element remain equal to their initial values.

Each element is labeled by the vector $\mathbf{m} = (i, j, k)$, and the element nodes are numbered by the index $n = 1 - 8$ and characterized by the coordinates $\mathbf{r}_n(\mathbf{m})$ [3]. Within each element $\mathbf{m}$, the concentrations of the dissolved reagent $u(\mathbf{m})$, the oxidized metal-ion catalyst $v(\mathbf{m})$, and the volume fraction of polymer $\phi(\mathbf{m})$ are taken to be spatially uniform. The latter value is related to the volume of the element $V(\mathbf{m})$ as $\phi(\mathbf{m}) = \Delta^3 \phi_0 / V(\mathbf{m})$. Within each element, we introduce a local coordinate system (ξ, η, ζ) as shown in Figure 8.1a. We then calculate the coordinates within the element $\mathbf{m}$ in this local coordinate system through the values of the nodal coordinates $\mathbf{r}_n(\mathbf{m})$ and a set of "shape functions," as detailed in Refs. [37, 38]. We perform all the necessarily volume and surface integrations within each linear hexahedral element in this local coordinate system [3].

Since the polymer network dynamics is assumed to be purely relaxational, the velocity of node n of the element $\mathbf{m}$ is proportional to the force acting on this node $\mathbf{F}_n(\mathbf{m})$: that is [2, 3],

$$\frac{d\mathbf{r}_n(\mathbf{m})}{dt} = M_n(\mathbf{m})\mathbf{F}_n(\mathbf{m}) \tag{8.12}$$

where $M_n(\mathbf{m})$ is the nodal mobility that depends on the volume fractions of polymer in the adjacent elements [3].

The total force acting on each node contains contributions from the elastic and osmotic properties of the system. In other words, we have shown [3] that the total force acting on node n of the element $\mathbf{m}$ consists of two contributions: $\mathbf{F}_n(\mathbf{m}) = \mathbf{F}_{1,n}(\mathbf{m}) + \mathbf{F}_{2,n}(\mathbf{m})$. The first term, $\mathbf{F}_{1,n}(\mathbf{m})$, describes the neo-Hookean elasticity contribution to the energy of the system, and can be expressed as a

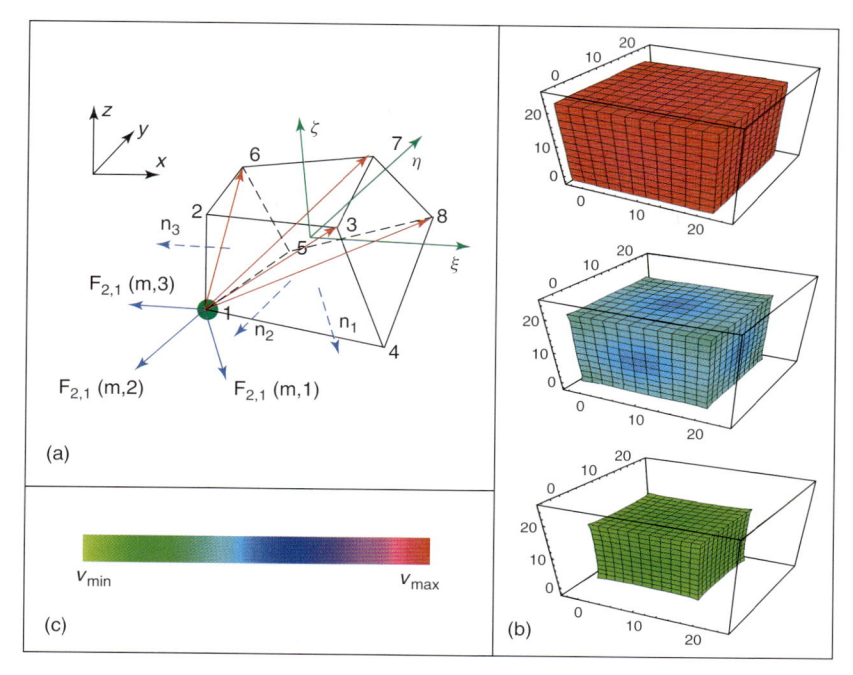

Figure 8.1 (a) Schematic of the 3D element. For each node, we provide its numbering within the element (1 : 8). Coordinate system local to this element (ξ, η, ζ) is marked by green arrows. Forces acting on the node 1 (marked by the green circle) of the element $\mathbf{m} = (i,j,k)$ are marked by the red and blue arrows. The red arrows inside the element mark the springlike elastic forces acting between the node 1 and the next-nearest and next-next-nearest neighbors within the same element $\mathbf{m}$ (as defined in Eq. (8.13)). The blue arrows outside of the element mark contributions to nodal forces from the isotropic pressure within this element (as defined in Eq. (8.14)). (b) Regular oscillations in sample of size $12 \times 12 \times 12$ nodes at simulation times $t = 1779, 1788, 1794$ (from top to bottom). Here, $f = 0.68$. The color represents the concentration of the oxidized catalyst v according to the color bar in (c). The minimum and maximum values in (c) are $v_{\min} = 8 \times 10^{-4}$ and $v_{\max} = 0.4166$, respectively.

combination of the linear springlike forces [3]:

$$\mathbf{F}_{1,n}(\mathbf{m}) = \frac{c_0 v_0 \Delta}{12} \left(\sum_{NN(\mathbf{m}')} w(n', n)[\mathbf{r}_{n'}(\mathbf{m}') - \mathbf{r}_n(\mathbf{m})] + \sum_{NNN(\mathbf{m}')} [\mathbf{r}_{n'}(\mathbf{m}') - \mathbf{r}_n(\mathbf{m})] \right) \tag{8.13}$$

Here, $\sum_{NN(\mathbf{m}')}$ and $\sum_{NNN(\mathbf{m}')}$ represent the respective summations over all the next-nearest neighbor nodal pairs and next-next-nearest neighbor nodal pairs belonging to all the neighboring elements $\mathbf{m}'$ adjacent to node n of the element $\mathbf{m}$. Above, $w(n', n) = 2$ if n and n' belong to an internal face, and $w(n', n) = 1$ if n and n' belong to a boundary face [3]. Unlike the situation for purely 2D deformations [2], there is no contribution from the interaction between nearest neighbors in Eq. (8.13).

The second contribution to the force acting on node n of the element $\mathbf{m}$, $\mathbf{F}_{2,n}(\mathbf{m})$, can be written as [3]:

$$\mathbf{F}_{2,n}(\mathbf{m}) = \frac{1}{4} \sum_{\mathbf{m}'} P[\phi(\mathbf{m}'), v(\mathbf{m}')] \left[\mathbf{n}_1(\mathbf{m}') S_1(\mathbf{m}') + \mathbf{n}_2(\mathbf{m}') S_2(\mathbf{m}') + \mathbf{n}_3(\mathbf{m}') S_3(\mathbf{m}') \right] \tag{8.14}$$

In the above equation, the summation is performed over all the neighboring elements $\mathbf{m}'$ that include node n of element $\mathbf{m}$. The pressure within each element, $P[\phi(\mathbf{m}'), v(\mathbf{m}')]$, is calculated according to Eq. (8.10). In Eq. (8.14), the vector $\mathbf{n}_l(\mathbf{m}')$ is the outward normal to the face l of element $\mathbf{m}'$, and S_l is the area of this face [3]. The vectors $\mathbf{n}_l(\mathbf{m})$ are shown in Figure 8.1a for element $\mathbf{m}$, which includes node $n = 1$ (here, faces 1, 2, and 3 correspond to $\zeta = -1$, $\eta = -1$, and $\xi = -1$ in the local coordinate system, respectively).

Both contributions to the force acting on node $n = 1$ of the element $\mathbf{m}$ from within this element are shown schematically in Figure 8.1a. The springlike forces between node $n = 1$ and the neighboring nodes are marked by red arrows, while the forces $\mathbf{F}_{2,n}(\mathbf{m})$ are depicted by the blue arrows. We emphasize that the total force acting on node n of element $\mathbf{m}$ includes similar contributions from each of the neighboring elements containing this node. If the nodal forces are known, the dynamics of polymer network can be described by Eq. (8.12).

The above expressions allow us to formulate the discretized equations for our model [3]. A detailed description of the 3D numerical approach is provided in Ref. [3], which also contains a discussion of how we validated this approach. In particular, we considered a limiting case that can be solved via an independent method and showed excellent agreement between the results obtained independently and those obtained with the gLSM [3].

To facilitate the ensuing discussion, we show the graphical output from our 3D gLSM simulations in Figure 8.1b. The gel sample in the images is $12 \times 12 \times 12$ nodes in size. These snapshots were taken during one period of oscillation at late times, ensuring that the simulations capture the regular, nontransient behavior. Within these images, the colors represent the concentration of oxidized catalyst v, with the color bar being given in Figure 8.1c, and the black lines mark the positions of the elements. We imposed the no-flux boundary conditions for u at the surface of the gel; unless specified otherwise, the ensuing examples also involve the no-flux boundary condition.

Figure 8.1b reveals the temporal synchronization of the chemical and mechanical oscillations for this sample, that is, the concentration of oxidized catalyst v attains its highest value (see color bar) when the gel is in the most swollen state (see top image in Figure 8.1b). As the concentration of oxidized catalyst decreases during the course of BZ reaction, the sample's degree of swelling also decreases, as shown in Figure 8.1b (two bottom images). This behavior is similar to the in-phase synchronization of the chemical and mechanical oscillations observed experimentally by Yoshida *et al.* for cubic gel pieces that were smaller in size than the characteristic length scale of the chemical wave [11]. The size of the gel in Figure 8.1b is sufficiently small, so the concentration of reagents within this sample at any instant of time is relatively uniform. For the larger sample sizes (see Ref. [3], as well as the simulations presented below), we observe traveling waves propagating throughout the sample.

8.3 Sensitivity to Mechanical Deformation

8.3.1 Capturing Effects of Local Mechanical Impact on Homogeneous BZ Gels

A vital function performed by skin is to send a chemical alarm signal throughout the system in response to irritation or damage. Our aim is to design a coating that can perform an analogous, biomimetic function by sending out an extensive chemical wave in response to a local mechanical impact. To study the effects of such a localized impact on a sample of the BZ gel, we apply a surface force along the $-z$ direction (see Figure 8.2a) to a square region of size $\delta \times \delta$ elements. To characterize the strength of the applied impact, we specify the value of the force in the $-z$ direction per element as F_e; this value is applied at time $t = t_{ON}$ and is kept constant until a time $t = t_{OFF}$. Below, we show that, depending on the size of the

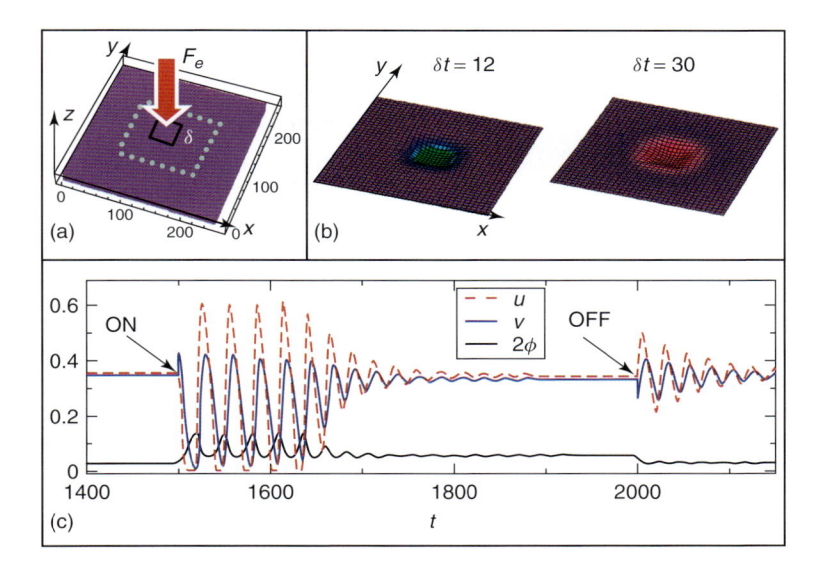

Figure 8.2 Transient oscillations generated by the impact. (a) Steady state before the impact. Here, $f = 0.64$ and the stationary values are $v_{st} = 0.3454$, $u_{st} = 0.35031$, and $\lambda_{st} = 2.14865$. (b) Transient oscillations after the impact with $F_e = 0.016$ and $\delta = 9$ elements at $t = 1512$ and $t = 1530$; here, $v_{min} = 0.006$, $v_{max} = 0.47$. (c) Evolution of ϕ, u, and v taken at the center of the impact (element (60, 60, 9)). Here, $t_{ON} = 1500$ and $t_{OFF} = 2000$. The size of the sample is $120 \times 120 \times 10$ nodes.

impacted area (δ), the force can induce waves that are either transient and spatially localized, or nondecaying and global.

The gel sample in Figure 8.2 consists of $120 \times 120 \times 10$ nodes. We initially set the swelling of the sample to the stationary value $\lambda_{st} = (\phi_0/\phi_{st})^{1/3}$ and set the values of u and v to have small random fluctuations around their respective stationary values u_{st} and v_{st} (choosing the standard deviations for these values to be 5%), where ϕ_{st}, u_{st}, and v_{st} are defined in Eq. (8.11). Taking into account that for these parameters $\lambda_{st} = 2.149$, the initial size of the coating is approximately $255.7 \times 255.7 \times 19.3$ dimensionless units, which corresponds to a dimensional size of roughly 10 mm $\times$ 10 mm $\times$ 0.8 mm. Because of the fluctuations in initial conditions, the sample undergoes small-amplitude, transient oscillations that decay until the sample reaches its stationary state (see Figure 8.2a) and remains in this state until we apply an external force F_e at $t = t_{ON}$. (In this way, we ensure that the stationary state before the impact is stable with respect to the small random fluctuations.)

We apply the force F_e within a relatively small region with $\delta = 9$. The dimensionless unit of pressure in our simulations corresponds to $\sim 10^8$ Pa, provided that the molar volume of solvent is taken equal to that of water. Hence, the total external force acting on the sample is $\sim$0.4N. The snapshots in Figure 8.2b reveal the transient oscillations caused by this impact; the values of $\delta t = t - t_{ON}$ at the top of each image are the corresponding simulation times (after the moment of impact). The colors represent the concentration of v, with the scale bar given in Figure 8.1c. The images in Figure 8.2b are enlargements of the affected region within the section marked by white dotted lines in Figure 8.2a; the mesh represents positions of the elements within the top layer and allows us to more clearly visualize the indentation caused by the impact.

The evolutions of u, v, and ϕ at the center of the affected region are shown in Figure 8.2c. (Since the absolute values of ϕ are relatively small, we plot the value of 2ϕ in Figure 8.2c to better illustrate the evolution of this parameter.) Before the impact, the latter three parameters are equal to their stationary values (as given in Eq. (8.11)). The force at $t = t_{ON}$ (see the arrow marked "ON" in Figure 8.2c) leads to a relatively rapid increase in ϕ, which in turn causes a rapid decrease in v and u. As the system relaxes after the moment of impact (while the applied force is held constant), the oscillations caused by the sudden increase in ϕ decay and the system reaches a new stationary state, where the external force is balanced by the response from the gel within the impacted region. This new stationary state is characterized by the values of v_{st_imp} and ϕ_{st_imp} within the top layer of the coating, at the center of the impact; note that $\phi_{st\_imp} > \phi_{st}$ because of the local compression (see Figure 8.2c). Additional simulations showed that, as anticipated, the value of $(\phi_{st\_imp} - \phi_{st})$ decreases with decrease in F_e. Hence, weaker applied forces lead to transient oscillations with smaller amplitudes in time and also smaller distortions away from the region of impact.

With the release of the external force at $t = t_{OFF}$ (see the arrow marked "OFF" in Figure 8.2c), we again observe transient oscillations, but now the amplitude is much smaller. Relatively quickly, the sample relaxes back to its stationary state,

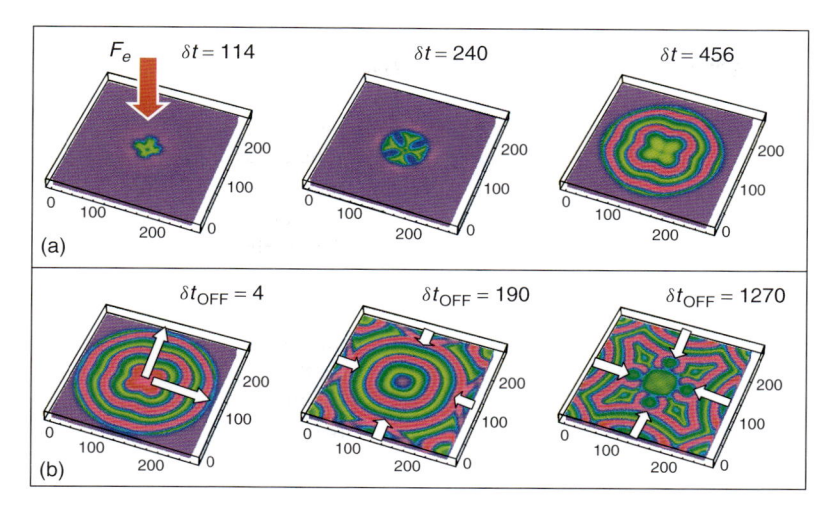

Figure 8.3 Nondecaying oscillations generated by the impact. (a) Oscillations at $t_{ON} < t < t_{OFF}$ with the time after the impact, $\delta t = t - t_{ON}$, provided on the top of each image. (b) Oscillations at $t > t_{OFF}$ with the time after the impact was lifted, $\delta t_{OFF} = t - t_{OFF}$, provided on the top of each image. Here, $\delta = 13$ elements, and the rest of parameters are the same as in Figure 8.2.

which is characterized by v_{st}, u_{st}, and ϕ_{st} throughout the entire sample, including the previously affected region.

In Figure 8.3, the size of the impacted region is increased to $\delta = 13$, while the applied force F_e and the time Δt_{imp} are the same as in Figure 8.2. As seen in Figure 8.3a, chemical waves are nucleated in the region of the local impact and propagate outward (the corresponding simulation time after the moment of impact δt is specified at the top of each image). These variations in chemical concentration produce propagating "ripples" on the surface of the coating. In contrast to the situation in Figure 8.2, which displayed only transient oscillations, this system remains in the oscillatory regime. Hence, the applied force causes a transition between the stationary and oscillatory states of this bistable system.

Another remarkable feature is that the system in Figure 8.3 continues to oscillate even after the release of the impact. The images in Figure 8.3b show the evolution of this sample at later times and after the external force was lifted. On the top of each image, we provide the corresponding simulation time after the impact was lifted, $\delta t_{OFF} = t - t_{OFF}$. These images reveal a complex dynamics that arises from the traveling waves reaching and interacting with the sidewalls. In particular, we observe traveling waves that (i) propagate *outward* from the former point of the impact and (ii) propagate *inward* from the sidewalls to the center of the sample (as marked by white arrows). At earlier times, the region encompassing waves traveling outward is much larger than the region involving the waves originating at the specific areas of the sidewalls and propagating inward (see arrows pointing inward of the sample). With time, the waves traveling from the sidewalls prevail and occupy the largest part of the sample (see the right image in Figure 8.3). Thus, the waves generated by the impact and initially traveling from the center to the sidewalls appear to be reflected from the sidewalls and travel inward to the center of the sample. At late times, we observe rhythmic oscillation at the center of the sample, where the waves traveling from the sidewalls collide with each other.

At fixed δ, the response of the sample to the impact depends on both the magnitude of the applied force and the time interval Δt_{imp} for which this force is applied [5]. In additional simulations, we fixed the parameters as those in Figure 8.3, but varied F_e and Δt_{imp}. For example, when we choose smaller values of F_e [5], the applied force only triggers transient oscillations (similar to those observed in Figure 8.2), which die off quickly. At higher values of F_e, the stronger impact generates nontransient traveling waves at the region of impact more quickly than in the case in Figure 8.3; hence, these waves reach the sidewalls at an earlier time. Summarizing information from the number of additional simulations [5], we find that, in order for the spatially localized impact to induce nontransient traveling waves that propagate throughout the sample, both the duration of the impact and the applied force should be larger than particular threshold values, which, in general, depend on the reaction parameters and on physical properties of the gel. On the basis of the above observations, as well as on a number of additional simulations, we conclude that the affected area can be detected on the surface of the coating as the region emitting traveling waves until these waves reach one or more sidewalls; then, however, the resulting pattern becomes more complicated because of the interaction of waves propagating in different directions, and hence, it becomes much more difficult to pinpoint the region where the impact originated.

In the next series of simulations, we exploit the above observations to control the traveling waves within the sample. Here, we consider a long, thin sample (consisting of $200 \times 40 \times 10$ nodes) and apply a force at its center. The parameters are the same as in Figure 8.3. Because of the large length-to-width ratio, the waves reach the walls in the y-direction relatively rapidly and trigger the oscillations in the respective sidewall areas of the gel, so that waves are generated at these areas and propagate inward to the center of the sample (see arrows pointing to the middle of the sample in Figure 8.4b,c). Thus, the waves generated by the impact appear to be reflected from the sidewalls in the y-direction. Along the x-direction, however, the waves continue to propagate from the center to the sidewalls (the white arrows pinpoint the position of the propagating front). The development of these traveling waves suggests an opportunity for using a mechanical impact to redirect the traffic of species or impurities within the system. In other words, the traveling pattern shown in Figure 8.4c, could transport species located at the regions schematically marked by red circles, and transfer them to the regions marked by green circles (see Figure 8.4c). We note that recently Yoshida *et al.* demonstrated the ability of traveling waves in chemoresponsive BZ gels to transport *millimeter*-sized objects [18]. In the scenarios considered herein, the external mechanical impact could generate the directed transport of such objects within initially stationary samples.

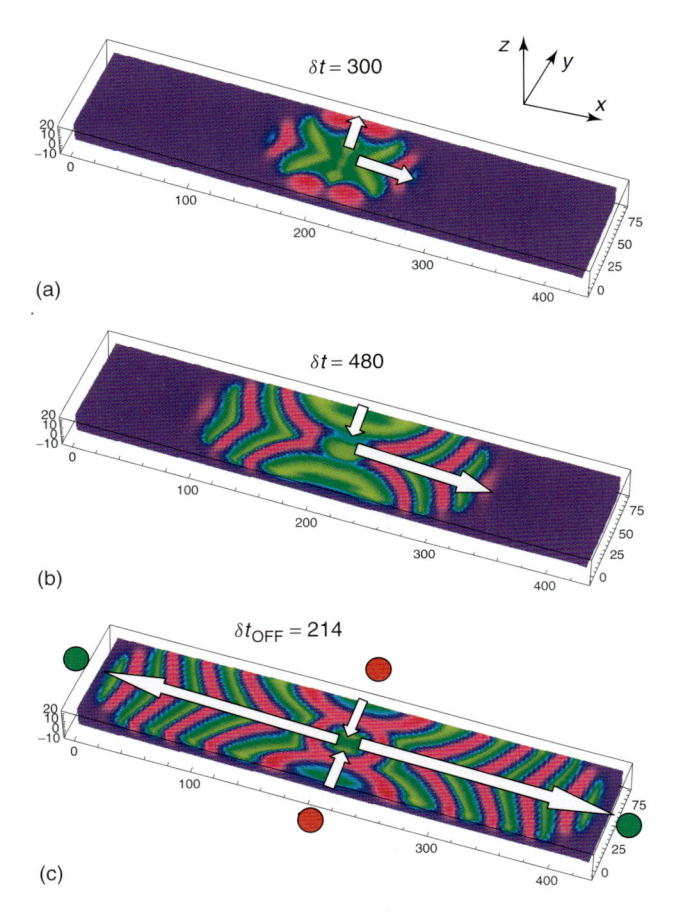

Figure 8.4 Response of a large-aspect-ratio sample to an impact [5]. (a,b) Oscillations after the impact at $\delta t = 300$ and $\delta t = 480$, respectively. (c) Oscillations after the impact was lifted at $\delta t_{OFF} = 214$. White arrows schematically mark propagation of traveling waves. All the parameters are the same as in Figure 8.3, except the sample size is now $200 \times 40 \times 10$ nodes.

8.3.2
Straining Heterogeneous BZ Gels

As discussed above, our results on the behavior of homogeneous BZ gels (which contain a uniform distribution of the polymer-tethered BZ catalyst) showed that this material can exhibit a biomimetic mechanochemical response to external mechanical action. Below, we describe our findings on the behavior of heterogeneous BZ gels, which encompass separate catalyst-containing regions and catalyst-free sections. These findings reveal that heterogeneous BZ gels can exhibit distinctive strain-dependent dynamical behavior. In the following, we use a simple model to

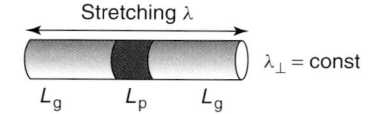

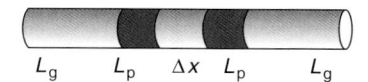

Figure 8.5 Symmetric 1D gels containing one and two patches. The BZ patches are marked in black. The total respective lengths of the nondeformed system were equal to $L_{ini} \equiv L_p + 2L_g$ and $L_{ini} \equiv 2(L_p + L_g) + \Delta x$ in the gels having one and two BZ patches. In the stretched state, Δx, L_p, and L_g are multiplied by the stretch ratio λ. In simulations, we set $\lambda_\perp = 1.1$ and $L_g = 10L_0$; L_p was taken equal to $3L_0$ and $5L_0$, and Δx was varied from $0.5L_0$ to $10L_0$.

examine how these dynamics depend on the arrangement of the BZ patches and the nature of the applied strain.

Our system consists of a swollen polymer gel where the polymer-tethered BZ catalysts are localized in specific patches (BZ patches) that are arranged along the X-axis (see Figure 8.5). In the nondeformed state, the gel sample has a length of L_{ini}. We assume that this sample is placed in a capillary tube (oriented along the X-axis), and becomes swollen so that its degree of swelling in the longitudinal direction is λ and the degree of swelling in the transverse direction is $\lambda_\perp$. The gel length L in the swollen state is $L = \lambda L_{ini}$. We then assume that a mechanical force is applied at the ends of the gel, so the longitudinal stretching λ is increased or decreased while the gel size in the transverse direction remains constant ($\lambda_\perp = $ const). Thus, the material can be treated as a purely 1D system.

We considered symmetric heterogeneous gels having one and two patches, as shown in Figure 8.5. The length of the catalyst-containing patches is L_p, and the distance from the patches to the sample ends is L_g. The lateral separation between patches, or the interpatch distance, is Δx. To solve the governing equations, Eqs. (8.1–8.3), we applied the 2D gLSM to a lattice, that is, only two nodes in height, kept the height constant, and prohibited the diffusion of the activator u across the boundaries in the latter direction. At the ends of the sample, we set the concentration of the activator to $u = 0$. The spatial discretization used in the simulations was $1/20\ L_0$. For the BZ reaction parameters (see Eqs (8.2–8.5)), we set $\varepsilon = 0.354$, $q = 9.52 \times 10^{-5}$. The polymer gel was characterized by $\phi_0 = 0.139$ and $c_0 = 1.3 \times 10^{-3}$, as in the studies described earlier. We simulated the dynamics of both nonresponsive and responsive BZ gels by setting $\chi^* = 0$ and 0.105, respectively. (In contrast to the chemoresponsive gels characterized by a positive value of χ^*, the systems with $\chi^* = 0$ are referred to as *nonresponsive* since the gels do not undergo a mechanical swelling or deswelling with the oscillations in the chemical reaction.) The parameters f and λ were used as variables.

We found that changing the lateral stretching λ of a heterogeneous gel having one BZ patch can result in a switching between the oscillatory and nonoscillatory

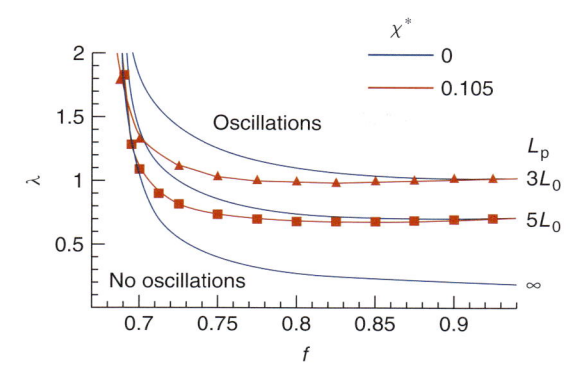

Figure 8.6 The location of the Hopf bifurcation points in a heterogeneous gel having one patch at the patch length of $L_p = 3L_0, 5L_0$, and ∞. Blue: obtained by the linear stability analysis of the nonresponsive gel [28]. Red: the symbols are the data points obtained by the numerical simulations of the responsive gel at $\chi^* = 0.105$.

regimes of the BZ reaction. Figure 8.6 shows that the lateral size at which the switching between the reaction regimes occurs depends on the value of f and on the length L_p of the BZ catalyst-containing region. The blue curves in Figure 8.6 correspond to the nonresponsive gel ($\chi^* = 0$) and show the position of the Hopf bifurcation point in the $\lambda - f$ plane as obtained through the linear stability analysis described in Ref. [28]. The domain of the oscillatory regime of a system is located above the corresponding curve. The red solid symbols present the results of numerical simulations of the responsive BZ gel ($\chi^* = 0.105$); the data points are connected by the red lines as a guide for the eye. The latter data points were obtained by starting the simulations at sufficiently low values of λ, at which the system exhibits no oscillations. Then, the longitudinal strain λ was gradually increased until the steady state ceased to exist. Specifically, the strain was increased in steps of $\delta\lambda = 0.01$; between two successive changes of λ, the simulations were run for 10^3 units of time. It is seen in Figure 8.6 that the domain of the oscillatory regime decreases with decreasing patch length L_p. In other words, gel samples having short BZ patches must be stretched to greater tensile strains in order to induce the BZ oscillations. It is also seen that the oscillatory domain is larger in the chemoresponsive gels than in the gels that do not respond to the BZ reaction by swelling and deswelling.

Further simulations revealed the existence of hysteresis; that is, for a certain range of the stoichiometric parameter f, the transition between the oscillatory and nonoscillatory regimes of the BZ reaction occurs at different values of λ depending on whether the transition is approached from within the oscillatory or steady-state domains. Figure 8.7 shows the presence of an area in which the oscillatory and nonoscillatory regimes coexist in the responsive (red symbols and lines) and nonresponsive (blue symbols and lines) heterogeneous gels at $L_p = 5L_0$.

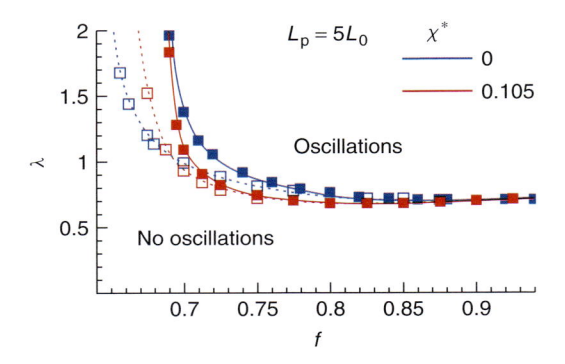

Figure 8.7 Coexistence of the oscillatory and nonoscillatory regimes in the nonresponsive (blue) and responsive (red) heterogeneous gels with one patch of length $L_p = 5L_0$. The regimes coexist in the area between the solid and dashed curves corresponding to the same χ^*. Symbols: the data obtained by the simulations.

In this figure, the data points marked by solid symbols and connected by the solid lines were obtained by increasing the stretching λ starting from the steady state as described earlier. The data points shown by the open symbols were obtained through the numerical simulations in which the strain λ was decreased starting from the oscillatory regime. The data points were obtained by varying λ in steps of $\delta\lambda = 0.01$, and are connected by the dashed lines. The oscillatory and steady-state regimes coexist in the area between the solid and the dashed lines. If the system is initially in the nonoscillatory regime and the dashed line is crossed from below, the system remains in the steady state until the solid line is reached. Correspondingly, if the system is initially in the oscillatory regime and the solid line is approached and crossed from above, then the oscillations continue to exist until the dashed line is crossed, and then the oscillations stop. These simulations also revealed that the area of coexistence depends on the value of the parameter f. As seen in Figure 8.7, the coexistence area in the both responsive and nonresponsive heterogeneous 1D gels disappears as the stoichiometric parameter f is increased. The observed hysteretic effect takes place because the Oregonator model exhibits the *subcritical* Hopf bifurcation at the values of the stoichiometric parameter of $f < 1$ [39]. Figure 8.7 also shows that the responsive gels exhibit smaller hysteresis than the nonresponsive samples.

Finally, the numerical simulations showed that the frequency of chemical oscillations within the heterogeneous BZ gels depends on whether the gel is responsive or not, and changes upon deformation of the sample. Figure 8.8 shows the oscillation frequency ω_0 as a function of the stretch λ for the responsive (black symbols and lines) and nonresponsive (red symbols and lines) gel having one patch of length $L_p = 5L_0$ at $f = 0.7$. The frequency of the chemomechanical oscillations in the responsive gel is seen to be lower than that of the purely chemical oscillations in the nonresponsive gel. In Figure 8.8, the open symbols mark the data obtained

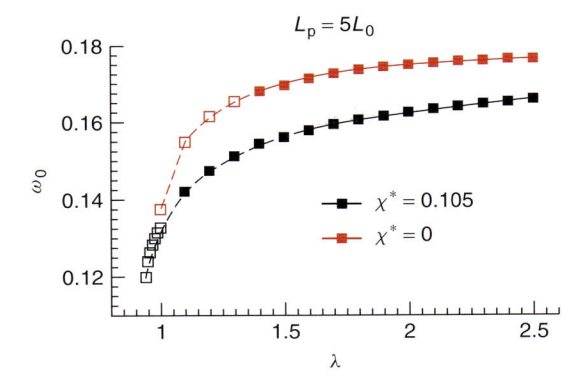

Figure 8.8 Frequency of oscillation in the nonresponsive (red) and responsive (black) heterogeneous BZ gel having one patch as a function of gel stretch λ at the patch length of $L_p = 5L_0$ and $f = 0.7$. The data were obtained by the simulations in the oscillatory regime (solid symbols) and in the area of coexistence of the oscillatory and nonoscillatory regimes (open symbols).

at λ within the area of coexistence of the oscillatory and nonoscillatory regimes, and the solid symbols correspond to the oscillatory regime. It is seen in Figure 8.8 that, in the oscillatory state, stretching the gel causes a slight increase in the frequency of oscillations on the order of several percent. The effect of deformations was found to be stronger in the coexistence area, to which oscillations within the gel could be driven by compressive strain.

The BZ reaction processes within the two catalyst-containing patches, which are placed next to each other, are correlated through the concentration field u in the gap between the patches. If the gap is sufficiently wide, the patches behave as independent chemical oscillators; in such cases, the effect of strain on the heterogeneous gel is similar to that described above. Altering the interpatch distance Δx, however, changes the strength of interaction between the BZ patches. In the following studies, we investigated the effect of varying both the interpatch distance Δx and the longitudinal stretch λ for nonresponsive and responsive heterogeneous gels that have two patches. Here, the length of each patch was set at $L_p = 5L_0$ and the stoichiometric parameter was set to $f = 0.7$. The numerical simulations revealed that in the oscillatory regime, the chemical oscillations within the two patches proceed with a constant phase difference equal to either 0 or to π (in-phase or out-of-phase synchronization, respectively). At a given chemomechanical coupling parameter χ^*, the observed mode of synchronization was found to depend on the interpatch distance Δx and the sample length λ in the deformed state. Furthermore, the simulations showed that the synchronization modes coexist at some Δx and λ, with the mode selection being dependent upon the initial conditions and the history of deformation.

Figure 8.9 presents the result of mapping the domains of the in-phase and out-of-phase modes of oscillation in the $\Delta x - \lambda$ coordinates for the nonresponsive

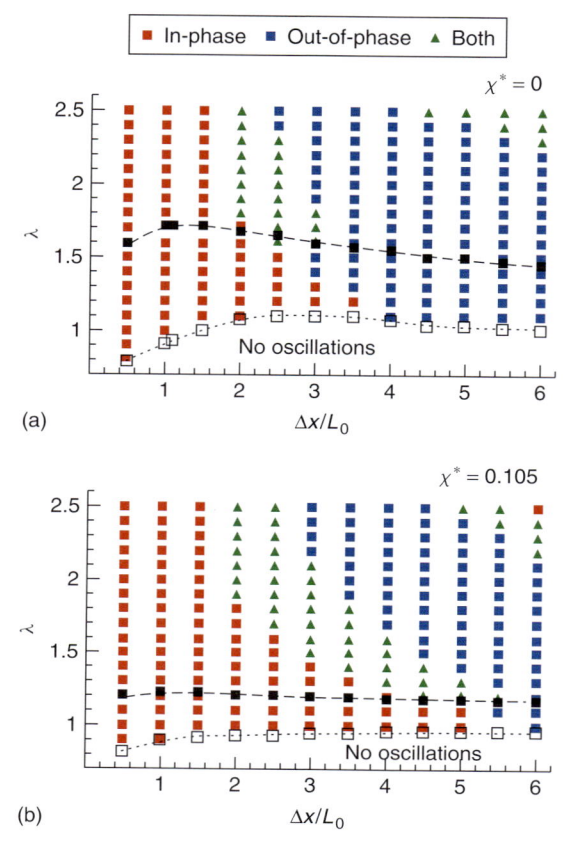

Figure 8.9 Domains of the in-phase and out-of-phase modes of synchronization in the (a) nonresponsive and (b) responsive heterogeneous gels with two patches as obtained by the numerical simulations at $L_p = 5L_0$ and $f = 0.7$. Also shown are the upper (solid black square symbols) and lower (open black square symbols) stability boundaries of the steady states.

(Figure 8.9a) and responsive (Figure 8.9b) gels. In Figure 8.9, the red and blue square symbols, respectively indicate the respective in-phase and out-of-phase oscillatory regimes, and the solid green triangular symbols show the points where both of these synchronization modes are observed. Figure 8.9 also shows the upper and lower stability boundaries of the steady state marked by the solid and open black square symbols, respectively, connected for convenience by dashed lines. To obtain the map in Figure 8.9, we initiated the simulations at the values of λ that correspond to the steady ($\lambda = 0.5$) or the oscillatory ($\lambda = 2.5$) states, and then respectively increased or decreased λ in a stepwise manner, with a step size of 0.1. For each point in the oscillatory regime, we considered both the in-phase and out-of-phase oscillations as the initial conditions and then observed which of these modes remained stable (or if both were stable). The simulations showed that, if a transition between the two modes of oscillation is possible, the transition might take as long as 2×10^3 units of time. To ensure that we captured such transition points, the simulations were run for 5×10^3 units of time at each value of λ.

As shown in Figure 8.9, the responsive gel exhibits a larger oscillatory domain, which is controlled mainly by the stretching λ, and a less pronounced hysteretic behavior than the nonresponsive gel. Closely spaced patches with $\Delta x \leq 1.5 L_0$ are seen to oscillate in phase at all the considered values of the applied stretch λ. The out-of-phase oscillations prevail over the in-phase mode if the patches are placed at greater distances and for higher elongations. The simulations also revealed that the longitudinal strain could cause a transition between the in-phase and out-of-phase modes at an interpatch distance in the intermediate range. The responsive gels were found to exhibit a notably larger area of coexistence of the synchronization modes than the nonresponsive gels (see Figure 8.9).

Further analysis of the data in Figure 8.9 revealed that the in-phase and out-of-phase oscillations exhibit different behaviors upon stretching the gel sample having two BZ patches. Moreover, the actual interpatch distance $\lambda \Delta x$ could be considered as a parameter that controls the mode selection. To demonstrate the latter two statements, in Figure 8.10 we show the normalized frequency for all of the data points *within the oscillatory domain* in Figure 8.9 as a function of $\lambda \Delta x$. In Figure 8.10, the frequencies ω for the system with two patches are normalized to the frequencies ω_0 for the system with one patch at the same values of λ and χ^* (see Figure 8.8). It is seen that the normalized frequencies corresponding to the in-phase and out-of-phase oscillations form distinct branches in the diagram for both the nonresponsive (Figure 8.10a) and responsive (Figure 8.10b) gels. In other words, the in-phase mode exhibits a weak dependence of the frequency on the actual interpatch distance, and exists only at sufficiently small ($\lambda \Delta x < 6 L_0$) or great ($\lambda \Delta x > 12 L_0$) interpatch spacings. In contrast, the out-of-phase mode exists at sufficiently high distances ($\lambda \Delta x > 3 L_0$), and its frequency is a notably decreasing function of $\lambda \Delta x$. Finally, the two modes of oscillation coexist at some interpatch distances (see Figure 8.10). It is also notable in Figure 8.6 that the mode frequencies in the nonresponsive and responsive gels exhibit qualitatively similar behavior as functions of $\lambda \Delta x$, although the frequency variation in the responsive gel is less prominent than in the nonresponsive one.

The above studies point to the fact that the heterogeneous BZ gels exhibit an ability to transmit information about mechanical strain. In particular, the above findings reveal that a tensile or compressive strain could induce transitions between the oscillatory and nonoscillatory regimes in gels having one and two BZ patches. Furthermore, in systems having two oscillating BZ patches, an applied strain can cause a switching between the in-phase and out-of-phase synchronization of the oscillations. Thus, in addition to the homogeneous BZ gels described in Section 8.3.1, heterogeneous BZ gels could also be used to create novel sensors for mechanical deformation.

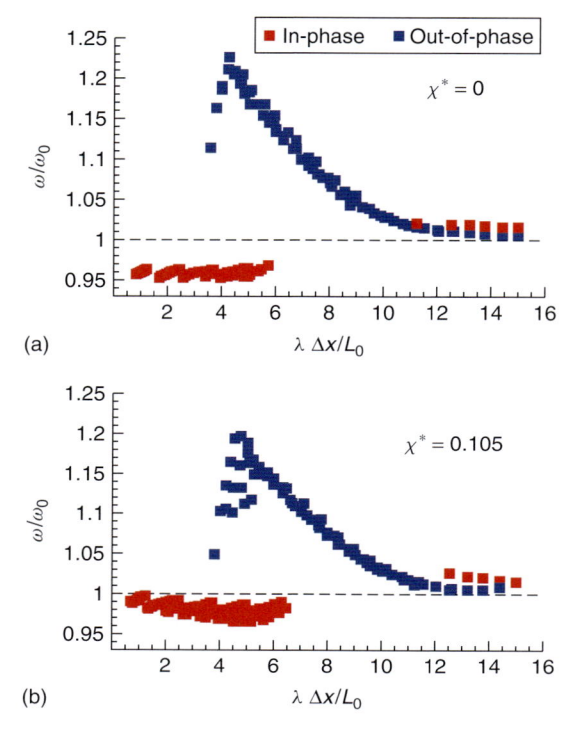

Figure 8.10 Normalized frequency of the in-phase (red symbols) and out-of-phase (blue symbols) oscillations as a function of the actual interpatch distance $\lambda \Delta x$ in the (a) nonresponsive and (b) responsive heterogeneous gels with two BZ patches at the patch length of $L_p = 5L_0$ and $f = 0.7$. Shown data points correspond to the oscillatory regime in Figure 8.5. ω_0 is the λ-dependent oscillation frequency in a single patch at $L_p = 5L_0$ shown in Figure 8.8.

8.4
Sensitivity to Light

In addition to mechanical impact and deformation, BZ gels are also sensitive to light of a particular wavelength [15, 40]. Thus, the dynamic behavior of BZ gels can be modulated by illuminating the sample with this light. The effect of this illumination is to suppress the oscillations [15, 40] and, as we show below, this phenomenon can be exploited to design BZ gel "worms" that exhibit autonomous, directed motion away from the light source.

To model the sensitivity of the BZ gels to light, we modify Eq. (8.5) to the following:

$$F(u, v, \phi) = (1-\phi)^2 u - u^2 - (1-\phi)\left[fv + \Phi\right]\frac{u - q(1-\phi)^2}{u + q(1-\phi)^2} \quad (8.15)$$

The variable Φ accounts for the additional flux of bromide ions due to the light [41]. With $\Phi = 0$ in the above equation, we recover our expression for the system in the absence of light, that is, Eq. (8.5) in Section 8.2.1. Our simulation box is $100 \times 10 \times 10$ nodes in the x, y, and z directions, respectively. We assume no-flux boundary conditions for u at the surfaces of the gel [2, 3].

In our first case (see Figure 8.11a), the light intensity is a step function, so that one region is kept in the dark and the other region is illuminated with a constant intensity. Initially, we place the gel sample along the x-direction so that one-quarter of the gel (on the left) is located in the dark and three-quarters (on the right) is within the illuminated region (see Figure 8.11a). For each element of the sample located within the dark, we set $\Phi = 0$, and within the illuminated region we set $\Phi = \Phi_c$, which is the critical value of the dimensionless flux of bromide ions. Above Φ_c, the chemical oscillations are completely suppressed [15]. Generally, Φ_c depends on the reaction parameters and the physical properties of the gel. By the simulations, we determined that for the system with the above parameters, $\Phi_c = 3.2 \times 10^{-4}$. The dimensionless value Φ_c can be related to the experimental value of the light intensity that is needed to suppress the oscillations within the BZ gel sample [15]. We initially set the swelling of the sample to its stationary value λ_{st} (taken at $\Phi = 0$) and set the values of u and v to have small random fluctuations around their respective stationary values, u_{st} and v_{st} defined in Eq. (8.11).

As the sample moves away from the stationary light, we set $\Phi = 0$ within those elements that are no longer illuminated. (Here, we neglect the attenuation of light within the gel and assume that the temperature is constant.) For the chosen parameters, the sample is in the oscillatory regime at $\Phi = 0$ and, thus, a wave of swelling and deswelling is generated at the nonilluminated left end and travels toward the right, ultimately giving rise to a net motion of the whole sample in the negative x-direction. Figure 8.11b shows the time evolution of the BZ gel at the early and late times; the gel is drawn within a larger box to more clearly illustrate the path of its motion; the nonilluminated region ($\Phi = 0$) is shaded in gray. The simulations revealed that, as the gel travels along the negative x-direction, it also bends in the x–z plane (Figure 8.11b). In the longer samples, this bending can occur in any direction, depending on the initial random seeds. In simulations involving shorter samples (i.e., $20 \times 10 \times 10$ nodes) under identical conditions, the bending was not observed, but the movement toward the dark remained robust.

In the absence of light, traveling waves appear in this sample and can cause a macroscopic displacement of the gel; however, there is no preferred direction for the wave propagation, so that the probabilities of the wave traveling to the left or to the right are equal. Hence, averaged over a large number of samples, the net displacement of this gel is approximately zero. The role of the spatially nonuniform illumination is to break the symmetry in the system. Now, oscillations always originate from the nonilluminated region, and thus, the traveling waves continually propagate from left to right.

Figure 8.12a shows the net displacement that the center of the gel attains as a function of time; the systematic decrease of this point's x-coordinate x_C reveals the net motion of the gel along the negative x-direction. Focusing on just one run (Figure 8.12b), we illustrate the variations in the position of the center of a gel

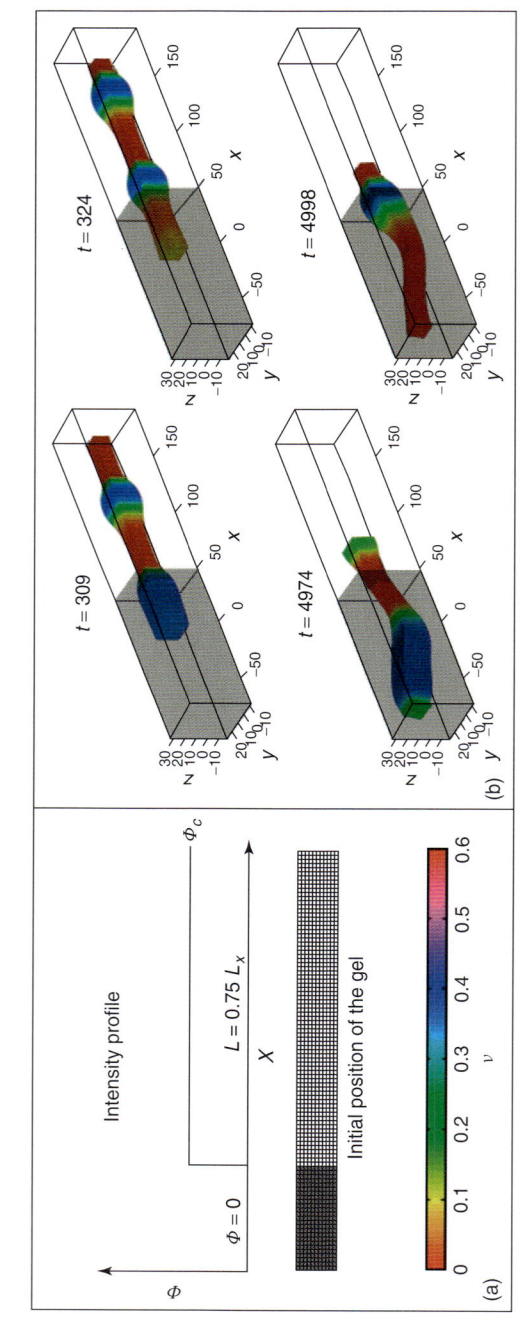

Figure 8.11 Motion of BZ gel under nonuniform light intensity. (a) Intensity profile of the incident illumination. The color bar represents the value of v. (b) Evolution of the sample during early ($t = 309$ and $t = 324$) and late ($t = 4974$ and $t = 4998$) stages. Here, we set $f = 0.7$ and choose the rest of the parameters as specified in the Section 8.2.

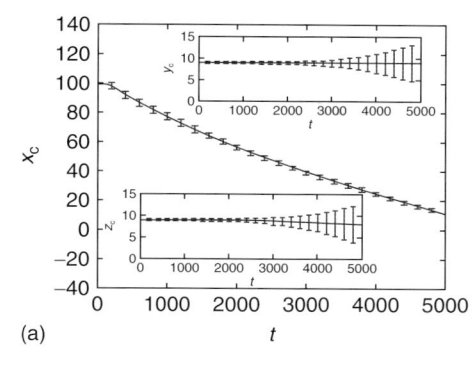

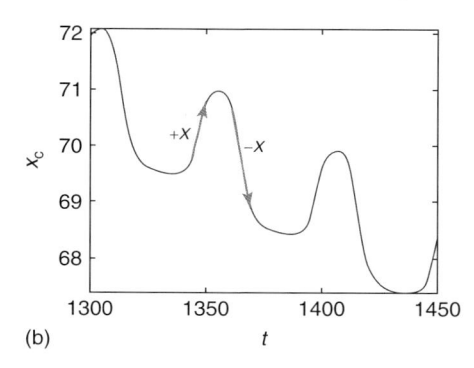

Figure 8.12 (a) Evolution of the coordinates of the center of the gel (x_C, y_C, z_C) for the intensity profile shown in Figure 8.11a. The gel's center is defined as a point within the center of the element (50, 5, 5). The vertical lines represent the error bars calculated over 10 independent runs. (b) The evolution of the x-coordinate for one such run; the arrows indicate the movement of the gel's center in the positive and the negative x-directions.

due to its periodic swelling and deswelling. As a chemical wave travels through a particular element within the sample, it causes the element to swell when the value of v is high and deswell when the value of v is low. The profile of the resulting oscillations within the gel is highly nonlinear and gives rise to a complex dynamics, where smaller amplitude motions to the right (see arrow marked "$+x$" in Figure 8.12b) alternate with larger amplitude motions to the left [6] (see arrow marked "$-x$"). Therefore, we observe that the direction of the net displacement of the gel is opposite to the direction of propagation of the traveling waves; this is consistent with our earlier observations for 2D samples of BZ gels for similar reaction parameters [2, 26]. In other words, while traveling waves that originate from the left drive the solvent concentration to oscillate, they ultimately push the solvent to the right; through the polymer–solvent interdiffusion, the gel is thereby driven toward the left.

By taking the slope of the curve for the evolution of the x-coordinate (Figure 8.12a) at relatively early times (but after the oscillations are fully developed within the sample), we find that gel's velocity along this direction is approximately 0.03 dimensionless units, which corresponds to $\sim 1.2\ \mu m\ s^{-1}$ using the characteristic length and time scales noted above. At later times, as the sample moves out of the illuminated region, the sample's velocity becomes even slower. Figure 8.12a also shows the plots for the evolution of the y and z coordinates (see upper and lower insets); these plots exhibit error bars that are larger than the actual displacements, indicating that the motion is random in these directions.

In addition to controlling the directionality of the motion, we can tailor the gel's velocity by varying χ^*, which is the parameter that characterizes the responsiveness of the gel to the BZ reaction. To obtain the plots in Figure 8.13a, we increased χ^* from 0.065 to 0.125 in steps of 0.010. Varying χ^*, however, results in a change of the stationary values for u, v, and ϕ. Thus, for each case, we must calculate the initial

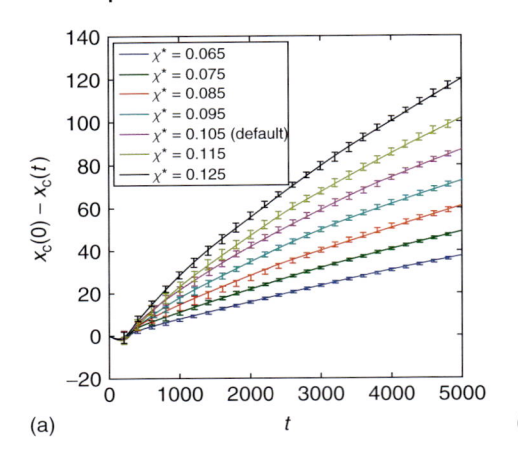

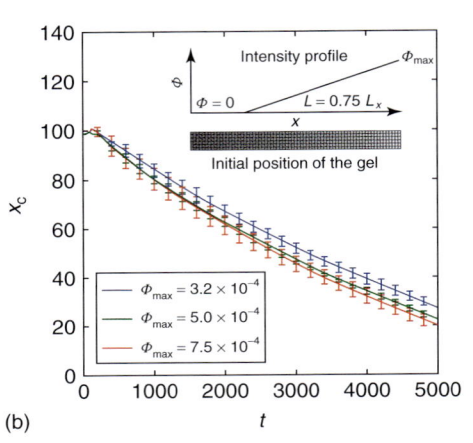

Figure 8.13 Controlling the velocity of motion. (a) Evolution of $x_C(0) - x_C(t)$ for different χ^*. The intensity profile is the same as Figure 8.11a. (b) Evolution of the x-coordinate of gel's center for different values of the light intensity gradient shown in top inset. Values of Φ_{max} are given in the legend. The vertical lines represent the error bars calculated over 10 independent runs.

values of u, v, and ϕ at that specific χ^*. Since samples with different values of χ^* exhibit different initial degrees of swelling (and hence, have different initial sizes), we characterize the directed motion by calculating the change in the x-coordinate of the gel's center $x_C(t)$ with respect to its initial value $x_C(0)$. Figure 8.13a shows the value of $x_C(0) - x_C(t)$ as a function of time for different χ^* and reveals that the slope of the curves, and hence the velocity, increases with increasing χ^*.

A step function in light intensity profile (as in Figure 8.11a) is not a necessary condition for directed motion. For example, in Figure 8.13b, the illuminated region now encompasses a gradient in the light intensity, which starts from zero and increases linearly to a value corresponding to Φ_{max} (see inset). The plot clearly reveals the gel's motion toward the left and further indicates that its velocity can be somewhat increased by increasing the gradient in Φ. (Similar behavior was found when the entire sample is exposed to this intensity gradient.)

By manipulating the intensity profile of the light, it is also possible to control the shape of the sample and effectively cause the gel to reorient with respect to its initial position. As illustrated in Figure 8.14a, we now illuminate the sample at the two ends while keeping the center of the gel in the dark. At early times (Figure 8.14b), the sample is aligned along the x-direction and oscillations are generated in the central dark region of the gel. These oscillations give rise to two waves traveling in opposite directions, toward the two ends of the sample. With the waves propagating to the sample's ends, the polymer is "pulled" toward the central dark region. This behavior is consistent with the cases above, where the gel exhibits autonomous motion away from the light. Here, the gel tries to "squeeze into" the central region, causing this portion to bend and rotate with respect to its initial position. In effect, the gel senses, reorients, and alters its shape in response to the light.

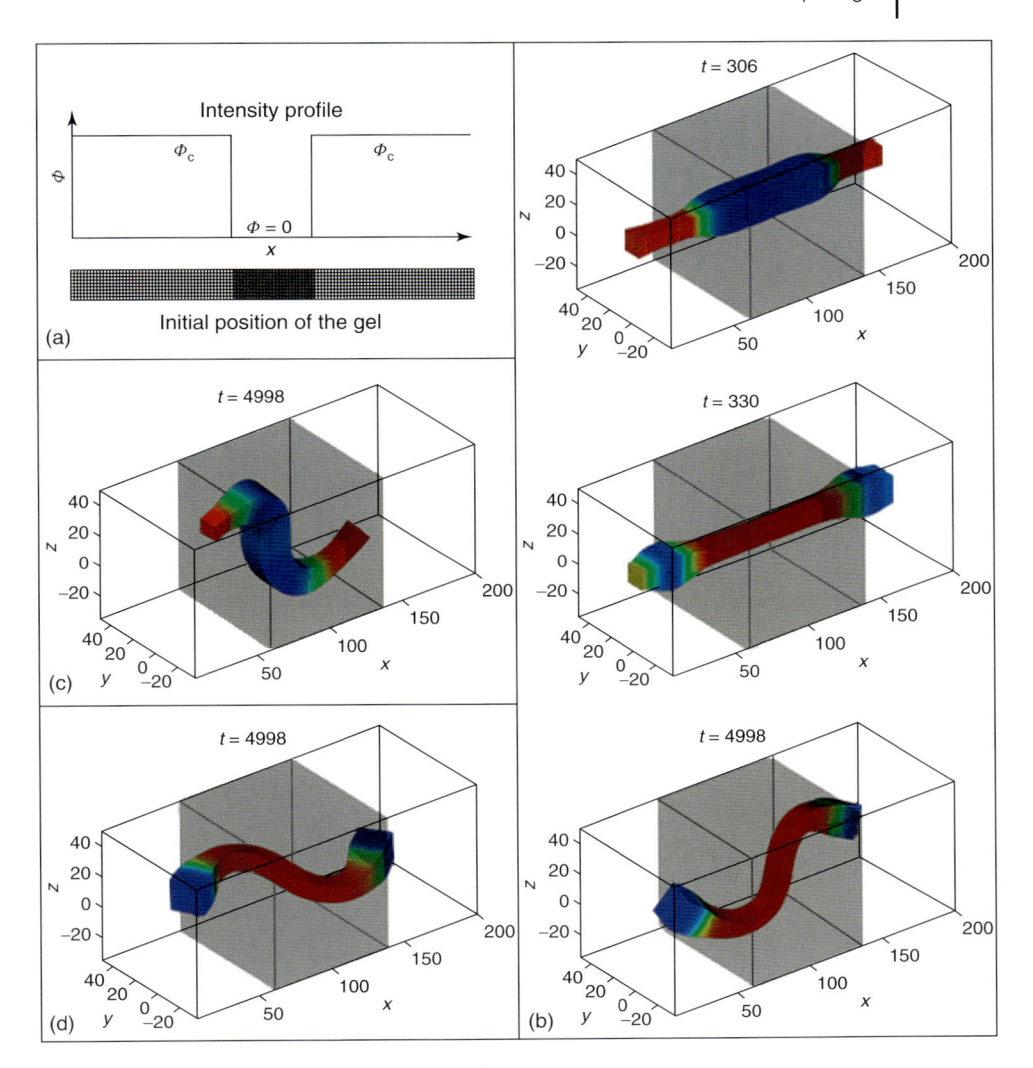

Figure 8.14 Shape changes and reorientation of the polymer gel. (a) The intensity profile of the light illumination; the length of the dark region is 40% of the gel's initial length. (b) Formation of an S-shape in the x–z plane. The images in (c) and (d) demonstrate the bending in different directions.

Figure 8.14b shows one such example in which the gel acquires an S-like shape located approximately in the x–z plane (at later times). Depending upon the initial random seed, this bending can take place in any of the planes; two other distinct examples are shown in Figure 8.14c,d. Additional simulations suggest that the degree of bending can be controlled by varying the total exposed region of the gel.

As noted in the Section 8.1, these synthetic BZ "worms" exhibit striking biomimetic behavior in their ability to move away from an adverse environmental condition, which in the context of the BZ reaction is the presence of light. It is noteworthy that the latter action mimics the adaptive behavior of the slime mold *Physarum*, which responds to nonuniform light illumination by moving toward the dark [42].

8.5
Conclusions

Our results provide the first predictions that local mechanical deformations can excite traveling chemical waves and widespread oscillations within BZ gels. The findings open up the possibility of harnessing BZ gels for a range of applications. Specifically, these materials could be used to create sensors that not only can transmit a signal in response to mechanical impact, but also transport reagents to address the aftereffects. Additionally, the BZ gels could be utilized to fabricate membranes that deliver reagents in response to an applied force; this feature provides a unique on–off mechanism for regulating the transport of compounds in the material. The coatings could ultimately provide a "synthetic skin" for robotics. By probing the relationships between the features of mechanical stimuli and the activated chemical waves, we can facilitate the development of these devices.

In addition to examining the response of homogeneous BZ gels to mechanical impact, we investigated the response of heterogeneous BZ gels to deformation. These studies can enhance our understanding of how chemical signals pass through complex media that encompasses reactive and nonreactive domains. In particular, our research is already revealing how BZ pieces that are separated by nonreactive polymer communicate with each other. Through such investigations, we can determine the role that the nonreactive network plays in the transfer of information throughout the system and the overall behavior of the material.

Our findings also indicate a new approach for directing the self-sustained motion of macroscopic samples of active materials. In other words, we have shown that by illuminating the sample with light of a particular wavelength, we can drive a gel sample to undergo directed movement away from the light source. Once guided along a particular path, the BZ gels will continue to move in that direction, even after the entire sample has left the illuminated region. We also found that the arrangement of the light and dark regions can be harnessed to manipulate the shape of the gel. Our findings provide guidelines for coupling chemoresponsive gels and photosensitive chemical reactions to design "soft robots" and devices that exhibit biomimetic functionality.

Acknowledgments

Financial support from the Army Research Office is gratefully acknowledged.

References

1. Yashin, V.V. and Balazs, A.C. (2006) Pattern formation and shape changes in self-oscillating polymer gels. *Science*, **314**(5800), 798–801.
2. Yashin, V.V. and Balazs, A.C. (2007) Theoretical and computational modeling of self-oscillating polymer gels. *J. Chem. Phys.*, **126**(12), 124707.
3. Kuksenok, O., Yashin, V.V., and Balazs, A.C. (2008) Three-dimensional model for chemoresponsive polymer gels undergoing the Belousov-Zhabotinsky reaction. *Phys. Rev. E*, **78**(4), 041406.
4. Zaikin, A.N. and Zhabotinsky, A.M. (1970) Concentration Wave Propagation in Two-dimensional Liquid-phase Self-oscillating System. *Nature*, **225**, 535–537.
5. Kuksenok, O., Yashin, V.V., and Balazs, A.C. (2009) Global signaling of localized impact in chemo-responsive gels. *Soft Matter*, **5**(9), 1835–1839.
6. Dayal, P., Kuksenok, O., and Balazs, A.C. (2009) Using light to guide the self-sustained motion of active gels. *Langmuir*, **25**(8), 4298–4301.
7. Yoshida, R., Takahashi, T., Yamaguchi, T., and Ichijo, H. (1996) Self-oscillating gel. *J. Am. Chem. Soc.*, **118**(21), 5134–5135.
8. Yoshida, R., Kokufuta, E., and Yamaguchi, T. (1999) Beating polymer gels coupled with a nonlinear chemical reaction. *Chaos*, **9**(2), 260–266.
9. Yoshida, R., Onodera, S., Yamaguchi, T., and Kokufuta, E. (1999) Aspects of the Belousov-Zhabotinsky reaction in polymer gels. *J. Phys. Chem. A*, **103**(43), 8573–8578.
10. Sakai, T. and Yoshida, R. (2004) Self-oscillating nanogel particles. *Langmuir*, **20**(4), 1036–1038.
11. Yoshida, R., Tanaka, M., Onodera, S., Yamaguchi, T., and Kokufuta, E. (2000) In-phase synchronization of chemical and mechanical oscillations in self-oscillating gels. *J. Phys. Chem. A*, **104**(32), 7549–7555.
12. Yoshida, R. (2005) Design of functional polymer gels and their application to biomimetic materials. *Curr. Org. Chem.*, **9**(16), 1617–1641.
13. Yoshida, R. (2008) Self-oscillating polymer and gels as novel biomimetic materials. *Bull. Chem. Soc. Jpn.*, **81**(6), 676–688.
14. Murase, Y., Maeda, S., Hashimoto, S., and Yoshida, R. (2009) Design of a mass transport surface utilizing peristaltic motion of a self-oscillating gel. *Langmuir*, **25**(1), 483–489.
15. Shinohara, S., Seki, T., Sakai, T., Yoshida, R., and Takeoka, Y. (2008) Photoregulated wormlike motion of a gel. *Angew. Chem. Int. Ed.*, **47**(47), 9039–9043.
16. Shen, J., Pullela, S., Marquez, M., and Cheng, Z.D. (2007) Ternary phase diagram for the Belousov-Zhabotinsky reaction-induced mechanical oscillation of intelligent PNIPAM colloids. *J. Phys. Chem. A*, **111**(48), 12081–12085.
17. Tateyama, S., Shibuta, Y., and Yoshida, R. (2008) Direction control of chemical wave propagation in self-oscillating gel array. *J. Phys. Chem. B*, **112**(6), 1777–1782.
18. Maeda, S., Hara, Y., Yoshida, R., and Hashimoto, S. (2008) Peristaltic motion of polymer gels. *Angew. Chem. Int. Ed.*, **47**(35), 6690–6693.
19. Maeda, S., Hara, Y., Yoshida, R., and Hashimoto, S. (2008) Control of the dynamic motion of a gel actuator driven by the Belousov-Zhabotinsky reaction. *Macromol. Rapid Commun.*, **29**(5), 401–405.
20. Suzuki, D. and Yoshida, R. (2008) Temporal control of self-oscillation for microgels by cross-linking network structure. *Macromolecules*, **41**(15), 5830–5838.
21. Suzuki, D. and Yoshida, R. (2008) Effect of initial substrate concentration of the Belousov-Zhabotinsky reaction on self-oscillation for microgel system. *J. Phys. Chem. B*, **112**(40), 12618–12624.
22. Sasaki, S., Koga, S., Yoshida, R., and Yamaguchi, T. (2003) Mechanical oscillation coupled with the Belousov-Zhabotinsky reaction in gel. *Langmuir*, **19**(14), 5595–5600.
23. Miyakawa, K., Sakamoto, F., Yoshida, R., Kokufuta, E., and Yamaguchi, T. (2000)

Chemical waves in self-oscillating gels. *Phys. Rev. E*, **62**(1), 793–798.

24. Maeda, S., Hara, Y., Sakai, T., Yoshida, R., and Hashimoto, S. (2007) Self-walking gel. *Adv. Mater.*, **19**(21), 3480–3484.

25. Suzuki, K., Yoshinobu, T., and Iwasaki, H. (2001) Induction of chemical waves by mechanical stimulation in elastic Belousov-Zhabotinsky media. *Chem. Phy. Lett.*, **349**(5-6), 437–441.

26. Kuksenok, O., Yashin, V.V., and Balazs, A.C. (2007) Mechanically induced chemical oscillations and motion in responsive gels. *Soft Matter*, **3**(9), 1138–1144.

27. Yashin, V.V. and Balazs, A.C. (2008) Chemomechanical synchronization in heterogeneous self-oscillating gels. *Phys. Rev. E*, **77**(4), 046210.

28. Yashin, V.V., Van Vliet, K.J., and Balazs, A.C. (2009) Controlling chemical oscillations in heterogeneous Belousov-Zhabotinsky gels via mechanical strain. *Phys. Rev. E*, **79**(4), 046214.

29. Kuhnert, L. (1986) A new optical photochemical memory device in a light-sensitive chemical active medium. *Nature*, **319**, 393–394.

30. Tyson, J.J. and Fife, P.C. (1980) Target patterns in a realistic model of the Belousov–Zhabotinskii reaction. *J. Chem. Phys.*, **73**(5), 2224–2237.

31. Yashin, V.V. and Balazs, A.C. (2006) Modeling polymer gels exhibiting self-oscillations due to the Belousov-Zhabotinsky reaction. *Macromolecules*, **39**(6), 2024–2026.

32. Scott, S.K. (1994) *Oscillations, Waves, and Chaos in Chemical Kinetics*, Oxford University Press, New York.

33. Barriere, B. and Leibler, L. (2003) Kinetics of solvent absorption and permeation through a highly swellable elastomeric network. *J. Polym. Sci., Part B: Polym. Phys.*, **41**(2), 166–182.

34. Atkin, R.J. and Fox, N. (1980) *An Introduction to the Theory of Elasticity*, Longman, New York.

35. Hill, T.L. (1960) *An Introduction to Statistical Thermodynamics*, Addison-Wesley, Reading, MA.

36. Hirotsu, S. (1991) Softening of bulk modulus and negative Poisson's ratio near the volume phase transition of polymer gels. *J. Chem. Phys.*, **94**(5), 3949–3957.

37. Smith, I.M. and Griffiths, D.V. (2004) *Programming the Finite Element Method*, John Wiley & Sons, Ltd, Chichester.

38. Zienkiewicz, O.C. and Taylor, R.L. (2000) *The Finite Element Method*, Butterworth-Heinemann, Oxford.

39. Mazzotti, M., Morbidelli, M., and Serravalle, G. (1995) Bifurcation-analysis of the Oregonator model in the 3-D space bromate malonic-acid stoichiometric coefficient. *J. Phys. Chem.*, **99**(13), 4501–4511.

40. Shinohara, S., Seki, T., Sakai, T., Yoshida, R., and Takeoka, Y. (2008) Chemical and optical control of peristaltic actuator based on self-oscillating porous gel. *Chem. Commun.*, **39**, 4735–4737.

41. Krug, H.J., Pohlmann, L., and Kuhnert, L. (1990) Analysis of the modified complete Oregonator accounting for oxygen sensitivity and photosensitivity of Belousov–Zhabotinsky systems. *J. Phys. Chem.*, **94**(12), 4862–4866.

42. Nakagaki, T., Iima, M., Ueda, T., Nishiura, Y., Saigusa, T., Tero, A., Kobayashi, R., and Showalter, K. (2007) Minimum-risk path finding by an adaptive amoebal network. *Phys. Rev. Lett.*, **99**(6), 068104.

9
Chemoelastodynamics of Responsive Gels

Jacques Boissonade, Pierre Borckmans, Patrick De Kepper, and Stéphane Métens

9.1
Introduction

Chemical reactions kept far from equilibrium by a permanent feed of fresh reactants can exhibit self-organization phenomena under the action of nonlinear kinetics [1]. These phenomena follow from instabilities that break the natural symmetries of the system. These instabilities result from the coupling of kinetic autoactivation processes to appropriate feedbacks. In well-stirred systems, they take the form of time-periodic, or more complex, oscillations of concentrations. In unstrirred systems, where a competition between reaction and mass transport is involved (mainly diffusion), they give rise to stationary concentration patterns (Turing patterns, labyrinthine patterns), traveling waves (target, spiral patterns, etc.), or more disordered spatiotemporal patterns. To sustain nonequilibrium conditions, it is necessary to resort to open reactors: for homogeneous systems, one generally uses a continuous stirred tank reactor (CSTR). For reaction–diffusion patterns, where convective transport must be impeded, the reactor is usually made of a piece of soft gel in which the reactants diffuse from boundaries that are kept in contact with one or several well-stirred reservoirs [2]. Concentration profiles and eventually patterns establish inside the gel. A CSTR, kept far from equilibrium by a short residence time, is commonly used as a unique reservoir providing all fresh reactants. Such one side fed reactors (OSFR) are very popular and used in this work. In rare cases, like in the well-known Belousov–Zhabotinsky (BZ) reaction, oscillations or patterns can transiently be observed in batch reactors during evolution of the reacting solution to equilibrium.

In standard reaction–diffusion experiments, the chemical nature and formulation of gels were chosen for their relative chemical inertness and mechanical stability. However, it is known that some gels, in particular polyelectrolyte gels, may undergo considerable swelling or shrinking under the influence of a wide range of stimuli [3]. We focus here on swelling (shrinking) resulting from the chemical composition of the solvent. In the latter case, the responses to pH and metal-ion concentration are most commonly studied. The responsiveness of these soft elastic materials to stimuli has led to propose numerous applications [4–6]. In this respect, it was

Nonlinear Dynamics with Polymers: Fundamentals, Methods and Applications.
Edited by John A. Pojman and Qui Tran-Cong-Miyata
Copyright © 2010 WILEY-VCH Verlag GmbH & Co. KGaA, Weinheim
ISBN: 978-3-527-32529-0

appealing to carry out the autoactivated reactions in such responsive gels to explore possible routes to autonomous mechanical motions and self-shaping phenomena in association with applications to soft actuating devices. A trend of research in this direction, mainly prompted by Yoshida [7–10], was soon followed by other groups [11–13].

This chapter is devoted to the spontaneous generation of mechanical oscillations by a responsive gel immersed in a reactive medium kept far from equilibrium. Two important cases will be considered. In Section 9.3, the chemomechanical instability is mainly driven by a kinetic instability leading to an oscillatory reaction. The approach is applied to the BZ reaction. In Sections 9.4 and 9.5, we elaborate on a mechanical oscillatory instability that emerges from the cross-coupling of a reaction–diffusion process and the volume or size responsiveness of the supporting material. In this case, there is no need for an oscillatory reaction. Bistable reactions, namely, the chlorite–tetrathionate (CT) and the bromate–sulfite (BS) reactions, are chosen as a support to this approach. Several theories have been developed to account for gel-swelling mechanisms and their coupling to reactants. Two different modeling processes will be used here. In Section 9.3, the gel swelling and reaction dynamics are described in terms of multicomponent nonequilibrium thermodynamics. In Section 9.4, a Maxwell–Stefan description applied to a two-fluid model will be used. Experimental observations are reported in Section 9.5.

The key points of these theories are the description of swelling dynamics and the way in which the concentration of the different species inside the medium influences the energy and/or the entropy of the system. A full presentation of this subject is far beyond the scope of this chapter. A brief survey of the related literature is given in Section 9.2. More details will be given in Sections 9.3 and 9.4 when needed.

9.2
Elastodynamics of Responsive Gels: a Brief Survey

Hydrogels consist of a crosslinked polymer network with hydrophilic moieties, pervaded by liquid water (the solvent) in which other ionized or neutral solutes can be present. Here, we wish to discuss the case where the solvated species are free to diffuse and react together. Depending on the chemical nature and the structure of the polymer matrix, changes in the gel environment (temperature, nature and concentration of the solutes, hydrophobicity, pH, ...) induce a concomitant swelling (or shrinking) process through the uptake, or expulsion, of the fluid solvent. After a step change in the gel environment, the whole system relaxes to a new thermodynamic equilibrium. If we are only interested in this final state, a thermodynamic approach is sufficient to describe the results [14].

The volume variation (swelling or shrinking) rate can be expressed as a balance between two forces. One is related to the osmotic pressure, which favors the swelling. The origin of this force is the interaction of the polymer chains with the solvent and solutes, and can be described in its simplest version, in terms of the Flory–Huggins polymer solution theory. In polyelectrolytes, there is an additional osmotic pressure due to free ions (see Section 9.4.3) and electrostatic interactions. Another force acts against swelling. It is induced by the elastic forces that are opposed to the extension of the macromolecular network. This contribution is most simply expressed by a model of rubber elasticity.

The volume changes, in response to specific external stimuli, can be quite dramatic. Depending on the formulation of the hydrogel matrices, the variation may occur continuously over a range of stimulus levels or discontinuously at a critical stimulus level. The possibility of such first-order volume phase transitions in gels was suggested by Dušek and Patterson in 1968 [15] and confirmed 10 years later by Tanaka [16]. This important aspect generated considerable interest, the results of which have been collected in various reviews [17, 18]. In some cases, such a phase transition could constitute an important ingredient but will not be considered in the theoretical developments that follow.

A description of the dynamical evolution of the system naturally requires to go beyond equilibrium thermodynamics. There are different approaches in the literature that are more or less equivalent. Differences and similarities will not be analyzed here. We briefly mention some of these descriptions and the related literature. A regular starting point lies in the derivation of the balance equations for the relevant quantities. To close this set of equations, transport models, based on some approximations, for example, the local equilibrium assumption and Onsager's linear assumption, must be added. They take into account osmotic pressure, hydrogel stress, species diffusive fluxes, and possibly electrical potential. The Tanaka–Fillmore [19] approach assumes that the deformation of the gel network takes place in an immobile viscous medium. In their model, the swelling rate is fixed by a balance between a driving force, that is, the osmotic pressure, and a friction described through a constant diffusion coefficient related to the bulk and shear modulus of the polymer network. Some extensions of this simple model have also been proposed [20]. More realistic models take the solvent and solute flows into account. Among the different approaches, a hydrodynamical theory of gels has been proposed in the framework of the elasticity of large deformations [21, 22]. It starts from a binary mixture including a variable describing the gel deformation. The balance equations allow for the calculation of the entropy production source. Then, the use of the equations of state and the application of Onsager's principle lead to a complete description (Section 9.3). An alternative starting point is to describe the relative motion of the gel and the solvent in terms of a generalized Darcy law, mechanical equilibrium, and incompressibility condition. This leads to a stress diffusion coupling model as proposed in Ref. [23]. A Lagragrian derivation of equivalent equations has been carried out in Refs. [24, 25]. The Maxwell–Stefan relations may also serves as a good starting point for a gel immersed in a multicomponent mixture. In this approach, both the diffusion coefficients and the friction coefficients depend on the composition of the system. The driving forces are expressed as the product of a friction coefficient and the relative velocities of the components expressed in an appropriate reference frame. The evolution equations are obtained through the balance of the driving and the friction forces [26, 27] (Section 9.4).

All these models require the values of diffusion coefficients and interaction parameters, which are, in most cases, not accurately known. Their measurement needs the implementation of some experimental methods as soft X-ray or neutron scattering, infrared spectroscopy, and application of rheological techniques [28]. In theoretical works, these coefficients are often empirically introduced.

9.3
Oscillatory Gel Dynamics Using an Oscillating Chemical Reaction

9.3.1
The Approach

We shall follow the principles of the method proposed in Refs. [21, 22], which are well adapted to treat the problem considered here. However, we will extend it to consider N chemical species besides the solvent [29–31]. The free energy is a central quantity. Indeed, on the one hand, it allows for the determination of the properties of the possible equilibrium states (these may be multiple, as the system is nonideal [17]). On the other hand, it plays a leading role in the derivation of the equations describing the space–time evolution of the gel in the framework of the local equilibrium assumption. For isothermal processes, the free energy f of the gel is the superposition of three contributions

$$f = f_{\text{mixing}} + f_{\text{elastic}} + f_{\text{ionic}} \tag{9.1}$$

where the first term is the free energy of the mixture of the polymer network, the solvent, and the (eventually reacting) solutes residing in the solvent. In its simplest form, it is given by the Flory–Huggins theory in terms of the volume fractions denoted by ϕ_i, where ϕ_p is that of the monomers,

$$f_{\text{mixing}} = \frac{Jk_BT}{v_1} \left(\sum_{i=1}^{N+s} \chi_{ip}\phi_i\phi_p + \frac{1}{2} \sum_{i \neq j}^{N+s} \chi_{ij}\phi_i\phi_j + \sum_{k=1}^{N+s} \frac{\phi_k}{N_k} \ln \phi_k \right) \tag{9.2}$$

where v_1 denotes the lattice site volume, that is, the volume of a monomer or of a solute molecule; k_B is the Boltzmann constant, T is the temperature, χ_{ip} and χ_{ij} are the Flory–Huggins energy parameters, respectively, between the polymer and the i molecules, and between the solute molecules i and j; N_k are the number of lattice sites occupied by molecules k. $J = \det(F_p^v)$ expresses the relation between the volume V of the deformed gel and its value V^0 in a reference state. F_p^v are the components of the deformation gradient tensor **F**.

There are many different expressions proposed to model the elastic part. They find their origins either in some statistical mechanical model like the Gaussian chains, the Flory [32] or the James and Guth models [33], or in some phenomenological laws tending to reproduce experimental observations [34]. In the case of the Gaussian chain approximation, the elastic contribution is given by

$$f_{\text{elastic}} = \frac{\nu_e k_B T}{2} \left(\text{Tr} \left(\mathbf{F} \mathbf{F}^T \right) - 3 - 2 \ln \left(\det \mathbf{F} \right) \right) \tag{9.3}$$

where ν_e is the number of partial chains per unit of reference volume, that is, the number of polymer chains between two cross-links.

We will postpone the discussion of the effects induced by f_{ionic}, describing the possible electrostatic interactions when ions are present, to a further section, as the problem we wish to present first is more related to hydrophobic effects than to ionic interactions.

From the free energy, we may express the equations of the state of the system in terms of the osmotic stress tensor $\mathbf{\Pi}$ and the chemical potentials making use of the Gibbs relation [21, 30, 31]. In terms of $\mathbf{\Pi}$, we have

$$(\Pi)_\nu^\lambda = \frac{F_p^\lambda}{J} \frac{\partial f}{\partial F_p^\nu} |_{T; J\phi_\alpha}$$
$$\frac{\widehat{\mu_\alpha^m}}{\overline{V_\alpha}} = \frac{\partial f}{\partial (J\phi_\alpha)} |_{T; F_p^\nu; J\phi_\beta \neq \alpha} \tag{9.4}$$

where $\widehat{\mu_\alpha^m}$ is the exchange chemical potential per unit mass.

Incompressibility of the gel is assumed here, implying that the total volume V of the system may be written as

$$V = M_p \overline{V_p} + M_s \overline{V_s} + \sum_{k=1}^{N} M_k \overline{V_k} \tag{9.5}$$

where the $\overline{V_a}$ denote the respective specific volumes. Dissipation effects due to internal friction of the viscous type are also not considered.

The details of the derivation of the dynamical equations are given in Ref. [31] and are too lengthy to be reproduced here. The natural experimental quantities are the permeation flux, which characterizes the solvent and solute transport in the gel matrix, and the diffusion coefficients of the solute species in the solvent. For these reasons, we introduce the fluxes $\mathbf{j}_k$ of species k

$$\mathbf{j}_k = \phi_k (\mathbf{v}_k - \mathbf{v}_{\text{mean}}) \tag{9.6}$$

where the mean velocity of the fluid is defined by $\mathbf{v}_{\text{mean}}(1 - \phi_p) = \sum_{k=1}^{N} \phi_k \mathbf{v}_k + \phi_s \mathbf{v}_s$ and $\mathbf{v}_s$ is the center of mass velocity of the solvent. The permeation flux is then given by

$$\mathbf{j}_{\text{perm}} = \left(\sum_{k=1}^{N} \phi_k + \phi_s \right) (\mathbf{v}_{\text{mean}} - \mathbf{v}_p) = \mathbf{j}_{\text{tot}} - \mathbf{v}_p$$
$$= (1 - \phi_p)(\mathbf{v}_s - \dot{\mathbf{r}}(R, t)) \tag{9.7}$$

where $\dot{\mathbf{r}}(R, t)$ is the velocity of a gel particle, labeled R, the movement of which we follow. From these quantities, the balance equations may be written as

$$\partial_t \phi_k = -\nabla \cdot (\phi_k \boldsymbol{v}_k) + G(\phi_k)$$

$$= -\nabla \cdot \left(\mathbf{j}_k + \mathbf{j}_{\text{perm}} \frac{\phi_p}{(1-\phi_p)} \right) + \phi_k \boldsymbol{v}_p + G(\phi_k)$$

$$= -\nabla \cdot \mathbf{j}_k - \mathbf{j}_{\text{perm}} \cdot \nabla \phi_k \frac{\phi_p}{1-\phi_p} - \frac{\phi_k \phi_p}{1-\phi_p} \nabla \cdot \mathbf{j}_{\text{perm}}$$

$$- \frac{\phi_k}{(1-\phi_p)^2} \mathbf{j}_{\text{perm}} \cdot \nabla \phi_p + G(\phi_k) \quad (9.8)$$

$$0 = \nabla \cdot \mathbf{j}_{\text{tot}} \quad (9.9)$$

where we have as usual $1 = \sum_{k=1}^{N} \phi_k + \phi_s + \phi_p$. We also assumed that $\phi_p < \phi_s$, $\phi_k \ll \phi_p$, and $\phi_k/\phi_s \ll 1$, as in the experiments [7, 8, 11, 13]. In addition, we also admit that the diffusion coefficients of the solutes inside the gel are independent of ϕ_p and may therefore be taken as constants. This equation can be easily expressed in terms of the concentrations of reacting species k, which is more commonly used, as

$$\partial_t c_k = D_k \triangle c_k + F(c_k) + \frac{\phi_p}{(1-\phi_p)} \mathbf{j}_{\text{perm}} \cdot \nabla \mathbf{c} + \frac{\phi_p}{(1-\phi_p)} c_k \nabla \cdot \mathbf{j}_{\text{perm}}$$

$$+ \frac{c_k}{(1-\phi_p)^2} \mathbf{j}_{\text{perm}} \cdot \nabla \phi_p \quad (9.10)$$

where $F(c_k)$ represents the reaction kinetics term.

This equation generalizes the usual reaction–diffusion equations (first two terms on the R.H.S.) in that it describes reaction–diffusion systems in gels that may undergo volume changes in response to some stimulus, (e.g., temperature, solvent composition, illumination). However, in order to furthermore describe the transduction of chemical to mechanical energy, we have to introduce a coupling of the chemistry with a property of the gel, so that chemistry may affect $\mathbf{j}_{\text{perm}}$.

This is not a generic feature and will depend on the particular nature of the gel composition and chemical kinetics at work. One should bear in mind that we are already working with a free energy density in the given frame of approximation that, as detailed in Section 2.4 of Ref. [17], while allowing for the description of the essential features of the volume phase transition, does not lead to a quantitative agreement between theory and experiments. We can introduce such couplings through the dependence of the Flory interaction parameter χ_{sp} on the concentrations of the reactive species. This method was first invoked in Ref. [35] to allow for the description of the volume phase transition of a neutral gel by an expansion of χ_{sp} in terms of the volume fraction of monomers such that $\chi_{sp} = \chi_1 + \chi_2 \phi_p + \ldots$. This procedure was generalized to include a dependence on the solute species [36]. In some experiments [37], such explicit dependence of χ_{sp} as a function of the volume fraction has even been measured. More examples may be found in Ref. [14]. We have successfully used such a treatment to test our method [30] and corroborate the results described in Section 9.4. For this purpose, we used a chemical system leading to spatial bistability, but different from those discussed in Section 9.4.

For the description of the experiments treated in the following subsection, we will rather include a supplementary contribution in the mixing free energy as in Ref. [38], because a main reacting group resides on the polymer.

9.3.2
Coupling to the Oscillating Belousov–Zhabotinsky Reaction

We now wish to test the above gel-reaction–diffusion approach (Eqs 9.7–9.10) using the BZ reaction, which consists in the oxidation of malonic acid by bromate ions in acidic medium. The reaction proceeds only when catalyzed by a suitable metal ion. In the experiments, the chosen catalyst is the ruthenium tris(2, 2′-bipyridine) that intervenes through its oxydo-reduction couple (Ru(bpy)$_3^{3+}$/Ru(bpy)$_3^{2+}$). In a batch reactor or a CSTR, this reaction exhibits well-documented periodic oscillations of concentrations in some region of the parameter space.

Using responsive gels (e.g., N-isopropylacrylamide) in which this catalytic ion is covalently bound to the polymer network–so that the oscillatory properties are confined to the gel contents–and the unique capacity of the reaction to exhibit oscillatory regimes over extended periods of time even in batch conditions, Yoshida [8] has obtained time-periodic modulations of the volume of the gel in a constant environment, contrary to the usual on–off switching by externally controlled stimuli.

An intuitive explanation of the oscillatory volume variations is given by Yoshida in Ref. [8]; it is based on hydrophobic effects that induce a difference of swelling rate between the oxidized and the reduced states of the catalyst. For a small piece of gel, the oscillations occur homogeneously [39]. But when the chemical wavelength is smaller than the system size, some inhomogeneous behavior in the form of waves may arise [8].

We restrict our presentation to the case of a *spherical* bead of gel submitted to the BZ reaction performed with feed and catalyst conditions analogous to the above-mentioned experiments (bonded catalyst). Each point of the sphere is defined by a triplet of variables (r, θ, ϕ), and furthermore, we shall assume that spherical symmetry is maintained during the volume oscillations, that is, the deformations are only radial as in Ref. [27, 40–42].

Starting from relations (9.8) and (9.9), we may derive the evolution equation for $\dot{r}(R, t)$. In the present geometry and conditions, this quantity represents the radial velocity for the motions of the particle labeled R, which is followed in its evolution

$$\dot{r}(R, t) = \frac{D_{\text{perm}}}{k_B T \nu_e} \frac{1 - \phi_p}{h(\phi_p)} \left(\frac{\partial \Pi_r^r}{\partial r} + \frac{1}{r}(2\Pi_r^r - \Pi_\theta^\theta - \Pi_\phi^\phi) \right) \quad (9.11)$$

where D_{perm} is the permeation coefficient that may depend on ϕ_p. A friction coefficient $\zeta(\phi_p)$ between polymers and solvent is also introduced

$$\zeta(\phi_p) = \frac{k_B T \nu_e}{D_{\text{perm}}} h(\phi_p) = \frac{k_B T \nu_e}{D_{\text{perm}}} \frac{\phi_p^{(3/2)}}{(\phi_p^0)^{(3/2)}(1 - \phi_p)^3} \quad (9.12)$$

The explicit forms of the chemomechanical Eq. (9.10) with the two-variable model for the BZ reaction [43] are then given by

$$\partial_t u = D_u \Delta u + \frac{\phi_p}{1-\phi_p} \left(j^r_{\text{perm}} \frac{\partial u}{\partial r} + u \left(\frac{\partial j^r_{\text{perm}}}{\partial r} + \frac{2}{r} j^r_{\text{perm}} \right) \right) + \frac{u j^r_{\text{perm}}}{(1-\phi_p)^2} \frac{\partial \phi_p}{\partial r}$$
$$+ f k_5 B v \frac{k_1 H^2 A - k_2 H u}{2(k_1 H^2 A + k_2 H u)} + k_3 H A u - 2 k_4 u^2 \tag{9.13}$$

$$\partial_t v = \frac{\phi_p}{1-\phi_p} \left(j^r_{\text{perm}} \frac{\partial v}{\partial r} + v \left(\frac{\partial j^r_{\text{perm}}}{\partial r} + \frac{2}{r} j^r_{\text{perm}} \right) \right)$$
$$+ \frac{v j^r_{\text{perm}}}{(1-\phi_p)^2} \frac{\partial \phi_p}{\partial r} + 2 k_3 H A u - k_5 B v \tag{9.14}$$

where u and v are reduced variables, respectively, proportional to the concentrations of the intermediate species $HBrO_2$ and $Ru(bpy)_3^{3+}$ (the oxidized form of the metal catalyst attached to the matrix) and where A, B, and H are reduced variables, respectively, proportional to the concentrations of the input species BrO_3^-, malonic acid, and H^+. The k_i are the kinetic constants. The concentrations A, B, and H are taken constant and uniform (pooled variables [1]) and are used as control and bifurcation parameters. We have kept D_v equal to zero as the transport of that group is effected solely through the permeation of the gel matrix represented by the non-reaction–diffusion type terms.

As stated previously, we now have to decide on the coupling between the chemistry at play and the polymeric part. In the absence of experimental determination of the χ parameters (that are now quite numerous), there is a great deal of arbitrariness in the choice of the values to be attributed. We choose the following $\chi_{up} = \chi_{us} = \chi_{vs} = 0$. The main chemomechanical contribution arises from the control of hydrophobicity by $Ru(bpy)_3^{3+}$. To account for this effect, we describe the interaction of $Ru(bpy)_3^{3+}$ with the polymer network by introducing a Flory–Huggins term containing the factor $\chi_{vp} \neq 0$. This is similar to the hypothesis found in Ref. [38]. As the catalyst consists of a large molecular group, we may assume that $\phi_v \gg \phi_u$ and then neglect $\chi_{sp}\phi_u\phi_p$ in comparison with $\chi_{sp}\phi_v\phi_p$. We further assume that each molecule occupies only one site in the network, that is, $N_s = N_i = 1$.

According to these hypotheses, the osmotic stress tensor components are [30, 31]

$$\Pi^r_r = \frac{k_B T}{v_1} \left(\phi_p \left\{ \chi_{sp}\phi_p + 1 + \beta v + \frac{v_e v_1}{\phi_p^0}[(F^r_r)^2 - 1] \right\} + \ln(1-\phi_p) \right)$$
$$\Pi^\theta_\theta = \frac{k_B T}{v_1} \left(\phi_p \left\{ \chi_{sp}\phi_p + 1 + \beta v + \frac{v_e v_1}{\phi_p^0}[(F^\theta_\theta)^2 - 1] \right\} + \ln(1-\phi_p) \right) \tag{9.15}$$

We chose $\beta = \overline{V}_v(\chi_{sp} - \chi_{vp})$ as the control parameter in the numerical simulations ($\overline{V}_v$ denotes the molar volume of v). The first term describes the usual polymer–solvent interactions that may depend on ϕ_p, as discussed previously, while the second one accounts for the effect of the oxidation state of the catalyst on the swelling properties of the gel. Its chemical potential is

$$\frac{\widehat{\mu_v}}{V_v^m} = \frac{k_B T}{v_1}\left(\ln \phi_v - \chi_{sp}\phi_p - \chi_{vp}\phi_p\right) \tag{9.16}$$

where the logarithmic term is dominant for dilute reactants.

At the interface between the gel and the bath, we have

1) Continuity of the chemical potential that ensures continuity for concentrations.

$$\phi_{u,\text{int}} = \phi_{u,\text{ext}} \tag{9.17}$$
$$\phi_{v,\text{int}} = \phi_{v,\text{ext}} = 0 \tag{9.18}$$

2) The osmotic stress tensor is equal to zero at the interface

$$\Pi_r^r|_{\text{int}} = 0 \tag{9.19}$$

9.3.3
Numerical Integration Results

Equations (9.11, 9.13, 9.14) have been numerically integrated for a spherical bead of gel using the above boundary conditions. Here, we only present a short summary. More details may be found in Refs. [29–31].

The values of the following parameters were kept fixed:

$B = 0.0035$ M
$k_2 = 3 \times 10^6$ M^{-2} s^{-1}
$k_3 = 42$ M^{-2} s^{-1}

$H = 0.3$ M
$D_u = 10^{-5}$ cm^2 s^{-1}
$k_4 = 3 \times 10^3$ M^{-1} s^{-1}

$k_1 = 2$ M^{-3} s^{-1}
$f = 0.7$
$k_5 = 6$ M^{-1} s^{-1}

We have checked that, with this model and in the range of A values used in the computations that follow, these conditions give rise to oscillations of u and v in a sphere of nonresponsive gel.

The temporal evolution of the radius R_{sphere} of a spherical chemoresponsive gel is represented in Figure 9.1(a) for a given value of A and of the coupling parameter β. Initially, the gel is in some chosen equilibrium (reference) state, $\phi_p(t=0)$, of radius $R_{\text{sphere}}(t=0)$ and volume V^0. This is the volume reached when the reactants' solution has imbibed the gel in the absence of coupling. At $t = 0$, the coupling is turned on ($\beta \neq 0$). The reactions set in and the concentrations start oscillating, as does the radius of the sphere. After a short transient, the mechanical oscillations become periodic. The amplitude of the radius oscillations grows as β increases. However, we have not found any dependence of the oscillation period on β. The volume variations remain small, but this may result from the particular choice of β. This choice is somewhat arbitrary, as previously mentioned.

Likewise, Figure 9.1(b) exhibits the temporal periodic oscillations of the concentration of the oxidized form of the catalyst $v(0, t)$ at the center of the sphere. They are similar to the relaxation oscillations observed in a passive gel, but no detailed comparison has yet been undertaken. As in the experiments of Yoshida et al. [39], the oscillations of R_{sphere} and $v(0, t)$ exhibit the same period with a small phase

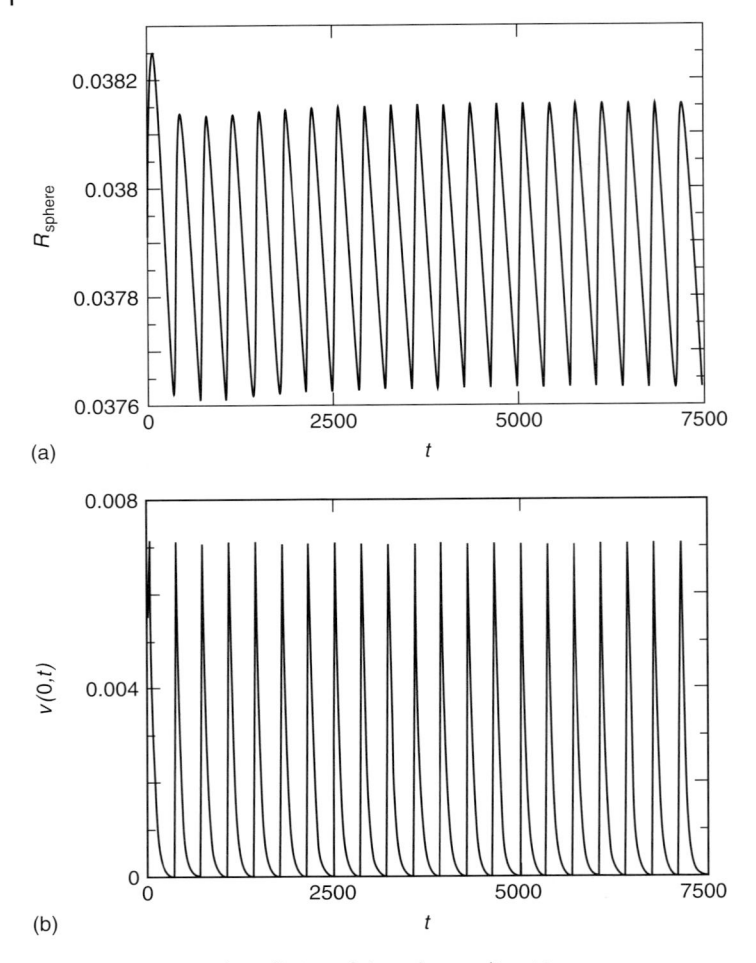

Figure 9.1 Temporal oscillation of the sphere radius (a) and of $v(0, t)$ (b) with $\chi_{sp} = 0.54$, $\phi_0 = 0.1$, $\phi_p(t = 0) = 0.1$, $R_{\text{sphere}}(t = 0) = 0.038$ cm, $A = 0.085$ M, $\beta = 0.98$.

shift. Although any value of $v(r, t)$ could be chosen as a norm to measure the frequence, $v(0, t)$ was selected because, here, it corresponds to the position where the maximum amplitude of change is observed.

In Figure 9.2, we show the evolution of the polymer volume fraction $\phi_p(r, t)$ over one oscillation period. We observe only weak spatial and temporal changes in the core of the sphere in accordance with the weak spatial variations of u and v in this region. The temporal changes become larger close to the boundary of the gel. In this region, the spatial gradient of $v(r, t)$ is large since, due to boundary condition (Eq. 9.18), there is no catalyst outside the gel. This behavior was expected since one of the driving terms in the radius equation (Eq. 9.11) is proportional to $\partial_r \Pi_r^r$, that is, also to $\partial_r v(r, t)$. The temporal evolution of the concentration profiles $v(r, t)$ and

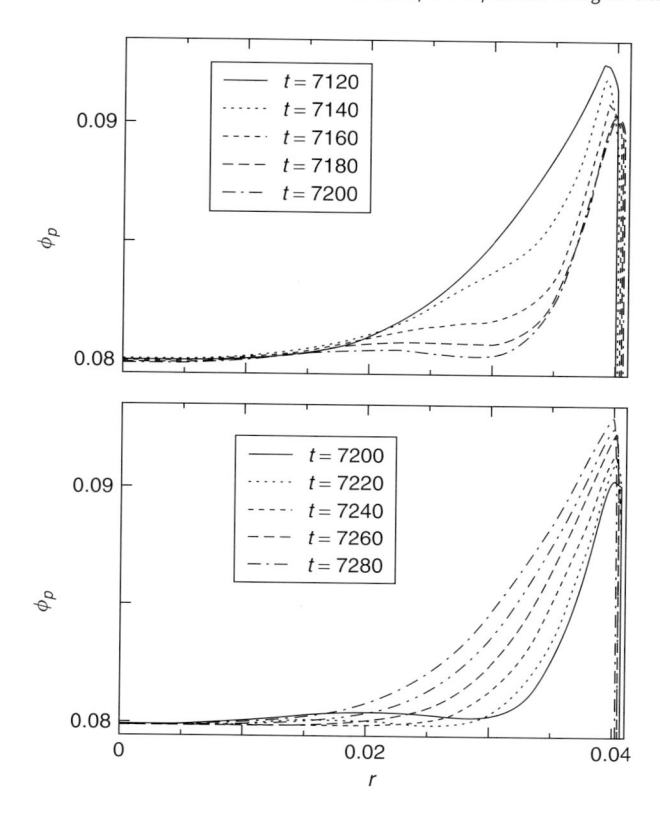

Figure 9.2 Temporal evolution of the radial profile of monomers' volume fraction ϕ_p during one period of oscillation; $\chi_{sp} = 0.54, \phi_0 = 0.1, q\phi_p(t=0) = 0.1, R_{\text{sphere}}(t=0) = 0.038$ cm, $A = 0.085$ M, $\beta = 0.98$.

$u(r, t)$ are only weakly affected by the coupling with the gel. Such a result might, however, be different in the case of larger volume variations or for other choices for the coupling constant.

The experimental determination of such profiles presents numerous challenges already found in the study of reaction–diffusion patterns in inert gels. In active gels, the problem is even more acute as shown in the experimental Section 9.5.

Another set of simulations was carried out, at fixed β, while varying the concentrations of the substrates. For instance, it is found that $v(0, t)$ increases monotonously with A. This induces a corresponding enlargement of the maximum and mean radii of the sphere. On the other hand, it is well known, for a BZ system, that the dependence of the period of oscillations, T, on the concentration of the substrates (bromate ion, malonic acid, and free protons) provides useful information [1]. In our simulations, we have obtained that T increases with the decrease in concentrations of the substrates following power laws with negative exponents. Similar behaviors are obtained for a sphere of passive gel, where the catalyst is also fixed to the matrix. However, the exponents differ. On the experimental side, in a millimetric size active gel, this power law dependence can show significant differences, in particular the exponent for the dependence on the proton concentration is positive [39]. Yet, when the reaction is performed in a suspension of nanogel beads, all the exponents are again negative [44].

From the differences that arise between the dynamics of the BZ reaction in an active and a passive gel, one may conclude that the dynamics of the active gel–BZ system does not result from a simple forcing of the gel by the chemistry. Numerical results clearly show that our three-variable model (including volume fraction ϕ_p) constitutes a dynamical system different from the two-variable BZ reaction–diffusion system in a passive sphere.

Nevertheless, the inherent oscillatory property of the BZ reaction, present in both experiments and modeling, introduces an intrinsic time periodicity at the root. In the following sections, we show that an oscillating gel may be obtained when neither the gel nor the chemical reaction ever exhibits oscillatory behavior.

9.4
Chemodynamic Oscillations Induced by Geometric Feedback

When a gel is immersed in a stationary reacting medium kept far from equilibrium, a distribution of concentrations settles in the gel as a result of the reaction–diffusion process. In a responsive gel, this can induce changes in size, which in turn modifies this distribution. In some cases, these mutual feedbacks are able to destabilize the whole system and create both mechanical and chemical oscillations even when the reaction has no oscillatory capacities. Among possible mechanisms for such oscillatory instabilities are the association of a bistable system and an appropriate feedback. Siegel and coworkers have exploited the hysteresis of the volume transition of a gel responsive to pH to create oscillations in a system of two compartments [45–47]. Here, we consider the case where the bistability and the associated hysteresis loop result from the chemical kinetics. We shall first recall the basic principles that rule spatial bistability, before elaborating on theoretical predictions with different models and concluding with a few experimental data.

9.4.1
Spatial Bistability and Related Chemomechanical Instabilities

Consider an inert gel immersed in a CSTR and a reaction with an autocatalytic path driven by species X. The result of the autocatalysis is that the reaction rate considerably increases when the amount of X becomes sufficient, so that the state of the reaction is essentially either an almost unreacted state F before this amount has been reached or a state T where the chemical conversion is almost complete after it has been reached. The CSTR contents are maintained in the unreacted state by imposing a short residence time τ. This fixes the gel state at the CSTR/gel boundary. The distribution of concentrations in the depth of the gel is controlled by the distance to the feeding edge, which defines the characteristic time for fresh unreacted species to reach the different points. Thus, the state of the gel is governed by $t_D \sim L^2/D$, where L is the characteristic size of the gel in its smaller dimension (the radius for a sphere or a cylinder) and D is the typical diffusion coefficient. If L is small, t_D is smaller than the induction time and the gel contents remain in the F state everywhere. If L is large, t_D is larger than the induction time and the core of the gel switches to state T. Under the effect of diffusion and autocatalysis, this state expands almost to the whole contents of the gel except for a boundary layer that insures concentration continuity along the feeding edges. This defines a mixed FT state. At intermediate L values, where t_D is only slightly smaller than the induction time, the final state depends on the initial state. If there is a sufficient initial amount of X present, the system will be pushed to an FT state. If not, it will remain in a F state. Thus, both are stable in a finite range of sizes [L_{inf}, L_{sup}]. This defines a spatial bistability and an hysteresis loop when L is alternatively increased beyond L_{sup} and decreased below L_{inf} [48].

When the inert gel is replaced by a chemoresponsive gel that swells in state F and shrinks in state FT, it is possible to find an appropriate CSTR composition and an initial gel size L so that the two limits L_{inf} and L_{sup} are crossed alternatively. Every time the system reaches one of these limits, the system jumps to the other state and reverses the direction of swelling/shrinking, giving rise both to mechanical (volume) oscillations and to chemical oscillations between F and FT. The process is made possible by the hysteresis loop and by the slow diffusion of the polymeric network that rules the swelling process.

9.4.2
Simple Models

The validity of this prediction has been checked on the basis of several simple models. They all rely on a Flory-type description of the gel and were applied to gel spheres of radius R_s. We again assume that no instability develops in nonradial directions and that the spherical symmetry is preserved. Different forms of the elastic contributions, of the chemical models, and of the mode of coupling between the chemical state and swelling properties have been used [30, 40, 41, 49]. These coupling modes are of energetic nature and are introduced by means of the energetic interaction terms χ or the χ_{ij}'s as mentioned in Section 9.3.1. In spite of these differences and the various approximations used in these computations, the results are qualitatively similar, and predict the emergence of oscillatory behaviors for some domains of parameters. Here, we shall assume that the volume fractions of the reagents are negligible in regard to those of the polymer matrix and the solvent (respectively, ϕ and $1 - \phi$) and present results obtained with a kinetic model that has been first proposed as a reduced renormalized model of the CT. Owing to numerous flaws, it must actually be taken as a simple toy model of reaction–diffusion system involving the concentration v of an autocatalytic species:

$$\frac{\partial u}{\partial t} = -u^2 v^2 + \nabla^2 u$$
$$\frac{\partial v}{\partial t} = \frac{12}{7} u^2 v^2 + \nabla^2 v \qquad (9.20)$$

When the nonequilibrium values ($u_0 = 1$, $v_0 = 0.05$) are imposed at the sphere boundary, the model exhibits spatial bistability for $4.48 < R_s < 5.42$ with $v \ll 1$ in the "unreacted" F state and $v \sim 1$ in the "reacted" state FT [40, 41]. In this example, one assumes that χ depends on v according to the sigmoidal law

$$2\chi(v) = \chi_{\min} + \chi_{\max} + (\chi_{\max} - \chi_{\min}) \tanh(s(v - v^*)) \qquad (9.21)$$

where $\chi_{\min}$, $\chi_{\max}$, v^*, and α are chosen in such a way that χ changes from 0.30 (for $v = 0$) to 0.53 (for $v \gg v^*$), so that the system swells in the F state and shrinks in the FT state. The contribution of elastic terms to the stress tensor σ is taken in agreement with the phenomenological theory of Bastide and Candau [50] as formulated by Barrière and Leibler [27]. The tensor can be decomposed in a mixing contribution and an elastic contribution according to

$$\sigma_{ij} = -\delta_{ij} \Pi_{\text{mix}} + \sigma_{ij}^{(\text{elas})} \qquad (9.22)$$

The mixing contribution to the osmotic pressure, given by the classical Flory–Huggins theory, is

$$\Pi_{\text{mix}} = -\left(\frac{RT}{V_S}\right)[\phi + \log(1-\phi) + \chi\phi^2] \qquad (9.23)$$

where V_S is the molar volume of the solvent. If r' and r are, respectively, the radial coordinates before and after deformation of an isotropic reference state of volume fraction ϕ_0, one shows that for this unidirectional swelling the contribution of elasticity to the stress tensor is

$$\sigma_{rr}^{(\text{elas})} = K_{\text{net}} \left(\frac{\phi}{\phi_0}\right)^{\frac{1}{3}} \left[1 + C_{\text{net}}\left(\lambda^2 - \frac{1}{\lambda}\right)\right] \text{ with } \lambda = \left(\frac{\phi_0}{\phi}\right)^{\frac{2}{3}} \left(\frac{r'}{r}\right)^2 \qquad (9.24)$$

where K_{net} is a constant that gathers network properties and C_{net} is a constant of order unity proportional to the shear modulus. The C_{net} term gives a contribution of the shear stress to the osmotic pressure as a function of the deformation factor λ. This term vanishes at equilibrium.

The relative motion of the polymer and the solvent is described in the frame of a Maxwell–Stefan approach by equating the internal forces with the friction forces of the solvent on the polymeric network [26]

$$\zeta(\phi)(v_P - v_S) = \nabla \cdot \boldsymbol{\sigma} \qquad (9.25)$$

Accounting for conservation of mass

$$\phi v_P + (1-\phi) v_S = 0 \qquad (9.26)$$

one gets

$$v_P = \frac{(1-\phi)}{\zeta(\phi)} \nabla \cdot \boldsymbol{\sigma} \quad \text{and} \quad v_S = -\frac{(\phi)}{\zeta(\phi)} \nabla \cdot \boldsymbol{\sigma} \qquad (9.27)$$

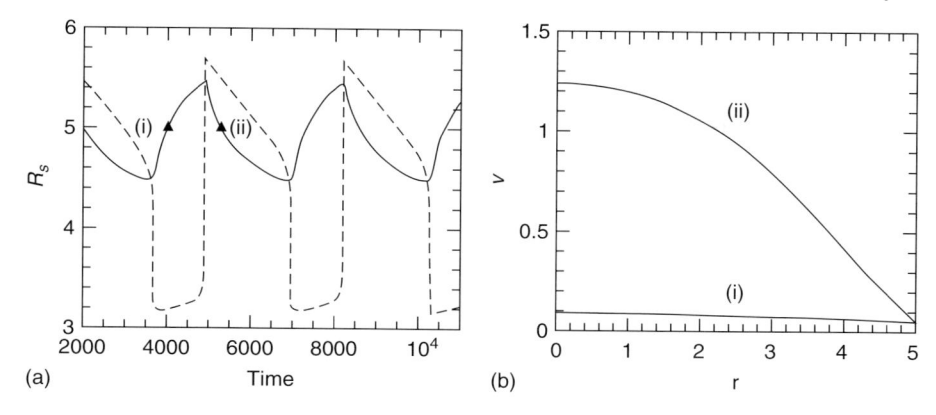

Figure 9.3 Self-oscillations induced by the system described by Eq. (9.20). (a) full lines: oscillations of the sphere's radius as a function of time; dotted lines: oscillations of concentration v at the sphere center. (b) profiles of concentration v as a function of r when $R_s = 5$ during the swelling phase (i) and the shrinking phase (ii). Parameters: $\chi_{min} = 0.2885$, $\chi_{max} = 0.53$, $v^* = 0.15$, $s = 10$ (gives $\chi(\phi_0) = 0.3$, $\eta = 5$, $D_0 = 1$, $C_{net} = 1$, $\phi_0 = 0.1$ (with $R_{s0} = 3$ for $\phi = \phi_0$).

There is no universally recognized form for the dependence of the friction coefficient $\zeta(\phi)$. A Ogston law [51]

$$\zeta(\phi) = \frac{RT}{V_S} \frac{1}{D_0} \phi\, e^{\eta\sqrt{\phi}} \qquad (9.28)$$

where η is a constant of a few units and D_0 has the dimension of a diffusion coefficient [26], was used. In the computations, the volume fraction was assumed to remain small ($\phi \ll 1$) so that the velocity of the solvent and the dependence of the diffusion coefficients on ϕ were neglected.

The motion of the polymeric network was obtained in numerical simulations by a multistep finite difference method from Eq. (9.26). In Figure 9.3(a), we show oscillations, obtained for appropriate parameters. In (b), we show the concentration distributions of the autocatalytic species for the same intermediate value of the sphere radius $R_s = 5$ taken, respectively, when the system is swelling (F state) or shrinking (FT state).

9.4.3
A More Realistic Model: The Polyelectrolyte Model

Despite the numerical support, the above model is quite unrealistic in regard to real systems. Actually, most hydrogels used in chemomechanics experiments are polyacids or polybases that swell/shrink as a function of the pH of the aqueous solution. In this case, the volume dependence cannot be accounted for on the same principles. The volume changes are associated with energetic effects due to the electrostatic interactions between ions attached to the polymer and with entropic effects due to an excess of free ions that exerts an additional osmotic pressure between the gel contents and the surroundings. Although the electrostatic energetic terms are sometimes introduced in complex models [52, 53], most authors consider the entropic effects to be dominant. At equilibrium, the swelling effect can be understood as resulting from a Donnan equilibrium between the gel contents and the surrounding as described by Rička and Tanaka [54]. Neglecting the electrostatic interaction, the additional osmotic pressure Π_{ion} can be computed easily. Assuming that the polymer is a polyacid with monofunctional groups HA that dissociates into H^+ and A^- with a dissociation constant K_g, the concentration c_a of the acid groups A^-, attached to the gel network, is given at equilibrium by

$$c_a = \frac{K_g c_{a0} \phi}{K_g + [H^+]} \tag{9.29}$$

where $c_{a0}\phi$ is the concentration of HA monomers in the absence of dissociation. If c_i is the concentration of *mobile* ion i inside the gel and and c'_i is the concentration of the same ion outside the gel, it results from the Boltzman law [54] that

$$c_i = c'_i K^{z_i} \tag{9.30}$$

where z_i is the charge number of species i and K is the Donnan ratio. For dilute systems, the osmotic pressure Π_{ion} created by the excess of mobile ion inside the gel is analogous to the pressure exerted by a perfect gas

$$\Pi_{ion} = RT \sum_i (c_i - c'_i) \tag{9.31}$$

The c_i's, c_a, and Π_{ion} can be expressed from Eq. 9.29–9.31 in terms of ϕ, K, and the fixed quantities c'_i's and c_{a0}. At equilibrium, the total osmotic pressure resulting from mixing, elasticity, and ionic pressure is null, so that, when accounting for electroneutrality, one gets a system of two nonlinear equations for the unknown variables ϕ and K,

$$\sum_i c_i z_i - c_a = 0 \quad \text{(electroneutrality)} \tag{9.32}$$

$$\Pi_{mix}(\phi) + \Pi_{elas}(\phi) + \Pi_{ion}(\phi, K) = 0 \quad \text{(mechanical equilibrium)} \tag{9.33}$$

the solution of which provides the volume ϕ_{eq} fraction at equilibrium and the c_i's within the gel by means of Eq. (9.30). Out of equilibrium, one can assume local equilibrium at the gel/surroundings interface and use these values as boundary conditions. Inside the gel, where a reaction is in progress, neutral and solvent molecules can be created or destroyed, so that the situation is more complex and cannot be generalized easily. A simple approach is to assume that the ionic osmotic pressure is proportional to the degree of ionization of the polymer, which can be written as

$$\Pi_{ion} = K_{ion} c_a \tag{9.34}$$

If the concentrations in the surroundings are constant, ϕ_{eq} can also be constant, a situation encountered in one-dimensional systems. In this case, K_{ion} can be obtained by equating the expressions of Π_{ion} obtained from Eqs (9.31) and (9.34) at the boundary [42]. If not, one must resort to more complex strategies to insure continuity at the boundary.

An improved model in regard to the simple model presented in Section 9.4.2 has been cut out for more realistic conditions [42]. Not only is the swelling process explicitly tied to an ionic pressure, but there are also other improvements: the transport description includes the solvent flow as well as the dynamics of charged species, the dependence of diffusion coefficients on the gel density is accounted for, and, finally, the kinetic toy model is replaced by a realistic model of a spatially bistable reaction. We consider that the functional unit HA of the polyacid has a unique weak-acid function.

For easier computations, a strictly unidimensional system of length L has been used. In these conditions, Eq. (9.24) remains valid with $\lambda = (\phi_0/\phi)^{2/3}$ and the elastic part of the stess tensor is reduced to an osmotic pressure $\Pi_{elas} = -\sigma_{rr}$. One end is assumed to be in contact with a CSTR, the composition of which can be computed easily [55]. The opposite end is an impermeable wall with no-flux boundary conditions. Assuming local equilibrium at the gel/CSTR boundary, one can compute the chemical composition and the volume fraction ϕ in agreement with the Donnan equilibrium as explained above. Since λ only depends on ϕ, this boundary condition is constant, so that a unique value of K_{ion} can be obtained and $\Pi_{ion}(r)$ can be determined at each point from the sole local chemical composition.

To describe the transport within the gel, one assumes that a free species i of charge number z_i is convected at the solvent velocity $\boldsymbol{v_S}$ and diffuses with a coefficient $D_i(\phi)$, which depends on the volume fraction occupied by the polymeric network, whereas the deprotonated form A^- of the monomer, attached to the network, is advected at velocity $\boldsymbol{v_P}$ and does not diffuse. To account for the permittivity and for the tortuosity of this network at $\phi \ll 1$, the following approximation $D_i(\phi) = D_{0i}(1 - 2\phi)$ was used [42], whereas for the friction coefficient of this network, the Ogston law, given by Eq. (9.28), was again employed. The fluxes of the free species and of the ions A^- are, respectively, given by the two equations

$$\boldsymbol{N_i} = \boldsymbol{N_{mi}} - D_i \nabla c_i + c_i \boldsymbol{v_S} \tag{9.35}$$

$$\boldsymbol{N_a} = c_a \boldsymbol{v_P} \tag{9.36}$$

where $\boldsymbol{N_{mi}}$ is the migration term of charged species under the effect of the local potential. Accounting for local electroneutrality and the absence of current, one can derive the following expression [42, 56]

$$\boldsymbol{N_{mi}} = \frac{t_i}{z_i}\left(\sum_{k \neq n} z_k(D_k - D_n)\nabla c_k - D_n z_a \nabla c_a - z_a c_a(\boldsymbol{v_P} - \boldsymbol{v_S})\right) \tag{9.37}$$

where $t_i = (z_i^2 D_i c_i)/(\sum_i z_k^2 D_k c_k)$ is the so-called tranference number of ionic species i and where one nonreactive ion of index n could be eliminated (Na^+ in our case).

The expression for the total osmotic pressure $\Pi(\phi)$ is obtained from Eqs (9.23, 9.24, 9.31) with $\lambda = (\phi_0/\phi)^{2/3}$. Taking into account both the mass conservation of the polymer

$$\frac{\partial \phi}{\partial t} + \nabla \cdot (\boldsymbol{v_P}\, \phi) = 0 \tag{9.38}$$

and the transport equations for all the species, one gets the following set of partial differential equations which completely describes the dynamics of the gel and the reactants

$$\frac{\partial c_i}{\partial t} = -\nabla \cdot \mathbf{N}_i + R_i \tag{9.39}$$

$$\frac{\partial c_a}{\partial t} = -\nabla \cdot \mathbf{N}_a + R_a \tag{9.40}$$

$$\frac{\partial \phi}{\partial t} = \nabla \left(\frac{\phi(1-\phi)}{\zeta(\phi)} \nabla \Pi \right) \tag{9.41}$$

where the R_i's and R_a are the reaction rate terms.

The numerical computations were performed with a previously tested realistic kinetic model of the BS reaction, a system that looked extremely promising for chemomechanical experiments. The reactants are bromate, sulfite, and sulfuric acid. A description of this kinetic model and the dynamic properties of this reaction is beyond the scope of this chapter. Details, extensive discussions, kinetic constants, and diffusion coefficients can be found in Refs. [42] and [55]. The domain of computed spatial bistability in a nonresponsive gel is reported in the diagram ($[H_2SO_4]_0, L$) in Figure 9.4(a) for $[BrO_3^-]_0 = 25$ mM, $[SO_3^{2-}]_0 = 60$ mM and a CSTR residence time $\tau = 500$ s, where the subscript 0 holds for the concentrations in the CSTR input flows. To compute these limits, a complexing agent with a fixed concentration was introduced to approximatively simulate the kinetic role that the polyacid plays in the responsive gel.

The parameters for the responsive gels were chosen to fit reasonable typical values for the gels usually used in experiments. The reference state was the state of maximum possible contraction (low pH) and fixed to $\phi_0 = 0.05$, which also fixes K_{net} using Eq. (9.23) at isotropic equilibrium. The corresponding reference size of the system L_0 is always smaller than the real size L since a part of the gel close to the CSTR boundary is always in a swollen state. The parameter D_0 in Eq. (9.28) was fixed to 0.03 mm^2 s^{-1} in regard to the typical swelling times, the Flory parameter χ was fixed to 0.515, since values slightly larger than $1/2$ are optimal for swelling, and C_{net} and η were, respectively, fixed to 1 and 5 as in Section 9.4.2.

The results of the simulations, performed in the Lagrangian mode, are given in Figure 9.4. If one choses $[H_2SO_4]_0$ as the experimentally tunable parameter, for all values for which spatial bistability is possible, there is a small domain of L_0 that leads to oscillations. Three examples of such oscillations are given in Figure 9.4(b). In the phase diagram section (Figure 9.4(a)), we report, in addition, the full domain of L_0 values (gray region) which actually lead to oscillations. As shown by the amplitude bars of the oscillations there is a good agreement with the heuristic theory that predicts that they approximately fit to the corresponding hysteresis domains. As expected, the period increases with the size of the gel, approximatively scaling as L^2, since all processes are ruled by diffusion of the species and of the network. This suggests that, in experiments, sizes of order 1 mm or less would be advisable in order to get reasonable periods.

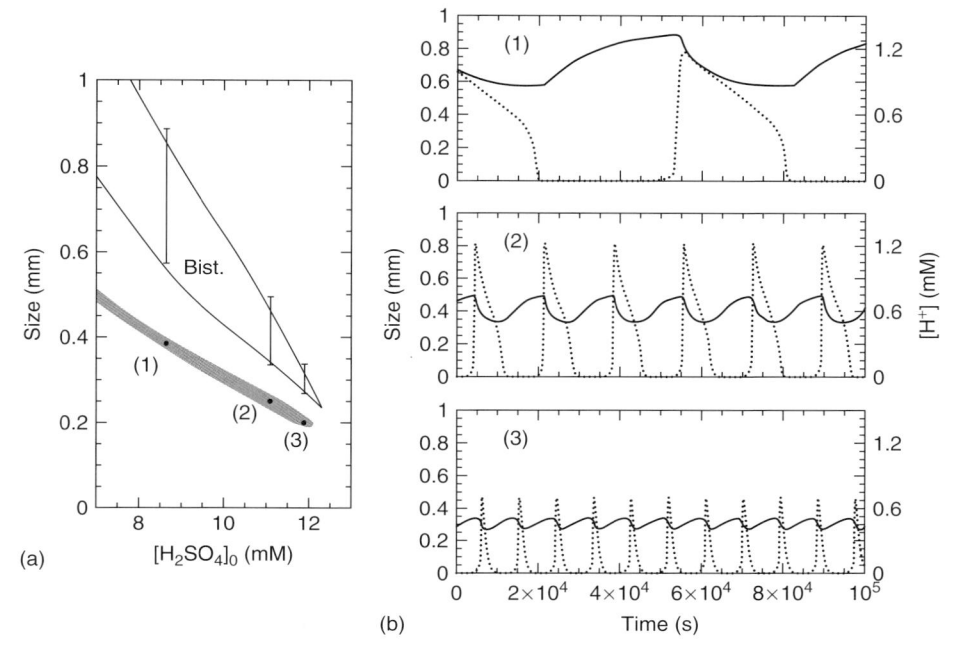

Figure 9.4 Self-oscillations induced in the polyelectrolyte model for the (BS reaction–responsive gel) system. (a) full lines: limits of bistability in a neutral gel for almost equivalent chemical conditions. Gray region: domain of reference sizes L_0 that gives rise to self-oscillations with the (black) test points labeled from 1 to 3. Vertical bars: amplitude of real sizes L covered by the corresponding oscillations. (b) Oscillations of size L (full lines) and of concentration $[H^+]$ at the impermeable wall (dotted lines) for the test points (1–3).

9.5
Experimental Observations

In the wake of the researches for oscillatory reactions more than a dozen pH-autoactivated reactions were shown to produce bistability when operated in a CSTR [57]. Theoretical calculations and experiments demonstrate that such systems readily give rise to spatial bistability when conducted in an OSFR. They would provide a large choice of reaction systems to test the chemomechanical instabilities theoretically described above. However, in our selection criteria, we have to take into account that many of these reactions can already exhibit kinetic oscillations over more or less wide ranges of feed parameters. Such complication can make it difficult to discriminate between kinetic and chemomechanic oscillatory instabilities. Furthermore, it has also been shown that in the case of proton-autoactivated system the natural faster diffusion of this species can lead to another source of oscillatory instability in an OSFR, the long range activation instability [58].

Typical pH differences between the steady-state branches of all those bistable systems range from one to several units within the pH = 10 to pH = 2 interval. Besides proton-autoactivated reactions, there are also a number of hydroxyl ion-autocatalyzed reactions where pH jumps from 6 to 11 are observed between the flow (F) and thermodynamic (T) branches [59]. Responsive polyelectrolyte gels have been reported to dramatically change size or volume over narrow pH domains (one or two units) within approximately the same spectrum of pH values than the above bistable system range. Many pH-responsive hydrogels are based on charged derivatives of N-isopropylacrylamide polymer networks. The most typical are copolymers of N-isopropylacrylamide and acrylic acid (poly(NiPAAm-co-AAc)), which exhibit quasi-first order volume collapse when the pH of the solvent drops below 5. This critical pH value can be increased (decreased) by increasing the hydrophobicity (hydrophilicity) of the network [60–62]. Gels that shrink in an alkaline medium and swell in neutral or acidic medium are obtained for networks with a basic moiety. In all cases, the game is to appropriately adjust the pH at which the gel swells or shrinks between the pH at which switching is obtained with the chemical bistable.

9.5.1
Experimental Results

Experiments were performed with long cylindrical or conical pieces of responsive gels hanging in a tubular CSTRs in which a strong recirculation of the fluid contents provided a uniform chemical composition around the gel [13]. These geometries minimize the mechanical constraints imposed by the anchoring of the gel inside the CSTR. However, the additional nonradial dimension unavoidably introduces additional dynamic complexity, and the spatiotemporal behavior of the system becomes richer, as we shall see.

The sources of chemomechanical instabilities described above have been tested in two quite different proton-autoactivated systems: the CT [63] and the BS [64] reactions.

9.5.1.1 Case of the Chlorite–Tetrathionate Reaction
During this reaction large amounts of protons are released in accordance with the following balance equation:

$$7ClO_2^- + 2S_4O_6^{2-} + 6H_2O \longrightarrow 7Cl^- + 8SO_4^{2-} + 12H^+ \tag{9.42}$$

In the range of reactant concentrations experimentally explored, the reaction is very sluggish if the initial pH of the solution is above 7, whereas when pH < 6, it very rapidly drops to pH $\sim$ 2. When operated in a CSTR, the reaction exhibits steady-state bistability, with pH differences between branches up to ΔpH $\sim$ 8), and does not lead to kinetic oscillations. However, in an OSFR, it was shown to display a diffusion-driven oscillatory instability because of the long-range activation of the free protons. But this instability can be quenched by weakly buffering the system with a low-mobility proton-binding species [58]. This condition is straightforwardly obtained in pH-responsive gels with large enough concentrations of caboxylic moieties on the polymer network.

Experiments were performed at 35 °C, that is, at a temperature slightly above the LCST of pure poly(NiPAAm), which is where maximum volume changes are obtained as a function of the pH of the solvent. Cylindrical gels (30 mm long with typical radius of 1 mm in the swollen state) were used. They are characterized by longitudinal size drop by about a factor of 2, in a 0.5 M solution of sodium chloride, when the pH of this salted solvent is decreased from 6 to 4.5. The gels are transparent above pH = 5 and become turbid below pH = 4.5, due to microphase separation. The changes in the size of the cylinders and the sharp variations of the polymer density inside are monitored by shadowgraphy while a direct view provides information (color, cloudiness) on the chemical state (F or FT) inside the gel.

In a non-chemoresponsive gel (e.g., agarose), spatial bistability is observed over a relatively large range of feed parameters. In this domain of parameters, the pH contents of the CSTR are high (10–9 units), except at the very end of the stability domain of the flow branch (F). At such high pH values, the reaction is extremely slow so that no spontaneous switch to the reacted state occurs in cylinders with a diameter size compatible with experimental times of a few days. In these conditions, the ratio R_{sup}/R_{inf} exceeds the value of the swelling factor of our responsive gels. Transitions from the swollen F state to the shrunken mixed state FT can be obtained by an acid perturbation. As a result, the collapsed state always invades the swollen state and in most cases the gel remains in the collapsed state. The swollen and shrunken states can be both stable for the same set of boundary conditions. This is the mechanical imprint of the underlying chemical spatial bistability. However, in a narrow region of parameter at the limit of stability of the collapsed state, quite complex spatiotemporal patterns can develop and persist over several swelling–shrinking cycles. The phenomenon progressively vanishes most likely because of the "aging" of the gel. Many of these chemomechanical patterns seem to find their source in the nonradial extended geometric component of the cylinders and in nonuniform local radial size deformations. These aspects are beyond the one-dimensional theories developed above.

Nonetheless, one observation is accounted for by a simple empirical extension of the above theory, the undamped propagation of a high-density polymer pulse as illustrated in Figure 9.5. In this experiment, the whole cylinder is initially in the swollen F state. An acid perturbation is made at the bottom end of the cylinder. This induces a transition to the collapse state of the gel, which starts to invade the swollen upper part. Yet, after some time, the collapsed part of the cylinder slowly reswells and relaxes back to the stable swollen state. The phenomenon can be explained by a chemomechanical version of the propagation of an excitability wave [13]. After the initial acid perturbation, the core of the cylindrical OSFR switches to the acid-core reacted state. The acid core contaminates the neighboring unreacted swollen part by diffusion, which, in turn, shrinks, following the acid wave. But the collapse of the gel undershoots the critical size R_{inf} below which the reacted acid state is stable. Diffusion from the boundary overtakes the reaction, the composition

(a) (b) (c) (d)

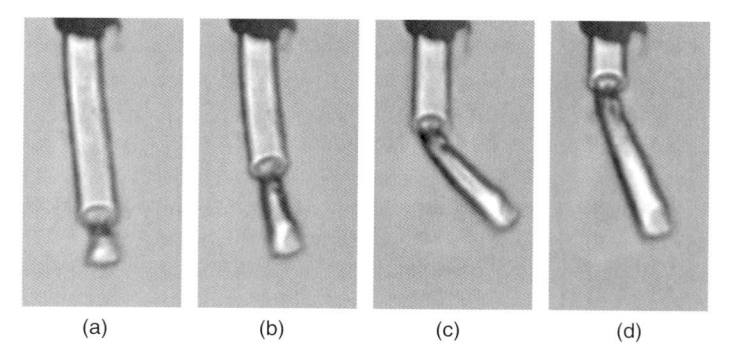

Figure 9.5 Chemomechanical excitability wave ("bottle-neck" shape) obtained after an acid perturbation made at the lower end of a cylindrical gel. The OSFR is fed with the CT reaction. Swollen gel diameter: 2.8 mm. Wave velocity: 5.6 mm h^{-1}. Time interval between snapshots (from (a) to (d)): 21, 30, 20 min. From Ref. [13]

returns to the unreacted alkaline state and the gel reswells. It remains in this stable state until another perturbation is applied or if some heterogeneity acts as a pacemaker for a new acid wave.

9.5.1.2 Case of the Bromate–Sulfite Reaction

This is a more mildly acid-autoactivated reaction. The gross stoichiometry of the activation process is given below:

$$BrO_3^- + 3HSO_3^- \longrightarrow Br^- + 3SO_4^{2-} + 3H^+ \qquad (9.43)$$

When operated in a 1 mm thick annular OSFR, this reaction shows spatial bistability [55], in good agreement with the numerical results reported in Figure 9.4(a), and contrary to the CT reaction no long-range activation diffusional instability is observed, but, in very large excess of bromate, a very narrow region of kinetic oscillations can develop. The latter is shown to be quenched in the presence of immobilized carboxylic functions at concentrations equal or above 0.01 M. As seen in Figure 9.4, chemomechanical oscillations are predicted over a wide range of size but in very narrow windows of [H_2SO_4]$_0$. To easily explore the predicted capacity, experiments were performed in narrow conical OSFRs made of pH sensitive gels similar to those used in the CT experiments. The concentration of carboxylic groups in the network is typically 0.05 M in the swollen state, a value much above that for which the kinetic oscillations are quenched. Preliminary results are reported in Figure 9.6.

The 21-shadowgram snapshots cover 4 h of experiment. In the figure, the swollen states appear brighter than the shrunken states, because the surface of the former is smooth while that of the latter is crumpled. A dark pulse of collapsed polymer gel is seen to propagate from top to bottom, down to a minimal position. The phenomenon repeats a few times. To guide the eyes, the locations of the denser polymer pulses are demarcated by red lines. Behind the dark pulse, the gel reswells

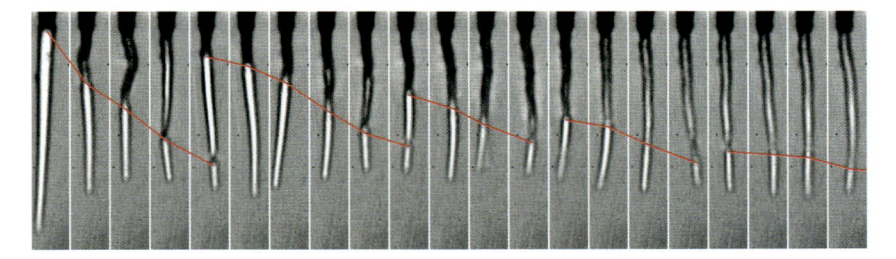

Figure 9.6 Periodic traveling polymer gel contraction in conical geometry. The OSFR is fed with the BS reaction. The sequence of shadowgrams is taken every 12 min. Initial length of the cone is 30 mm. Diameters: large base (top) 2.2 mm, small base (bottom) 1.0 mm. Red line: see text. (Unpublished results by courtesy of J. Horváth).

more slowly leading to the growth of a clearer area in the shadowgram. As the time increases, the pulses start from a lower position on the cone. Pulsations stop after the fourth one. It is presumed that above the red lines, the gel does not recover to the F state but stays in the FT state. The slow reswelling in this area is due to a chemical degradation inducing a decay in the pH responsiveness of the gel. Despite their limited number, these are genuine chemomechanical oscillations. In gel networks with weak size responsiveness, no oscillations are observed for the same feed and initial OSFR size conditions. Details will be provided in a forthcoming publication.

9.6
Conclusions and Perspectives

Theoretical and experimental studies of chemomechanical instabilities in gels are still in their infancy. In experiments, a strong limitation to refined studies is the "aging" of the responsiveness of the gels. Either more chemically resistant polymer will have to be tested or less aggressive reaction will have to be used. One of the most appealing aspect of combining reactions and responsive material is the capacity for emerging time and space self-organization phenomena. The fact that the responsive material evolves on timescales independent (most often slower) of those of the chemical processes makes it possible to experimentally explore systems with more than two time and space scales. In this respect, it would be interesting to investigate the coupling of a Turing bifurcation to stationary pH patterns, as the one recently discovered in the TuIS reaction [65] with pH size responsive gels. The development of such patterns in thin gel-disk OSFRs has been shown to be very sensitive to the thickness of the disk. Once the local stationary pH pattern develops, the slow local collapse can oppose, with delay, the settling of the patterns. Traveling or blinking spot patterns can be foreseen in this case. This brings out the problem of matching gel and reaction properties, which is unfortunately not related to a universal property. From the theoretical point of view, the consideration of systems of higher dimensionality than those examined above introduces no conceptual problem, but will lead to more advanced categories of numerical simulations, specially in the presence of charged species. In this case, an alternative approach would be to use discretized systems with local phenomenological laws as proposed by Balazs *et al.* [66]. The volume variations modeled in our simulations are quite small, and therefore the condition where phase transitions would occur is not attained. In the experimental observations, the core of the shrunken gels becomes turbid, a sign that this part of the gel enters the spinodal region, since one would expect that a totally collapsed gel would become clear again. This poses the general problem of describing the interaction of dynamical phenomena with a phase transition. This problem has been scarcely examined using oversimplified models of swelling/shrinking in the absence of any chemical process [24]. The study of this class of problem, for which other examples arise in hydrodynamics [67] or Ultra-High Vacuum surface chemistry [68], still lacks a true development.

References

1. Epstein, I.R. and Pojman, J. (1998) *An introduction to Nonlinear Chemical Dynamics*, Oxford University Press.
2. Lázár, A., Noszticzius, Z., Fösterling, H.-D., and Nagy-Ungvárai, Z. (1995) *Phys. D*, **84**, 112.
3. Horkay, F. and Mc Kenna, G.B. (2009) in *Physical Properties of Polymers Handbook* (ed. J.E. Mark), Chapter 29, Springer, New York.
4. Peppas, N.A., Bures, P., Leobandund, W., and Ichikawa, H. (2000) *Eur. J. Pharm. Biopharm.*, **50**, 27.
5. Calvert, P. (2008) *MRS Bull.*, **33**, 207.
6. Gerlach, G. and Arndt, K.-F. (2009) *Hydrogel Sensors and Actuators*, Springer Series on Chemical Sensors and Biosensors, vol. 6. Springer, New York.
7. Yoshida, R., Ichijo, H., Hakuta, T., and Yamaguchi, T. (1995) *Macromol. Rapid Commun.*, **16**, 305.
8. Yoshida, R., Kokufuta, E., and Yamagushi, T. (1999) *Chaos*, **9**, 260.
9. Yoshida, R. (2009) in *Chemomechanical Instabilities in Responsive Materials*, Springer NATO series A (eds P. Borckmans, P. De.Kepper, A. Khokhlov, and S. Métens), p. 39.
10. Yoshida R., (2010) Gels coupled to oscillatory reactions, *Nonlinear Dynamics with Polymers*, Wiley-Vch Verlag GmbH.
11. Crook, C., Smith, A., Jones, R., and Ryan, A. (2002) *Phys. Chem. Chem. Phys.*, **4**, 1367.
12. Swann, J.M.G. and Ryan, A.J. (2009) *Polym. Int.*, **58**, 285.
13. Labrot, V., De Kepper, P., Boissonade, J., Szalai, I., and Gauffre, F. (2005) *J. Phys. Chem. B*, **109**, 21476.
14. Wu, S., Li, H., Chen, J.P., and Lam, K.Y. (2004) *Macromol. Theory Simul.*, **13**, 13.
15. Dušek, K. and Patterson, D. (1968) *J. Polym. Sci. A-2*, **6**, 1209.
16. Tanaka, T. (1978) *Phys. Rev. Lett.*, **40**, 820.
17. Shibayama, M. and Tanaka, T. (1993) *Adv. Polym. Sci.*, **109**, 1.
18. Khokhlov, A.R., Starodubtzev, S.G., and Vasilevskaya, V.V. (1993) *Adv. Polym. Sci.*, **109**, 123.
19. Tanaka, T. and Filmore, J. (1979) *J. Chem. Phys.*, **70**, 1214.
20. Komori, T., Takahashi, H., and Okamoto, N. (1998) *Colloid Polym. Sci.*, **266**, 1181.
21. Sekimoto, K. (1991) *J. Phys.II (France)*, **1**, 19.

22. Sekimoto, K. (1992) *J. Phys.II (France)*, **2**, 1755.
23. Doi, M. (2008) Modeling of Gels (slides for IMA Tutorial) at *http://www.ima.umn.edu/matter/fall/t1.html*.
24. Tomari, T. and Doi, M. (1995) *Macromolecules*, **28**, 8334.
25. Tomari, T. and Doi, M. (1994) *J. Phys. Soc. Jpn.*, **63**, 2093.
26. Bisschops, M.A.T., Luyben, K.Ch.A.M., and van der Wielen, L.A.M. (1998) *Ind. Eng. Chem. Res.*, **37**, 3312.
27. Barrière, B. and Leibler, L. (2003) *J. Polym. Sci.*, **41**, 166.
28. Han, C.D. (ed) (2007) in *Rheology and Processing of Polymeric Materials*, Vol.1, Oxford University Press.
29. Villain, S. (2007) Comportement mécanique de gels soumis à des réactions autocatalytiques, Ph.D thesis, Université Paris 7 - Denis Diderot.
30. Métens, S., Villain, S., and Borckmans, P. (2010) *Phys. D*, doi:10.1016/J.Physd.2009.06.002.
31. Métens, S., Villain, S., and Borckmans, P. (2009) in *Chemomechanical Instabilities in Responsive Materials*, Springer NATO Series A (eds P. Borckmans, P. De Kepper, A. Khokhlov, S. Metens), p. 39.
32. Flory, P.J. and Erman, B. (1982) *Macromolecules*, **15**, 800.
33. James, H.M. and Guth, E. (1949) *J. Polym. Sci.*, **4**, 153.
34. Bastide, J., Candau, S., and Leibler, L. (1981) *Macromolecules*, **14**, 719.
35. Erman, B. and Flory, P.J. (1986) *Macromolecules*, **19**, 2342.
36. Orofino, T.A. and Flory, P.J. (1957) *J. Chem. Phys.*, **26**, 1067.
37. Chiu, H., Lin, Y., and Hsu, Y. (2002) *Biomaterials*, **23**, 1103.
38. Yashin, V.V. and Balazs, A.C. (2006) *Macromolecules*, **39**, 2024.
39. Yoshida, R., Tanaka, M., Onodera, S., Yamaguchi, T., and Kokufuta, E. (2000) *J. Phys. Chem. A*, **104**, 7549.
40. Boissonade, J. (2003) *Phys. Rev. Lett.*, **90**, 188302.
41. Boissonade, J. (2005) *Chaos*, **15**, 023703.
42. Boissonade, J. (2009) *Eur. Phys. J. E*, **28**, 337.
43. Tyson, J. (1977) *J. Phys. Chem.*, **66**, 905.
44. Sakai, T. and Yoshida, R. (2004) *Langmuir*, **20**, 1036.
45. Misra, G.P. and Siegel, R.A. (2002) *J. Control. Release*, **81**, 1.
46. Dhanarajan, A.P., Misra, G.P., and Siegel, R. (2002) *J. Phys. Chem.*, **A106**, 8835.
47. Siegel, R.A. (2010) Oscillatory systems created with polymer membranes, *Nonlinear Dynamics with Polymers*, Wiley-Vch Verlag GmbH.
48. Blanchedeau, P., Boissonade, J., and De Kepper, P. (2000) *Phys. D*, **147**, 283.
49. Boissonade, J. and De Kepper, P. (2009) in *Chemomechanical Instabilities in Responsive Materials*, Springer NATO Series A (eds P. Borckmans, P. De Kepper, A. Khokhlov, and S. Métens), p. 95.
50. Bastide, J. and Candau, S.J. (1996) in *Physical Properties of polymeric Gels* (ed. J.P. Cohen Addad), John Wiley & Sons, Ltd, Chichester, p. 143.
51. Ogston, A.G., Preston, B.N., and Wells, J.D. (1973) *Proc. R. Soc. Lond. A*, **333**, 297.
52. Achilleos, E.C., Christodoulou, K.N., and Kevrekidis, I.G. (2001) *Comput. Theor. Polym. Sci.*, **1**, 63.
53. Dolbow, J., Fried, E., and Ji, H. (2004) *J. Mech. Phys. Solids*, **52**, 51.
54. Rička, J. and Tanaka, T. (1984) *Macromolecules*, **17**, 2916.
55. Virányi, Z., Szalai, I., Boissonade, J., and De Kepper, P. (2007) *J. Phys. Chem. A*, **111**, 8090.
56. Newman, J. and Thomas-Alyea, K.E. (2004) *Electrochemical Systems*, Chapter 11, John Wiley & Sons, Inc., New-York.
57. Rábai, Gy. (1998) *ACH-Models Chem.*, **135**, 381.
58. Szalai, I., Gauffre, F., Labrot, V., Boissonade, J., and De Kepper, P. (2005) *J. Phys. Chem. A*, **109**, 7843.
59. Kovacs, K., McIlwaine, R., Gannon, K., Taylor, A.F., and Scott, S.K. (2005) *J. Phys. Chem. A*, **109**, 283.
60. Philippova, O.E., Hourdet, D., Audebert, R., and Khokhlov, A.K. (1997) *Macromolecules*, **30**, 8278.
61. Siegel, R.A. and Firestone, B.A. (1988) *Macromolecules*, **21**, 3254.
62. Mujumdar, S.K., Bhalla, A.S., and Siegel, R.A. (2007) *Macromol. Symp.*, **254**, 338.

63. Nagypál, I. and Epstein, I.R. (1986) *J. Phys. Chem.*, **90**, 6285.
64. Szántó, T.G. and Rábai, Gy. (2005) *J. Chem. Phys.*, **109**, 5398.
65. Horváth, J., Szalai, I., and De Kepper, P. (2009) *Science*, **324**, 772.
66. Kuksenok, O., Yashin, V.V., and Balazs, A.C. (2008) *Phys. Rev. E*, **78**, 041406.
67. Assenheimer, M., Khaykovich, B., and Steinberg, V. (1994) *Phys. A*, **208**, 373.
68. Imbihl, R. and Ertl, G. (1995) *Chem. Rev.*, **95**, 697.

10
Oscillatory Systems Created with Polymer Membranes
Ronald A. Siegel

10.1
Introduction

While much of the study nonlinear chemical dynamics over the past several decades has focused on reactions with small molecules [1, 2], usually of the redox kind, there have been significant parallel developments in understanding biochemical oscillators [3] and in oscillations in which a membrane plays a critical mechanistic role [4–6]. Interest in membrane oscillators stems, in part, from the observation that cellular processes do not occur in a homogeneous, well-stirred medium. Cells, besides being separated from their immediate environment by a lipid bilayer/protein membrane, are compartmentalized into organelles that are also bounded by membranes. Membranes regulate transfer of reactants and products, and they confine and compartmentalize enzymes. By this means, reactions within the cell become highly heterogeneous, and reaction networks exhibit a wealth of complex behaviors that would not be possible in a homogeneous medium, where kinetics are constrained by intrinsic reaction rates.

The initial motivation for studying membrane oscillators was to understand biological processes [6–8]. Model synthetic systems have been constructed that are simpler than the biological systems they were intended to mimic, and the goal has been to isolate key principles using "pared down" designs. As often occurs with model systems, however, synthetic membrane oscillators exhibit their own peculiar behaviors, leading to new lines of study.

Introduction of membranes may, in some cases, lead to more flexibility in the design and study of chemical oscillators. The continuous-stirred tank reactor (CSTR) configuration, which is often used to study chemical oscillators because it maintains reaction and product concentrations away from equilibrium [1, 2], controls the transport of reactants, intermediates, and products by fluid flow, and does not discriminate among species. Membrane selectivity between chemical species can provide a basis for selection of dynamical behaviors that are unavailable with a CSTR.

For some years, we have entertained the possibility that autonomous membrane oscillations can be used to drive rhythmic release of drugs and hormones in an

Nonlinear Dynamics with Polymers: Fundamentals, Methods and Applications.
Edited by John A. Pojman and Qui Tran-Cong-Miyata
Copyright © 2010 WILEY-VCH Verlag GmbH & Co. KGaA, Weinheim
ISBN: 978-3-527-32529-0

implantable system. Since the 1970s, there have been numerous demonstrations of episodic endogenous secretion of hormones modulating disparate physiological processes such as reproduction, growth, energy metabolism, fluid balance, and stress [9, 10]. A partial list of such hormones and their endogenous periodicities is provided in Table 10.1 [10]. Of particular interest is gonadotropin-releasing hormone (GnRH) [11–14]. This hormone is normally secreted in rhythmic pulses from the hypothalamus, and it triggers the release of the gonadotropins luteinizing hormone (LH) and follicle-stimulating hormone (FSH) in the anterior pituitary gland. The latter hormones modulate secretion of other hormones from the gonads in both males and females. Hypogonadotropic hypogonadism (HH) is an endocrine condition in which GnRH release is absent or suppressed [15–18]. Treatment of HH requires that GnRH be administered with the normal endogenous pulse pattern [19]. In fact, steady administration of GnRH has long been known to inhibit

Table 10.1 A variety of hormones with varied physiologic functions endogenously released as a train of pulses. gonadotropin-releasing hormone (GnRH) is bolded here since its release from the hypothalamus drives anterior pituitary release of luteinizing hormone (LH) and follicle-stimulating hormone (FSH), which send signals leading to release of gonadal hormones, including estradiol, progesterone, and testosterone, and since it was the hormone of interest in the hydrogel/enzyme oscillator featured in this work. frequencies listed are averages, and separate entries correspond to different studies of the same hormone.

Hormone	Frequency (pulses/day)
Growth	9–16, 29
Prolactin	4–9, 7–22
Thyroid-stimulating hormone	6–12, 13
Adrenocorticotropic hormone	15, 54
Gonadotropin-releasing hormone (GnRH)	**12–24**
Luteinizing hormone (LH)	7–15
Follicle-stimulating hormone (FSH)	4–16
Estradiol	8–16, 8–19
Progesterone	8–12, 6–16
Testosterone	3, 8–12
β-Endorphin	13
Melatonin	18–24, 12–20
Vasopressin	12–18
Renin	6, 8–12
Parathyroid hormone	24–139, 23
Insulin	108–144, 120
Pancreatic polypeptide	96
Somatostatin	72
Glucagon	103, 144
Aldosterone	6, 9–12
Cortisol	15, 39

Data taken from Ref. [10].

secretion of pituitary gonadotropins, shutting down reproductive function [19]. As a first step in developing an implantable system to treat HH, a hydrogel/enzyme construct has been designed that releases GnRH in rhythmic pulses, using glucose as a free energy source [20–28].

In this chapter, we provide a brief survey of membrane-based chemical and biochemical oscillators. A comprehensive review of such systems was written by Larter in 1990 [6]. The purpose of the present survey is to summarize some key systems already described by Larter, and to describe a few developments since her review. We will focus on the single-periodicity oscillations. More complex behaviors such as period doubling, multiple periodicity, chaos, and dissipative spatial structures will not be covered. Following the survey, we present results on our glucose-driven hydrogel/enzyme system. This system relies on hydrogel properties, in particular the volume phase transition, which were not available in the previous membrane systems.

The survey is not essential for the material regarding the glucose-driven hydrogel/enzyme system; it can be skipped at the reader's discretion. Other oscillating hydrogel systems, developed in Japan and France, are described in Chapters 7–9 of this book.

10.2
Survey of Synthetic Membrane Oscillators

10.2.1
Teorell Oscillator

The original motivation for studying synthetic membrane oscillators was to mimic rhythmic spiking behaviors (action potentials) seen in neurons. The first synthetic oscillator was devised by Teorell [7, 29], who had previously investigated the electrochemical properties of ionic membranes [30]. The original Teorell oscillator consisted of a fritted glass plate with negative charges fixed to the pore walls, set between two well-stirred NaCl solutions of differing concentrations. In this system, bulk hydraulic flow from one chamber to the other leads to changes in hydrostatic pressure head between the two chambers. Reversible electrodes, connected to each other by high-impedance circuitry, permit determination of the electrical potential between the two solutions. Under normal conditions, with water levels the same on both sides, solute diffuses from the high concentration side to the low concentration side, and a steady, relatively small diffusion potential arises as a result of differences in ion mobilities and/or Donnan potentials at the membrane/chamber interfaces.

To produce transients, a steady electrical current is introduced across the porous frit, using a second pair of electrodes. Below a threshold current value, the trans-frit potential and hydrostatic pressure change monotonically until they attain stationary values. In this regime, the frit acts as a slightly nonlinear electrical resistor, whose resistance depends on the partially distorted internal mobile ion concentration profile. Electro-osmotic flow due to the presence of fixed charges on the frit occurs until it is cancelled out by pressure-induced back flow. For an interval of current values above the latter threshold, approach to steady state is preceded by decaying oscillations. Beyond a second threshold current, however, oscillations in potential and pressure differences across the frit are sustained indefinitely. On further increasing the current, voltage and pressure waveforms transition from smooth sinusoids to more switchlike morphologies characteristic of relaxation oscillators [31]. This sequence of behaviors is reminiscent of other systems that undergo a Hopf bifurcation with variation of a critical parameter [1–3, 32].

The glass frit in the Teorell oscillator can be replaced by nanoporous filter membranes [33–35], or even capillary electrophoresis tubes [36]. By these means, important parameters such as salt and hydraulic permeability, salt concentration, pore width, and fixed charge relevant to electro-osmotic flow (which may in turn depend on pH) can be varied systematically.

Teorell originally explained the system's oscillatory behavior by postulating that current flow through the charged membrane leads to electro-osmotic flow of solvent followed by development of hydrostatic pressure between the chambers [29]. Electro-osmotic flow also distorts the salt concentration profile in the membrane's pores, leading to changes in membrane resistance, and hence electrical potential. With sufficient pressure buildup, convective backflow overcomes the electro-osmotic flow, redistributing the intramembrane salt ions. A simple model accounting for electro-osmosis, hydrostatic pressure buildup and backflow, and nonlinear flow-dependent resistance, with a finite relaxation time for changes in resistance, semiquantitatively reproduced for many of the observations.

In Teorell's model, the relaxation time for resistance was taken to be a free fitting parameter, and the *ad hoc* nature of this assumption was criticized. Later researchers sought to clarify the nature of the oscillations by working with mechanistic models that included structural assumptions about the membrane. Modeling the membrane channels as straight, cylindrical capillaries lined with fixed charges, Kobatake and Fujita [37, 38] pointed out that electro-osmotic flow depends on the radial distribution of ions in the capillaries, which in turn depends on local salt concentrations in accord with the Gouy–Chapman theory for the electrical double layer. They also noted that while pressure-induced convection flow has a parabolic Poiseuille radial flow profile, electro-osmotic flow is pluglike, and the total flow through the capillary is a superposition of these two components, with a complex flow profile that could reverse direction depending on radial position. By solving the relevant electrostatic, hydrodynamic, and mass transfer equations in the capillary model, Kobatake and Fujita demonstrated that above a critical transmembrane pressure, relations between transmembrane voltage and salt flow exhibit bistable, or "flip-flop," behaviors. Such bistability, coupled with the hydrostatic relationship between transmembrane flow and pressure, was taken to explain the dynamical behavior of the Teorell oscillator.

Although Kobatake and Fujita did not compare their predictions with experiments, their work showed that Teorell's assumption of a finite relaxation time for membrane resistance was unnecessary. Subsequent studies by Meares and Page [33, 39], using a very similar modeling perspective, confirmed that the flip-flop approach to the Teorell oscillator has broad explanatory value, by conducting a range of experiments controlling various parameters. In particular, oscillations are supported by very thin membranes, in which ion concentration relaxations are almost instantaneous, demonstrating that a finite relaxation rate of the membrane's resistance is not a requirement for oscillations. On the other hand, there have been reports of oscillations in thicker membranes in which the flip-flop relations are not present [34]. For such membranes, finite relaxation times for electrical resistance may play a role in producing oscillations. The reader is referred to other theoretical papers concerning the Teorell oscillator [40, 41].

10.2.2
Polyelectrolyte Membrane-Based Oscillators

In a second membrane oscillator described by Shashoua [42], a very thin polybase layer (homopolymer of *N,N*-diethylaminoethyl acrylate) was reacted with a thin polyacid layer (a copolymer of acrylamide and acrylic acid). The resulting composite membrane, with polycationic and polyanionic outer layers and a neutral polyelectrolyte complex interfacial layer in the middle, showed periodic voltage spiking when stimulated by a threshold DC voltage, even when the two sides of the membrane had identical electrolyte (NaCl) concentrations and were at the same hydrostatic pressure. Katchalsky and Spangler [5] suggested that the Shashoua membrane oscillates as a result of Donnan-potential-based selectivity of Na^+ and Cl^- transport through the polyanion and polycation layers, respectively, leading to buildup of salt inside the polycomplex midlayer. Salt buildup is followed first by osmotic swelling of the midlayer and then by salt-induced contraction of the polyelectrolyte layers, until a critical internal pressure is reached. Salt solution is then ejected from the center, returning the membrane to its initial state and allowing the chain of events to restart. A similar and more quantitative elaboration of this model was presented by Huang and Spangler [43], who studied periodic electrical spiking in thin polyglutamate membranes that were effectively neutralized on one side by calcium ions.

Rhythmic spiking of potential driven by a steady ionic strength (KCl) gradient was observed by Minoura *et al.* [44] in a thin membrane formed from a triblock polymer with short poly(glutamatic acid) flanking blocks and a long poly(leucine) center block. In this construct, self-assembly leads to a structure containing poly(glutamic acid) channels spanning the membrane, embedded in a hydrophobic poly(leucine) continuum. Circular dichroism measurements showed that the poly(glutamic acid) blocks undergo a helix–coil transition with increasing ionic strength. It was suggested that permeabilities of salt ions, and therefore the diffusion potential, change at the helix–coil transition. Oscillations in membrane potential were interpreted as arising from periodic advance and retreat of a "front" in helicity in the membrane and its effect on ion gradients. The frequency of oscillations is maximal around pH 3, where it is expected that the helix and coil states are most sensitive to salt concentration. Notice that electrical spiking in this system is "passive" since neither current nor voltage is applied across the membrane.

In a recent development, Ito and Yamaguchi [45] plasma-grafted poly(N-isopropylacrylamide) (pNIPA) chains partially substituted with Ba^{2+}-selective crown ether side chains into the micropores of a polyethylene membrane. At ~40 °C and in the absence of Ba^{2+}, the pNIPA chains are relatively collapsed onto the pore walls, and the membrane is relatively permeable to water. When Ba^{2+} enters the pores and binds to the immobilized crown ethers, the polymer becomes a polyelectrolyte and swells, substantially reducing hydraulic permeability. On the other hand, fixed charge present on the swollen polymer excludes further entry of Ba^{2+} because of buildup of a Donnan potential.

The membrane is placed between an aqueous salt-free compartment containing unpolymerized crown ether and an aqueous compartment containing a higher $BaCl_2$, with the latter compartment having a higher osmotic pressure. When a hydrostatic pressure head is created with a column of $BaCl_2$ solution in the latter compartment, pressure-mediated flow of solution across the membrane is punctuated by short and small amplitude quasiperiodic reversals. Such oscillations do not occur when $BaCl_2$ is replaced with $CaCl_2$, presumably because Ca^{2+} does not bind to the crown ether.

Oscillations in this system are explained in terms of switchovers between hydraulically and osmotically dominated flows. With the polymer initially collapsed against the pore walls, pressure-driven flow is dominant, and $BaCl_2$ enters the pores by convection. Then, Ba^{2+} binds to the polymer, causing it to swell, filling the pore and reducing convective flow. At this stage, the membrane also acts as a barrier to $BaCl_2$ transport, and osmotic flow from the compartment containing free crown ether overcomes and reverses the pressure flow. Eventually, Ba^{2+} is stripped off the polymer by the free crown ether, the polymer reswells, and the cycle is poised to restart.

10.2.3
Thermofluidic Oscillator

The $BaCl_2$ driven, pNIPA based fluidic oscillator functions in the absence of an electrical current or voltage gradient. Instead, it operates by "playing off" gradients in pressure, $BaCl_2$ concentration, and free crown ether concentration. Gu et al. [46, 47] provide another example involving the interplay of gradients. In this system, shown in Figure 10.1a, a cross-linked pNIPA hydrogel is polymerized in a crosscut glass wafer. The wafer is scored on each side with parallel trenches cut more than halfway into the glass, and the trenches on the two sides are perpendicular. A continuous network of trenches results, with voids crossing the whole slide at the right angle intersection points. The pNIPA hydrogel is polymerized in the trenches, forming an interlock with the wafer. At low temperatures (<33 °C), the hydrogel swells and filled the trenches, effectively blocking water flow, while at higher temperatures (>33 °C) the hydrogel retracts from the trench walls, opening up pathways for out-of-plane water flow across the slide through the right angle trench junctions.

The slide is connected on one side to a fluid line leading to an ice water source with positive pressure head, and the other side carries water to a balance, where

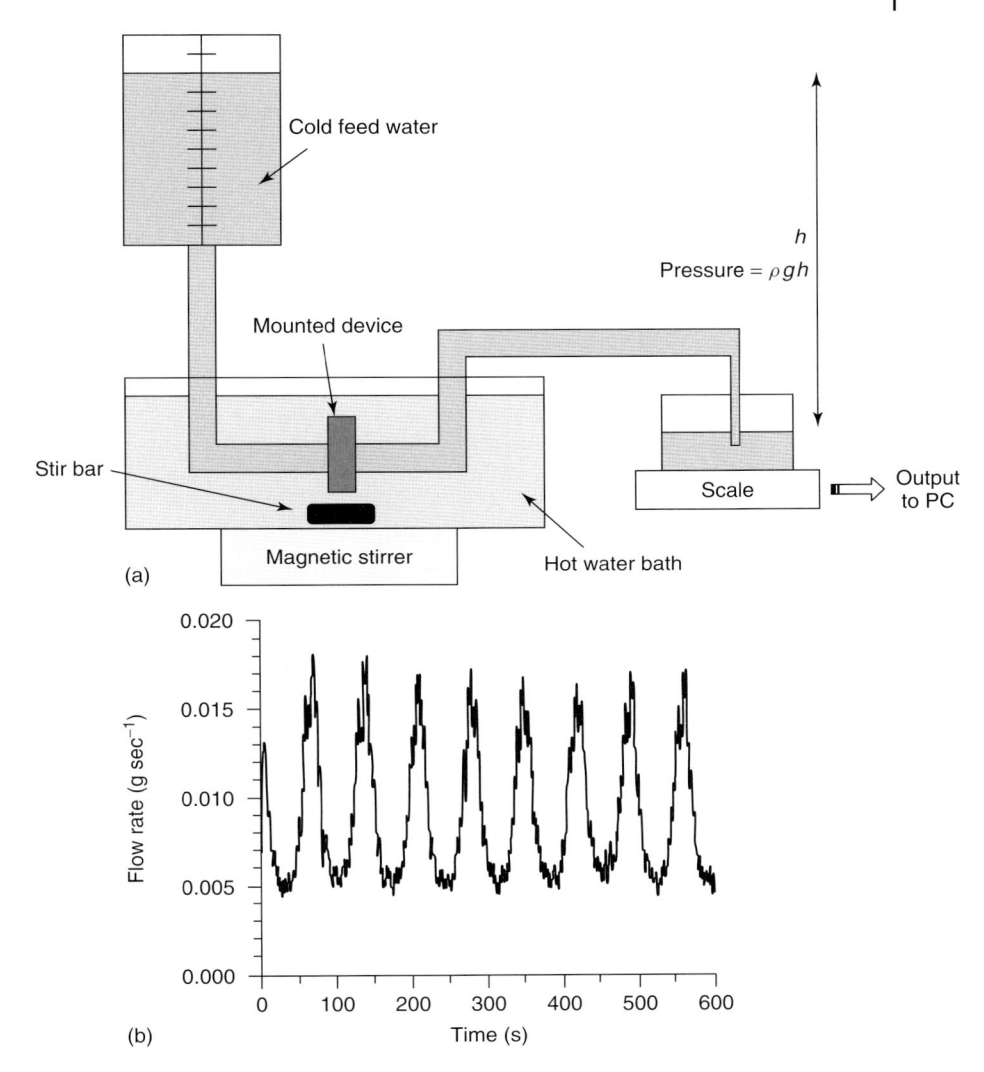

Figure 10.1 (a) Schematic of thermofluidic oscillator. The "mounted device" is a glass wafer with crosscut trenches containing poly(N-isopropylacrylamide) (pNIPA) hydrogel. (b) Recorded flow oscillations. Reproduced with permission from Ref. [47]; © 2009 Elsevier.

water accumulation is measured. The slide and adjacent feed tubing are immersed in a hot (50 °C) water reservoir. Cold water flowing through the tubing causes the hydrogel to swell and plug the slide, reducing flow. Heat transfer across the inlet tubing then warms the water in the tubing, ultimately leading to shrinkage and retraction of the hydrogel from the trench walls and reestablishment of fluid flow through the trench junctions. Eventually, cold water replaces warm water in the inlet tubing, and the swelling/shutdown/warming/shrinking cycle is repeated. A typical flow oscillation is shown in Figure 10.1b. In this thermofluidic system, chemical reaction plays no role, and oscillations arise from feedback between the pressure gradient across the hydrogel-gated slide and the temperature gradient across the tubing.

10.2.4
Lipid/Organic Membrane Analogs

Another class of membranes exhibiting periodic voltage spikes in the presence of constant current relies on the periodic modulation or breakdown of an aqueous/organic barrier [6, 48–50]. Breakdown may be due to buildup of capacitive charge separation across the membrane, which then seeks a lower free energy state by forming conducting channels. Current flows through these channels until the charge is relaxed. The channels then collapse, and the charging/discharging cycle is restarted. This mechanism seems particularly favorable when the organic phase is bounded by a surfactant, which participates in channel formation. Lipids incorporated into porous filter membranes have been studied intensively. It has been shown that such membranes are inclined toward threshold and oscillation phenomena in response to tonic electrical stimulation when they are close to their gel–liquid phase transitions, which are characterized by large changes in dielectric constant and conductivity.

In an elegant modeling study of lipid bilayers, Yagisawa *et al.* showed that oscillations can be induced by a transmembrane pH and salt gradient, with no electrical stimulation or pressure gradients [51]. Briefly, the pH difference leads to a transmembrane dipole and electrical stress on the nonpolar interior of the bilayer, triggering a gel/liquid crystal transition. Following this transition, permeability to salt increases and there is a relaxation of electrical stress, followed by reversal of the lipid transition, restoration of membrane potential, and reinitiation of the cycle.

10.2.5
Membrane/Enzyme Oscillators

In the hormone delivery oscillator considered below, neither imposed electric fields or currents, nor pH or salt gradients are used, since they are difficult to impose in a physiological environment. Instead, feedback interactions between an enzyme-catalyzed chemical reaction and transport across the membrane are key. Previous efforts coupling enzymes and membrane transport in artificial membranes, leading to oscillations, will now be reviewed.

The earliest example of a membrane–enzyme oscillator was presented by Naparstek, Caplan, and coworkers [52]. This oscillator consists of papain immobilized in a porous collodion membrane that is permeable to water, substrate, and ions. The membrane is cast as a thin film against a pH electrode and is exposed to an alkaline (pH 10) external solution of benzyl arginine ethyl ester (BAEE), which is a substrate for papain. While stationary values of pH are recorded from the electrode under most circumstances, particular combinations of external BAEE concentration and membrane thickness lead to pH oscillations at the electrode surface. Presence or absence of stirring also affects whether the system oscillates. In a complementary study, Naparstek *et al.* showed that the steady-state pH of an electrode coated with immobilized papain exposed to constant concentration of BAEE varies with external pH in a hysteretic manner [53]. Thus the diffusion–reaction system exhibits bistability, which, as we have already pointed out, can be exploited to generate oscillations.

Papain cleaves BAEE into an acidic product, which dissociates and lowers the local pH in the membrane. Papain's activity exhibits a bell-shaped curve as a function of pH [54]. Activity increases with decreasing pH in the alkaline region, so BAEE cleavage and acid production is autocatalytic. Under proper conditions, local BAEE depletes rapidly as pH drops. The reaction then shuts down, and OH^- ions diffusing in from the external solution realkalinize the region. Once BAEE is replenished, also by diffusion, the cyclic process can start anew. It should be emphasized that, for this mechanism to produce oscillatory behavior, the diffusion and reaction fluxes must play off each other so as to prevent the appearance of a stationary state. When one flux dominates, a steady state is reached without sustained oscillations.

The immobilized papain system has been subjected to much theoretical analysis using diffusion–reaction and partial differential equation models that take into account the pH-sensitivity of papain's activity [55–58]. The models predict a sharp pH front that moves back and forth across the membrane. Comparison of model predictions with experiment has been disappointing, however [58]. The models predict much sharper oscillations than are attained in the experiments.

The original measurements of Naparstek *et al.* have not been reproduced, perhaps because of difficulties in preparing stable experimental systems. Yang and coworkers substituted a polyacrylamide gel for collodion, and searched for oscillatory conditions based on theoretically calculated "phase diagrams" [58]. These diagrams identified a very narrow window in which oscillations could occur. In their experiments, the best results showed slow fluctuations in pH, but nothing resembling rhythmic behavior.

In a review of their work, Caplan and Naparstek pointed out that a simpler system might have been an enzyme-free membrane separating the alkaline BAEE solution from a small chamber containing the papain [56]. The chamber could be treated as homogeneous, and quasi-steady-state ordinary differential equations could account for transport of substrate, acid, and base across the membrane as well as for the enzyme-catalyzed reaction. By fixing the external concentrations of BAEE and H^+ (and hence OH^- since the dissociation product is constant), it was shown that conditions exist under which diffusional and reaction fluxes that balance each other are unstable, and the system is directed to a limit cycle. This simplified membrane–chamber system was further investigated theoretically by Ohmori *et al.* [59], who identified regions of parameter space that are predictive of pH oscillations for compartmentalized papain and other proteolytic enzymes. To our knowledge, no experiments have been reported with this simplified system.

Membrane-mediated oscillations in phosphofructokinase (PFK) were studied by Hervagault and coworkers [60, 61]. This enzyme is a critical component in natural glycolytic oscillators [3, 62], and its kinetics is complex [63]. Concentrations of reaction intermediates in solutions containing purified PFK may oscillate in a CSTR as a result of product activation by adenosine diphosphate (ADP) followed by substrate depletion, ultimate loss of ADP, and resetting of enzyme activity and reactant concentrations due to flow [64]. To suppress this mode of oscillation in the membrane system, Hervagault *et al.* used a preparation containing enzymes that recycle ADP to adenosine triphosphate (ATP). This preparation was placed in a small chamber separated by a glutaraldehyde cross-linked protein membrane from a reservoir containing constant concentrations of fructose-6-phosphate (F6P) and ATP.

With ADP action suppressed, the reaction velocity of PFK shows a peaked profile with respect to ATP concentration, with the peak shifting upward and to the right in a saturated manner as a function of F6P concentration. Beyond the peak, the profile has an inflection point. In order to arrive at a steady state, fluxes of both substrates across the membrane must exactly balance the consumption of substrates by the enzyme. The complex enzyme kinetics of PFK is such that this balance is unstable over particular regions of parameter space, and the system is driven toward limit cycle oscillations. In other cases, two stable steady states are available, and hysteresis with respect to changes in system parameters is possible. In Hervagault's work, oscillations were observed when the ratio of membrane permeabilities to F6P and ATP was below a threshold value, while stationary behavior was observed above that threshold, signaling a Hopf bifurcation. It was pointed out, importantly, that for this system the frequency of oscillation is determined primarily by geometrical factors such as membrane thickness and the membrane surface to reaction chamber volume ratio, instead of intrinsic reaction kinetics.

The reader is also referred to works in which uricase or acetylcholinesterase are immobilized in membranes, with evidence for oscillatory behavior when exposed to appropriate substrates [65–68]. Catalase-mediated conversion of H_2O_2 to O_2 has also been shown to exhibit oscillations when exposure to H_2O_2 is controlled by a membrane [69]. Finally, there have been theoretical speculations regarding the oscillatory potential of more general systems, including membrane-delimited systems with two enzymes with different pH–activity relationships [70, 71], though no corroborating experimental evidence has been presented to this author's knowledge.

10.2.6
General Discussion

For the membrane-mediated systems described in this survey, oscillations are clearly of dissipative nature, and they operate when external conditions are away from equilibrium. As usual, they operate as a result of positive and negative feedbacks between components, which may include reactants, products, enzymes, and the membrane itself. These feedbacks are unstable in the sense that they forbid the system from reaching a stationary state. Two general ideas exist as to the source of instability. The first scheme assumes smooth relationships between state variables, but feedback occurs after a delay, mediated by chains of reactions or processes [4, 72, 73]. The original model of the Teorell oscillator is of this type, with extra delay provided by relaxation of membrane resistance. The other conception assumes bistability of system states, with feedback causing the system to flip between states before they can reach stationary points [5, 38]. The second notion underlies the work on the Teorell oscillator with very thin membranes, and also appears to be the general analytical scheme on which the membrane-mediated enzyme oscillators are based. For the Teorell oscillator, either conception may hold depending on system specifics such as membrane thickness.

Changes in membrane resistance and electro-osmotic properties as salt redistributes play a critical role in the Teorell oscillator, so the membrane is an active player in the oscillation mechanism. Changes in membrane permeabilities to various species (including solvent and current carriers) also play a role in most of the nonenzymatic oscillators discussed. We also showed that the membrane can act simply to limit transport into and out of a reactor, with the membrane's own properties remaining constant – the PFK system is exemplary of this limit. Here, the membrane's selectivity to different reactants contributes to oscillatory behavior. In the discussion of the hydrogel–enzyme system in the next section, the membrane and enzyme behaviors are seen to be mutually coupled, and the most significant transitions occur inside the membrane.

We close this survey by pointing out that, in most cases described, the membrane serves as a transport barrier. Another possible role of a membrane was described by Schenning *et al.* [74], who dissolved manganese porphyrins in phospholipid membranes, into which lipidized rhodium complexes were incorporated. It was shown that this construct catalyzes the reaction $HCOOH + O_2 \rightarrow O_2 + H_2O_2$, and that the oxidation state of manganese oscillates in time, as detected by absorbance at 435 nm. Here the membrane colocalizes reactive centers and controls their reactions rates. These vesicle systems are therefore regarded as oscillatory enzyme mimics.

10.3
Hydrogel–Enyzme Oscillator for Rhythmic Hormone Delivery

In order to produce medically useful oscillators that deliver drugs or hormones in a rhythmic manner, it is necessary to work with substances that are endogenous, or at least nontoxic. The oscillator must also function under physiological conditions such as normal body temperature (37 °C), pH 7.4, 155 mM ionic strength, ~5 mM glucose, etc. Finally, the oscillator itself must be biocompatible. The oscillator to be described in this section represents a step toward these goals [20–28].

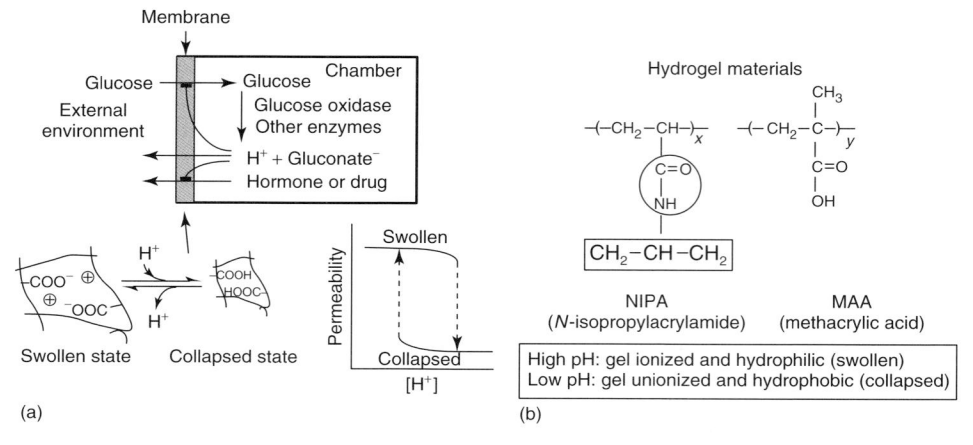

Figure 10.2 (a) Schematic of hydrogel/enzyme oscillator, with enzymes confined in chamber delimited by a pH-sensitive membrane (top). Bistable, hysteretic relationship between membrane permeability to glucose and hydrogen ion concentration, and therefore pH (bottom). Reproduced from Ref. [25] with permission; © 2002 Wiley-VCH Verlag GmbH & Co. KGaA. (b) Components of hydrogel membrane. N-isopropylacrylamide (NIPA) has both hydrophobic (rectangle) and hydrophilic (circle) components, while methacrylic acid (MAA) is ionizable. Temperature- and pH-dependent ionization state of MAA controls hydrophobic/hydrophilic balance of NIPA, and ionization of MAA also leads to swelling due to Donnan osmotic pressure. Reproduced from Ref. [28] with kind permission of Springer Science and Business Media; © 2009.

10.3.1
General Scheme

A schematic of the hydrogel/enzyme oscillator is shown in Figure 10.2 [25, 28]. The oscillator consists of a chamber in contact with an external physiological medium (blood or interstitial fluid) containing a constant level of glucose. The chamber is delimited by a hydrogel membrane that is swollen and permeable to glucose at low pH values inside the chamber, but becomes relatively collapsed and impermeable to glucose at higher pH values. The chamber contains the drug or hormone to be delivered, and the membrane's permeability to that species increases and decreases with membrane swelling and shrinking.

The chamber contains the enzymes glucose oxidase (*GluOx*), gluconolactonase (*GluLac*), and catalase (*Cat*). *GluOx* and *GluLac* catalyze the conversion of glucose to gluconic acid according to the following schemes [75–78]:

$$\text{Glucose} + O_2 + H_2O \xrightarrow{GluOx} \text{Gluconolactone} + H_2O_2$$

$$\text{Gluconolactone} \xrightarrow{GluLac} \text{Gluconic Acid}$$

Catalase eliminates H_2O_2 according to [79]

$$H_2O_2 \xrightarrow{Cat} H_2O + \tfrac{1}{2}O_2$$

so the net reaction is

$$\text{Glucose} + 1/2 O_2 + H_2O \xrightarrow{Enzymes} \text{Gluconic Acid}$$

Gluconic acid dissociates into hydrogen ion and gluconate anion:

$$\text{Gluconic Acid} \leftrightarrow H^+ + \text{Gluconate}^-$$

with $pK_a \sim 3.6$. Therefore, over the range of pH values to be considered below, it is reasonable to state that one equivalent of H^+ is produced for every equivalent of glucose consumed by *GluOx*. It will be seen, however, that the gluconic acid/gluconate$^-$ buffer system can have important effects on oscillator behavior.

The hydrogel membrane is a network of polymer chains consisting primarily of *N*-isopropylacrylamide (NIPA), substituted up to 10 mol% by *n*-alkylacrylic acids, usually methacrylic acid (MAA), and cross-linked with small amounts of ethylene glycol dimethacrylate (EGDMA). The hydrogel is cast as a thin sheet between glass plates in water/ethanol, followed by removal of unreacted materials and several conditioning steps during which the solvent is converted to aqueous. Details are available elsewhere [23, 24].

10.3.2
Bistability of Hydrogel Membrane Permeability

NIPA/MAA hydrogels with 10 mol% MAA substitution were chosen for initial study because of their apparent capacity to undergo a discrete, hysteretic transition in their permeability to glucose as a function of intrachamber pH, with external pH held constant at 7.4. This property was demonstrated in a thermostated (37 °C) side-by-side diffusion cell with independent pH control of the two sides, as illustrated in Figure 10.3a [80]. The hydrogel membrane was mounted and clamped between the two sides, or cells. "Cell I" was charged with ^{14}C-glucose and maintained at pH 7.4, while "Cell II" was subjected to pH staircases. Aliquots were collected at time intervals from Cell II and subjected to liquid scintillation counting to determine glucose flux across the membrane. Both cells were vigorously stirred during the experiments.

Results are shown in Figure 10.3b [80]. Starting at pH 5.3, downstream pH was stepped down by 0.1 unit every hour. Glucose flux, as determined from the slope of the accumulation curve in Figure 10.3b, was essentially constant until pH 4.9, at which point glucose flux became sharply attenuated. The membrane remained relatively impermeable to glucose as pH was stepwise decreased to 4.7, and then increased stepwise until pH reached 5.2, at which point the initial glucose flux was restored. The difference in the transition pH values going down and up, that is, hysteresis, is evidence for membrane bistability. To check that hysteresis was not due to kinetic lags, the downstream medium was challenged with various pH programs that would detect such kinetic effects [80, 81]. Transitions always occurred at pH 4.9 and 5.2. Hysteretic behavior has also been established with ramp pH programs [27].

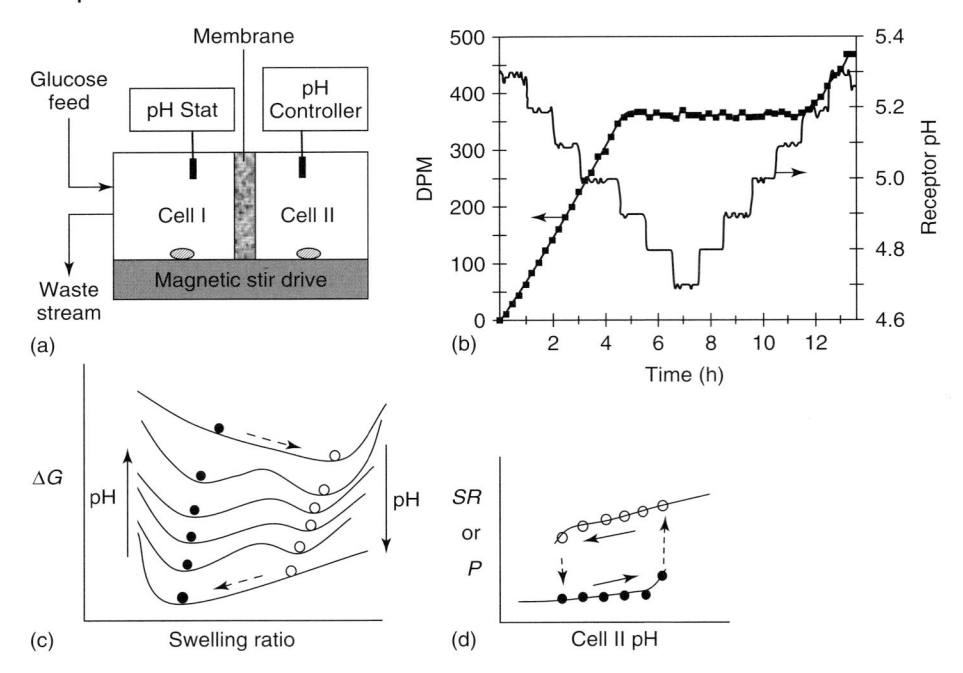

Figure 10.3 (a) Schematic of side-by-side diffusion cell used to determine ^{14}C-glucose flux across a hydrogel membrane as a function of pH in Cell II. pH in Cell I was maintained at 7.0, while pH in Cell II was stepped up and down. Hatched ellipses are magnetic stir bars. Reproduced from Ref. [28] with kind permission of Springer Science and Business Media; © 2009. (b) Accumulation of ^{14}C-glucose in Cell II (receptor) as pH in Cell II is stepped up and down. Reproduced from Ref. [80] with permission; © 1996 Wiley-VCH Verlag GmbH & Co. KGaA. (c) Schematic of changes in free energy landscape for hydrogel with change in pH, illustrating presence of two minima at intermediate pH values. Curves are translated vertically for clarity. (d) Hysteresis in swelling ratio (SR) or permeability (P) of hydrogel resulting from presence of two free energy minima.

To explain the membrane's hysteretic switchlike behavior with respect to glucose permeability, a brief exposition of the forces determining hydrogel swelling and permeability is warranted. According to standard Flory–Rehner–Donnan theory [82, 83], three forces dominate swelling behavior of polyelectrolyte gels, namely, mixing, polymer elasticity, and Donnan potential. Mixing refers to the translational entropy gain due to dispersion of polymer chains in solvent, plus the enthalpy and entropy changes resulting from polymer/solvent contacts. The NIPA/water system is unusual in this respect, as there is a strong inverse temperature relationship. pNIPA and water become less miscible with increasing temperature due to a change in dominance of hydrophobic over hydrophilic interactions, which appears to be cooperative along poly(NIPA) chains [84–88]. When MAA is in its neutral, protonated form at low pH, it too is hydrophobic, which reinforces the tendency of p(NIPA-co-MAA) chains to segregate from water. (Hydrogen bonding interactions between NIPA and MAA have also been proposed [88].) When MAA is ionized at higher pH values, it polarizes water in its vicinity and substantially reduces hydrophobic interactions, thus encouraging mixing of water and polymer chains.

The elastic force is due to distortion of polymer chains from their "unperturbed" random walk configurations, and to other constraints placed on their positions by the cross-links [89]. This force does not depend on pH in the present system. Increased incorporation of cross-linker reduces the swelling range, and may even prevent a discrete, hysteretic transition from occurring when swelling is free. On the other hand, when the hydrogel is constrained, the elastic force, which is transmitted throughout the material by the cross-links, may engender some unusual effects, as will be discussed shortly.

The Donnan potential refers to screened electrostatic repulsion between fixed charge groups on the hydrogel, usually MAA^- in the present system. Since these groups cannot leave the hydrogel, their charge must be balanced by an excess of mobile counterions and a smaller deficiency of coions compared to the external solution. The net accumulation of mobile species inside the hydrogel leads to water ingress in order to increase translational entropy of the microions, analogous to osmotic pressure that is observed in classical Donnan equilibria [82]. For hydrogels containing pendant acid groups, this force increases with increasing pH, as the fraction of fixed MAA groups that are ionized increases.

Swelling transitions in hydrophobic, polyacid hydrogels are normally attributed to shifts in the balance between hydrophobic forces that dominate at low pH and the Donnan potential that dominates at high pH, with polymer elasticity ultimately limiting swelling and collapse. Hysteresis may be due to subtle factors causing the free energy landscape of the hydrogel to possess more than one minimum as a function of water content, with sizable free energy barriers between the minima [90–92], as illustrated in Figure 10.3c. The free energies of these minima are pH dependent. A hydrogel that is in a swelling state corresponding to its global free energy minimum at one pH may find itself trapped in that state following a change in pH, since it is now at a local free energy minimum. It will remain at the local minimum until the pH change is large enough to remove the free energy barrier, allowing the system to relax to its new global free energy minimum [90]. Trapping in local free energy minima is manifested in bistability and hysteresis in swelling and permeability, as illustrated in Figure 10.3d.

A factor complicating the interpretation of hysteretic behavior in the system under study is the imposition of two broken symmetries on the hydrogel. First, the hydrogel is clamped along its perimeter, so swelling and shrinking are anisotropic. The elastic force becomes tensorial in character, and lateral tension could lead to phase separation and pattern formation [27, 93–95]. Simple conceptions based on single variables such as degree of swelling break down in this case. Second, imposing a pH gradient across the membrane leads to a gradient in swelling forces inside the membrane. Overall permeability is probably controlled by a relatively thin skin layer facing the solution in Cell II where pH is variable. Hydrogel facing the solution in Cell I should always be swollen since proximal pH is 7.4, well above the pK_a

of MAA (∼5). The swollen and collapsed skin layers are conformal at their interface owing to the polymer cross-links, and therefore they exert stress on each other.

The two symmetry breaks may interact strongly. Hydrogel swelling is best understood in the isotropic, free swelling case, and there is much to be learned about systems subjected to mechanical constraints and gradients in swelling/modulating environmental variables. In particular, the degree to which clamping and imposition of a pH gradient across the hydrogel introduces or modulates hysteretic behavior is not known.

For hydrophilic solutes such as glucose, permeability through the hydrogel depends on solute size, availability of connected pools of free water in the hydrogel [96–98], and reduced presence of obstacles from polymer chains that accompanies swelling [99–104]. (Hydrophobic solutes may actually be more permeable when the hydrogel is collapsed, however [105].) Thus switching and hysteresis in swelling are mirrored by similar changes in glucose permeability. In fact, permeability changes may be even more readily observed than swelling/shrinking when the latter is confined to a thin skin.

Before returning to our discussion of the hydrogel/enzyme oscillator, we note in passing that several researchers have tried to couple pH-sensitive hydrogels with *GluOx/GluLac/Cat* to produce glucose-sensitive insulin delivery systems [106–111]. Typically, hydrogels incorporating weak basic groups such as N,N-dimethylaminomethacrylate (DMAEMA) have been used. Such hydrogels include monomers such as hydroxyethyl methacrylate or polyethylene glycol methacrylate as comonomers, which are always hydrophilic and permit glucose diffusion under all circumstances. Swelling affects insulin permeability more strongly, however, because of its size. Higher levels of glucose (hyperglycemia) lead to increased acid production by the enzymes, increased swelling of the hydrogel membrane caused by ionization of the DMAEMA sites by acid protons, and increased permeability to insulin. At lower glucose levels (normoglycemia), acid production is reduced and the membrane becomes less ionized and shrinks, leading to reduced insulin delivery.

10.3.3
Oscillator Operation

The hydrogel/enzyme oscillator considered in the present work is distinguished from glucose-sensitive systems in that the membrane shrinks with increasing acid production by the enzymes, and in doing so glucose permeability is substantially reduced because of the presence of the hydrophobic NIPA comonomer. A negative feedback loop is established between the membrane and the enzyme. This feedback, combined with the observed hysteresis (i.e., bistability) of the hydrogel's glucose permeability response, can drive the system into an oscillatory instability under proper circumstances. This system operates at a constant concentration of glucose in the external environment, again in contrast to the glucose-sensitive responsive delivery systems. In the oscillator, glucose (along with O_2) is a free energy source that is provided at constant rate through the feed stream, while the lower free energy products, the most important of which is H^+, are removed in the waste stream.

Oscillations occur as follows (refer to Figure 10.2): Starting with a swollen membrane at high intrachamber pH, glucose diffuses into the chamber and is converted to gluconic acid, which dissociates and lowers the intrachamber pH. Acidic protons diffuse into the hydrogel and neutralize the charged MAA groups in the skin layer. The skin layer collapses and becomes relatively impermeable to glucose. Enzymatic production of acid is then reduced, and pH inside the chamber rises. Eventually pH reaches a point where the skin is sufficiently ionized that it reswells. Acidic protons that were bound to the membrane diffuse into the external medium, where they are diluted and carried away. The system is now poised to repeat the cycle.

The oscillatory mechanism is expected to operate over a bounded range of external glucose concentrations. When glucose concentration is too low, acid production by the enzymes is insufficient to collapse the membrane. When glucose concentration is too high, residual permeability of the collapsed membrane to glucose will permit enough acid production to maintain the collapsed state. Oscillations will occur only in an intermediate glucose concentration region, in which acid production places the membrane in its bistable region [22, 112].

10.3.4
Oscillator Prototype

The hydrogel/enzyme oscillator concept has been tested in the same side-by-side cell as was used to study hysteretic pH effects on the hydrogel membrane's glucose permeability [23–25, 112]. Figure 10.4a illustrates the setup, with the NIPA/MAA hydrogel membrane mounted between Cell I and Cell II. The cells each contain 75 ml of solution, and the area of the membrane supporting transport is 3.14 cm². Glucose solutions of constant concentration flow at 1.37 ml min^{-1} through Cell I, which is pH-stated at 7.4. The flow outlet provides a waste stream. *GluOx*, *GluLac*, and *Cat* are introduced into Cell II, either as free enzymes or immobilized in polyacrylamide microparticles. A pH electrode is placed into Cell II. Thus Cells I and II correspond, respectively, to the physiological environment and the chamber in Figure 10.2. Both cells are well stirred and exposed to the atmosphere, providing adequate O_2 levels such that the enzyme reactions are not limited by oxygen.

As a final element, a piece of marble (solid $CaCO_3$) is placed in Cell II. The role of this mineral is to provide a "shunt" pathway for elimination of H^+ from Cell II, via the heterogeneous reaction [113]

$$H^+ + CaCO_3 \rightarrow Ca^{2+} + HCO_3^-$$

followed by

$$H^+ + HCO_3^- \rightarrow H_2CO_3 \rightarrow H_2O + CO_2$$

To compensate for rapid removal of H^+, it is necessary to raise the feed glucose concentration, increasing the rate of glucose presentation to the enzyme. Accelerated acid production removal leads to faster pH changes in Cell II. Without marble,

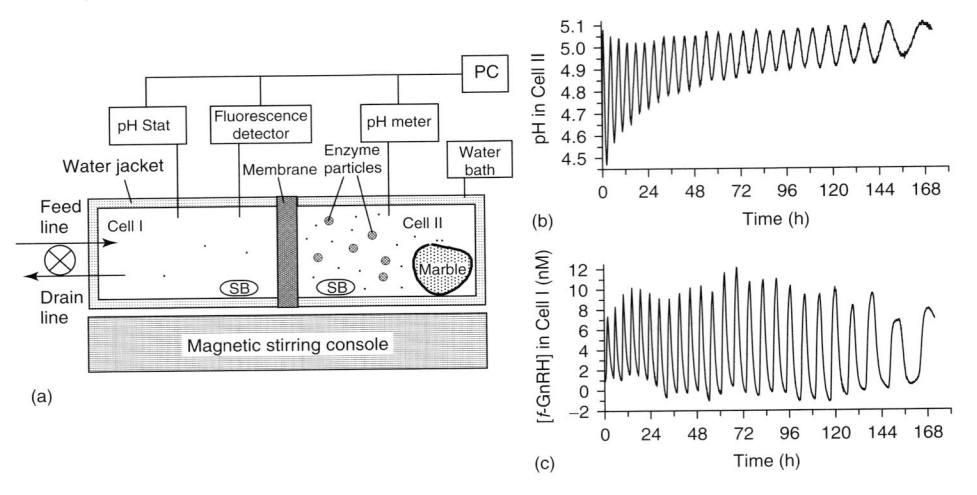

Figure 10.4 (a) Schematic of prototype oscillator setup. Cells I and II, each containing 75 ml fluid, are well stirred with magnetic stir bars (SBs). Glucose at constant concentration is continuously fed into Cell I at 1.37 ml min^{-1}, and fluid is drained from Cell I at the same rate. pH in Cell I is maintained at 7.0 by pH stat. Enzyme particles in Cell II contain GluOx, GluLac, and Cat. Fluorescently labeled gonadotropin-releasing hormone (f-GnRH) is introduced at time zero into Cell II. A water jacket maintains the temperature in both cells at 37°C. Fluctuating pH is monitored in Cell II, and f-GnRH is monitored by circulation through a fluorescence monitor (return time ~1 min). (b) pH oscillations measured in Cell II. (c) Concentration–time profile of f-GnRH in Cell I, which is the convolution of the oscillating delivery rate of f-GnRH through the membrane, and the fluid residence time distribution in Cell I. Assuming that the latter is exponentially decaying with time constant ~55 min, the f-GnRH delivery rate profile is much sharper. Reproduced from Ref. [24] with permission; © 2002 Elsevier.

pH in Cell II changes very slowly, and the hydrogel membrane is coaxed into an intermediate, stationary glucose permeability state, and sustained oscillations do not occur [22]. Apparently, a minimal slew rate for pH is needed to support oscillations.

To demonstrate the potential for pulsed hormone delivery, fluorescently labeled gonadotropin-releasing hormone (f-GnRH) was introduced into Cell II, and its concentration in Cell I was monitored as a function of time. Figure 10.4b,c shows results when glucose concentration in Cell I was 50 mM, and the hydrogel membrane contained 10 mol% MAA [24]. Sustained pH oscillations were observed in Cell II for 1 week as shown in Figure 10.4b, while Figure 10.4c shows coherent oscillations in f-GnRH concentration in Cell I. Note that this concentration profile reflects the convolution of the f-GnRH delivery rate into Cell I with the molecular residence time distribution in that cell. Assuming Cell I to be well stirred with volume, this distribution is a decaying exponential with time constant 75 ml/(1.37 ml/min) ≈ 55 min. The actual delivery rate profile, which can be obtained by deconvolution, has much sharper peaks than the f-GnRH concentration profile [24].

Figure 10.4b,c can be regarded as a proof of principle, whereby a constant level of glucose presented to the enzyme–membrane system can drive rhythmic delivery of a hormone. However, the present system is far from ready for testing as an implantable device for hormone treatment. The size of the system, its reliance on marble to produce oscillations, the low intrachamber (Cell II) pH range at which the system operates, and the high glucose level used to produce oscillations (~10-fold higher than normal blood glucose level) are indicators that much work is needed to make the system physiologically compatible. Since *f*-GnRH is not relevant to fundamental questions about the system, it has not been included in subsequent mechanistic studies.

10.3.5
Analysis of Factors Affecting Oscillations Over Time

The pH oscillations portrayed in Figure 10.4b are not quite uniform. Amplitude decreases during the first several periods and there is a steady increase in period (decrease in frequency) with time, with oscillations ultimately halting after 1 week. On the basis of several lines of experimental evidence, we believe that these phenomena are due to accumulation of gluconate in Cell II and not, say, to depletion of enzyme activity. Enzyme was included in great excess, and the time constant for enzymatic catalysis was established to be about 15 min (unpublished), much less than the period of pH oscillation, which was at least 2 h. Catalase in the preparation removed H_2O_2, a primary cause of *GluOx* deactivation [79]. In several experiments, the solution contents of Cell II were removed and dialyzed, and the remaining enzyme was reconstituted with gluconate-free "starter" solution. Oscillations resumed with the initial period, suggesting that enzyme activity was uncompromised [114].

In a more detailed set of experiments, samples were removed at selected intervals from Cell II and subjected to titration with HCl [112]. At the beginning of the run, the titration curve was sharp and was affected weakly by the enzyme. Titration curves became shallower with each subsequent sample, and careful data-fitting suggested a single buffer species with $pK_a \approx 3.6$, with buffer concentration increasing with time. This pK_a value is very close to that of gluconic acid/gluconate$^-$. Fits were not improved by including bicarbonate (produced from the marble reaction) as a second buffering species in the calculations. It was also shown that spiking gluconic acid into Cell II could quench oscillations, and that restoring gluconate-free medium into Cell II permitted restart of oscillations, with a nearly identical amplitude–period pattern as was observed at the beginning of the run [114].

All of these experiments suggest that gluconate, which is of similar size to glucose but may be Donnan-excluded from the membrane because of its predominantly negative charge over the pH range of interest, is a player in the dynamics of pH oscillations. Accumulation of gluconate readily explains the slowing of pH oscillations in Cell II, due to its buffering capacity.

A second potential player is Ca^{2+}, which is generated by the marble reaction [112]. This species should also accumulate with time. It also may affect the membrane's swelling properties as a result of favorable Donnan partitioning and reduction in Donnan osmotic pressure. The dynamics and effect of Ca^{2+} have not been adequately explored to date, however. Note that Ca^{2+} probably diffuses more rapidly than gluconate through the hydrogel membrane, because of its favorable size and charge.

Reduction in pH oscillation amplitude observed at early stages may also be due in part to buffering by gluconate. Another potential explanation for amplitude reduction is a finite relaxation time of the hydrogel's skin layer between its swollen glucose-permeable state and its collapsed glucose-impermeable state. Changes between these states may not be able to follow rapid pH swings, so overshoots may occur. As pH swings slow down, the skin layer's permeability state presumably becomes more in line with pH in Cell II. It should be emphasized that the skin layer is probably dynamic in both time and space, growing from and shrinking toward the interface with the Cell II solution as the latter's pH waxes and wanes.

As indicated above, oscillations terminate after seven days. It has been hypothesized [24] that as pH swings become slower, an intermediate permeability steady-state attractor becomes available, similar to that seen in slow systems without marble. The collapsed skin of the hydrogel develops stress because of its perimeter being clamped. Analogous to the cracking of mud, it seems quite plausible that, with enough time, the skin can separate into dense and partially swollen domains, the latter being partially permeable to glucose, leading to a stationary pH gradient across the membrane, which stabilizes the surface pattern. Such surface effects are well established for pinned thermosensitive gels [94], and we have presented preliminary optical evidence for changes in surface morphology following initial collapse [27]. We have also shown that slow ramps in pH in Cell II lead to an intermediate permeability to glucose [27].

To summarize the present picture, gluconate accumulation leads to slowing of pH oscillations in Cell II. As long as the pH swings are fast enough, the oscillations drive the hydrogel's skin layer between states of high and low permeability to glucose. However, if pH swing becomes slow enough, then the hydrogel can select the intermediate glucose permeability state, which is self-sustaining. Precise verification of this mechanism has been elusive, as it is very difficult to observe and characterize the morphologies and chemical gradients inside large deformation-sustaining hydrogels when they are challenged with dynamic gradients of stimulating factors such as pH. Further complicating matters are the change in concentrations of buffering species, Ca^{2+} generated by the marble reaction, and the fluxes of species such as O_2, HCO_3^-, and CO_2.

10.3.6
Tuning pH Range of Oscillations

The low pH range ($\sim$pH 5) at which the oscillator operates is also problematic, as it is far from physiologic pH. The system does not oscillate at all when exposed to physiologic buffers containing bicarbonate or phosphate (unpublished). The low operating pH, along with the capture of acidic protons by marble, places heavy demands on the enzyme and glucose to produce H^+, and these demands are presently met using glucose concentrations that well exceed normal physiological levels. Improvement of oscillator performance will require, at the very least, that membranes be produced whose pH ranges of bistability and oscillation are below but close to physiologic pH, preferably without requiring marble. Presumably, raising the oscillator's pH range of operation will also lower the glucose concentration needed to fuel the oscillations, hopefully to normoglycemic levels.

To tune the pH operating range, recall that hydrogel swelling is determined by hydrophobicity of the polymer, Donnan potential and hence osmotic pressure, and polymer elasticity. Elasticity is unlikely to be a dominant factor in determining the transition pH's, although it may limit swelling and shrinking, and there may be subtle effects due to clamping. Donnan osmotic potential is controlled by the number density and extent of ionization of pendant MAA groups in the hydrogel, the latter being controlled by pH through deprotonation of MAA. Reasoning that the swelling transition will occur when the Donnan swelling pressure is high enough to overcome hydrophobic forces, and the reverse for the shrinking transition, it was postulated that pH values associated with the transitions could be increased by (i) replacing MAA with acid groups with higher pK_a [28, 114], (ii) reducing the concentration of MAA [23], or (iii) increasing hydrophobicity of the nonionic components of the hydrogel [115]. Glucose concentrations needed for oscillations would be correspondingly reduced since fewer acidic protons need to be produced at increased pH target ranges. Experimental approaches and results associated with each of these methods will now be described.

To raise the intrinsic pK_a of the pendant acid groups in the hydrogel, it is necessary to replace the carboxylic acids with weaker acid functionalities. Bae et al. [116–119] noted that sulfonamides, a readily available class of drugs, present a variety of pK_a values surrounding physiologic pH, depending on the electron-withdrawing capability of adjacent groups. They also showed how to derivatize sulfanomides with vinyl groups and incorporate them into hydrogels with dimethylacrylamide as the majority comonomer, obtaining sharp pH swelling transitions. By incorporating *GluOx* directly into the hydrogel, the transition range broadened somewhat, but they were able to show swelling sensitivity to external glucose concentration. Following their lead, we acryloylated sulfamethoxypyridazine (SMPA) and produced 3 : 1 : 96 mol% SMPA/MAA/NIPA hydrogel membranes. When incorporated into the oscillator prototype, slow and irregular oscillations were observed, indicating that, while this method is promising, much more work is needed [28, 114].

In Figure 10.5a, it is shown that reducing MAA content in the NIPA/MAA copolymer hydrogel from 10 to 5 mol% leads to an alkaline shift in the pH range of oscillations [23]. Furthermore, oscillations occurred at a reduced glucose concentration. Further reduction of MAA to 2 mol% leads to stable focuslike behavior, that is, pH oscillations decaying to a stationary value. It may be that with such low MAA doping of the membrane, there is no bistable pH range for swelling/collapse. Alternatively, the bistable range may very narrow and difficult to target by altering glucose concentration.

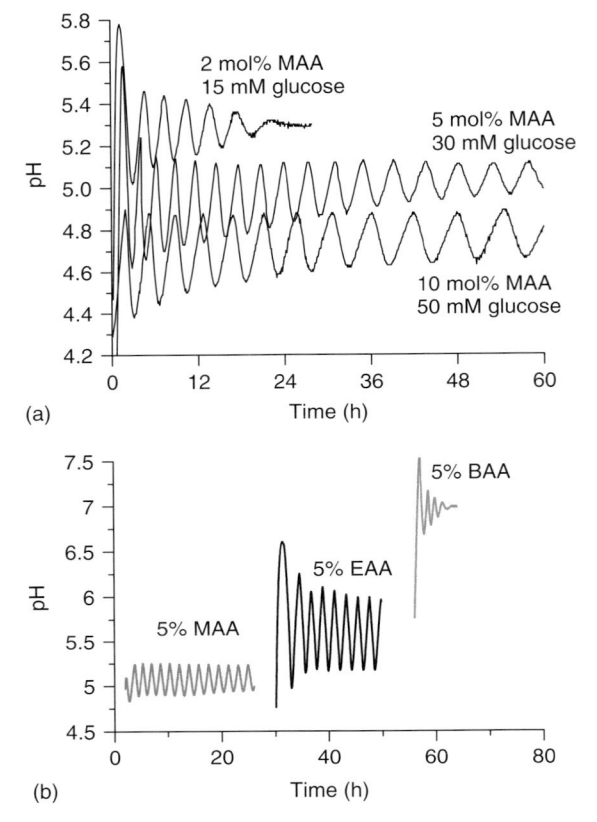

Figure 10.5 (a) Lowering methacrylic acid (MAA) content leading to alkaline shift in pH range of oscillations, and lower glucose concentrations in Cell II required to elicit oscillations. Reproduced from Ref. [23] with permission; © 2002 American Chemical Society. (b) Replacing MAA with more hydrophobic EAA and BAA also leads to alkaline shift in oscillation range. Glucose concentrations were 50, 70, and 20 mM for membranes containing 5 mol% MAA, EAA, and BAA, respectively. Reproduced from Ref. [115] with permission; © 2007 Wiley-VCH Verlag GmbH & Co. KGaA.

To study the effects of increasing hydrophobicity, MAA in the copolymer hydrogel membranes was replaced with a homologous series of α-alkylacrylic acids [115, 120]. The longer the α-alkyl side chain, the more hydrophobic the monomer, and it has been shown that there is a $\sim$0.5 pH unit alkaline shift in the free swelling transition for each added methylene unit on the side chain. Figure 10.5b shows the effects of such substitutions on oscillations, with 5 mol% α-alkylacrylic acid incorporation in each hydrogel membrane. Ethylacrylic acid (EAA) containing hydrogels oscillated over a more alkaline pH range than hydrogels containing MAA. Interestingly, butylacrylic acid (BAA) doped hydrogels exhibited oscillations very close to physiologic pH, but these oscillations were not sustained, perhaps because of improper targeting of glucose concentration.

Thus, each of the three methods to raise operating pH range of the oscillating system shows promise, but further work is needed to produce stable oscillations near physiologic pH.

10.3.7
Discussion and Conclusion

The work reviewed here on the glucose-driven hydrogel enzyme oscillator is clearly very preliminary with respect its targeted application, namely, rhythmic pulsatile delivery of hormones such as GnRH. Aside from the needs to clarify and mitigate the mechanism by which oscillations cease after a time, and to raise the pH range over which stable oscillations occur, it has already been noted that size must be reduced substantially to make the system implantable, and marble should not be required as a component. Marble was included to speed up oscillations, at the cost of requiring more glucose. Plausibly, marble could be eliminated by designing device geometries with larger surface/volume ratios (primarily by decreasing volume), which by itself would accelerate changes in pH. It is hoped that, if a suitable membrane that will oscillate steadily near physiologic pH can be produced, then the driving glucose concentration can be reduced to physiologic levels. Reduced external glucose concentration will also reduce gluconate accumulation, and perhaps lead to a limiting oscillation frequency that does not engender the putative stress-induced phase separation in the limiting skin layer. In the long rung, this frequency should match that of the endogenous hormone oscillation (1–2 h in the case of GnRH [13]).

We conclude with a general discussion about the assumptions that have been made regarding the mechanism of oscillation. In line with many of the membrane/biochemical oscillators described in the survey section, we have assumed that the hydrogel's bistability, coupled with negative feedback from the enzyme-mediated reaction in the chamber, is the probable mechanism. It is also concluded, on the basis some experimental evidence, that as the oscillations slow down as a result of buffer accumulation, the hydrogel's bistability is compromised and a third, intermediate permeability state is found, leading to stationary behavior. As already noted, asymmetries due to hydrogel clamping and pH gradients across the membrane make precise modeling very difficult, and models that treat the hydrogel as uniform throughout or as a one-dimensional continuum are probably too simplistic.

It is yet to be proved directly that pH swings are correlated with hysteretic (first-order) pH-driven transitions in skin permeability to glucose. An alternative is that the transitions are very sharp although not of first order and represent a strong negative feedback which, when coupled with a cascade of rate-limiting chemical or mechanical processes, can also undergo an oscillatory feedback instability [4, 21, 72, 73, 121, 122]. As noted earlier, such a combination may be operative in the Teorell oscillator for thick membranes. Generally speaking, however, such oscillations are not as robust as those driven by bistability/feedback mechanisms [4, 5, 21, 123]. On the basis of observations on hysteresis in membrane permeability as a function of pH, we believe that bistability/feedback is the root mechanism for oscillations in the hydrogel/enzyme driven oscillator.

Acknowledgments

This work was funded by Amgen, Inc., the National Science Foundation (CHE-9996223), and the National Institutes of Health (HD040366). Experimental work reported here was carried out by J.P. Baker, J.C. Leroux, G.P. Misra, A.P. Dhanarajan, A.S. Bhalla, and S.K. Mujumdar. Details of individual contributions can be found in the references.

References

1. Gray, P. and Scott, S.K. (1990) *Chemical Oscillations and Instabilities*, Clarendon, Oxford.
2. Epstein, I.R. and Pojman, J.A. (1998) *An Introduction to Nonlinear Chemical Dynamics*, Oxford University Press, New York.
3. Goldbeter, A. (1996) *Biochemical Oscillations and Cellular Rhythms*, Cambridge University Press.
4. Hahn, H.-S., Ortoleva, P.J., and Ross, J. (1973) Chemical oscillations and multiple steady states due to variable boundary permeability. *J. Theoret. Biol.*, **41**, 503–521.
5. Katchalsky, A. and Spangler, R. (1968) Dynamics of membrane processes. *Quart. Rev. Biophys.*, **2**, 127–175.
6. Larter, R. (1990) Oscillations and spatial nonuniformities in membranes. *Chem. Rev.*, **90**, 355–381.
7. Teorell, T. (1959) Electrokinetic membrane processes in relation to properties of excitable tissues. I. Experiments on oscillatory transport phenomena in artificial membranes. *J. Gen. Physiol.*, **42**, 831–845.
8. Rapp, P.E. (1979) An atlas of cellular oscillators. *J. Exp. Biol.*, **81**, 281–306.
9. Crowley, W.F. and Hofler, J.G. (1987) *The Episodic Secretion of Hormones*, John Wiley & Sons, Inc., New York.
10. Brabant, G., Prank, K., and Schöfl, C. (1992) Pulsatile patterns in hormone secretion. *Trends Endocrinol. Metab.*, **3**, 183–190.
11. Schally, A.V., Arimura, A., Kastin, A.J., Matsuo, H., Baba, Y., Redding, T.W., Nair, R.M.G., Debeljuk, L., and White, W.F. (1971) Gonadotropin-releasing hormone: one polypeptide regulates secretion of luteinizing and follicle-stimulating hormones. *Science*, **173**, 1036–1038.
12. Belchetz, P.E., Plant, T.M., Nakai, Y., Keogh, E.J., and Knobil, E. (1978) Hypophysical responses to continuous and intermittent delivery of hypothalamic gonadotropin-releasing hormone. *Science*, **202**, 631–633.
13. Crowley, W.F., Filicori, M., Spratt, D.I., and Santoro, N. Jr. (1985) The physiology of gonadotropin-releasing hormone (GNRH) secretion in men and women. *Recent Prog. Hormone Res.*, **41**, 473–527.
14. Conn, P.M. and Crowley, W.F. Jr. (1994) Gonadotropin-releasing hormone and its analogs. *Ann. Rev. Med.*, **45**, 391–405.
15. Nachtigall, L.B., Boepple, P.A., Pralong, F.P., and Crowley, W.F. Jr. (1997) Adult-onset idiopathic hypogonadotropic hypogonadism-a treatable form of male infertility. *N. Engl. J. Med.*, **336**, 410–415.
16. Hayes, F.J., Seminara, S.B., and Crowley, W.F. Jr. (1998) Hypogonadotropic hypogonadism. *Endocrinol. Metab. Clin. N. Am.*, **27**, 739–763.
17. Beranova, M., Oliveira, L.M.B., Bedecarrats, G.Y., Schipani, E., Vallejo, M., Ammini, A.C., Quintos, J.B., Hall,

J.E., Martin, K., Hayes, F., Pitteloud, N., Kaiser, U.B., Crowley, W.F., and Seminara, S.B. Jr. (2001) Prevalence, phenotypic spectrum, and modes of inheritance of gonadotropin-releasing hormone receptor mutations in idiopathic hypogonadotropic hypogonadism. *J. Clin. Endocrinol. Metab.*, **86**, 1580–1588.

18. Seminara, S.B., Hayes, F., and Crowley, W.F. Jr. (1998) Gonadotropin-releasing hormone deficiency in the human (idiopathic hypogonadotropic hypogonadism and kallmann's syndrome): pathophysiological and genetic considerations. *Endocrine Revs.*, **19**, 521–539.

19. Southworth, M.B., Matsumoto, A.M., Gross, K., Soules, M.R., and Bremner, W.J. (1991) The importance of signal pattern in the transmission of endocrine information: pituitary gonadotropin responses to continuous and pulsatile gonadotropin-releasing hormone. *J. Clin. Endocrinol. Metab.*, **72**, 1286–1289.

20. Siegel, R.A. and Pitt, C.G. (1995) A strategy for oscillatory drug release. general scheme and simplified theory. *J. Controlled Release*, **33**, 173–186.

21. Siegel, R.A. (1997) in *Controlled Release: Challenges and Strategies* (ed. K. Park), American Chemical Society, Washington, DC, pp. 501–527.

22. Leroux, J.-C. and Siegel, R.A. (1999) Autonomous gel-enzyme oscillator fueled by glucose--preliminary evidence for oscillations. *Chaos*, **9**, 267–275.

23. Dhanarajan, A.P., Misra, G.P., and Siegel, R.A. (2002) Autonomous chemomechanical oscillations in a hydrogel/enzyme system driven by glucose. *J. Phys. Chem.*, **106**, 8835–8838.

24. Misra, G.P. and Siegel, R.A. (2002) A new mode of drug delivery: long term autonomous rhythmic hormone release across a hydrogel membrane. *J. Controlled Release*, **81**, 1–6.

25. Siegel, R.A., Misra, G.P., and Dhanarajan, A.P. (2002) in *Polymer Gels and Networks* (eds Y. Osada and A.R. Khokhlov), Marcel Dekker, New York, pp. 357–372.

26. Dhanarajan, A., Urban, J., and Siegel, R.A. (2003) in *Nonlinear Dynamics in Polymeric Systems* (eds J.A. Pojman and Q. Tran-Cong-Miyata), American Chemical Society, Washington, DC, pp. 44–57.

27. Dhanarajan, A. and Siegel, R.A. (2005) Time-dependent permeabilities of hydrophobic, Ph-sensitive hydrogels exposed to pH gradients. *Macromol. Symp.*, **227**, 105–114.

28. Siegel, R.A. (2009) in *Chemomechanical Instabilities in Responsive Materials* (eds P. Borckmans, P. De Kepper, A. Khokhlov, and S. Mètens), Springer, Dordrecht, pp. 175–201.

29. Teorell, T. (1959) Electrokinetic membrane processes in relation to properties of excitable tissues. II. some theoretical considerations. *J. Gen. Physiol.*, **42**, 847–863.

30. Teorell, T. (1956) Transport phenomena in membranes. *Disc. Faraday Soc.*, **21**, 9–26.

31. Caplan, S.R. and Mikulecky, D.C. (1966) in *Ion Exchange* (ed. J.A. Marinsky), Marcel Dekker, pp. 1–64.

32. Murray, J.D. (1989) *Mathematical Biology*, Springer-Verlag, Berlin.

33. Meares, P. and Page, K.R. (1974) Oscillatory fluxes in highly porous membranes. *Proc. R. Soc. Lond. A.*, **339**, 513–532.

34. Rastogi, R.P., Mishra, G.P., Pandey, P.C., Bala, K., and Kumar, N. (1999) Bistability and electrokinetic oscillations. *J. Colloid Interface Sci.*, **217**, 275–287.

35. Langer, P., Page, K.R., and Wiedner, G. (1981) A teorell oscillator with fine pore membranes. *Biophys. J.*, **36**, 93–107.

36. Bala, K., Kumar, K., Saha, S.K., and Sristava, R.C. (2004) Single-capillary teorell oscillator–studies with nonelectrolytes. *J. Colloid Interface Sci.*, **273**, 320–323.

37. Kobatake, Y. and Fujita, H. (1964) Flows through charged membranes. Ii. flip-flop current vs voltage relation. *J. Chem. Phys.*, **40**, 2212–2218.

38. Kobatake, Y. and Fujita, H. (1964) Flows through charged membranes. II. Oscillation phenomena. *J. Chem. Phys.*, **40**, 2219–2222.

39. Meares, P. and Page, K.R. (1973) The teorell membrane oscillator as a mechano-electric transducer. *J. Memb. Biol.*, **11**, 197–216.
40. Franck, U.F. (1978) A quantitative treatment of oscillatory phenomena in coarse-grained ion-exchanger membranes. *Electrochim. Acta*, **23**, 1081–1091.
41. Gedalin, K. (1997) Electro-osmotic oscillations. *Phys. D*, **110**, 154–168.
42. Shashoua, V.E. (1967) Electrically active polyelectrolyte membranes. *Nature*, **215**, 846–847.
43. Huang, L.-Y.M. and Spangler, R.A. (1977) Dynamic properties of polyelectrolyte calcium membranes. *J. Memb. Biol.*, **36**, 311–335.
44. Minoura, N., Aiba, S.-I., and Fujiwara, Y. (1993) Spontaneous oscillations of the electrical membrane potential in tri-block copolypeptide membranes composed of L-glutamic acid and L-leucine. *J. Amer. Chem. Soc.*, **115**, 5902–5906.
45. Ito, T. and Yamaguchi, T. (2006) Nonlinear self-excited oscillation of a synthetic ion channel-inspired membrane. *Angew. Chem. Int. Ed.*, **45**, 5630–5633.
46. Ziaie, B., Baldi, A., Lei, M., Gu, Y., and Siegel, R.A. (2004) Hard and soft micromachining for biomems: review of techniques and examples of applications in microfluidics and drug delivery. *Adv. Drug Deliv. Revs.*, **56**, 145–172.
47. Gu, Y., Dhanarajan, A.P., Hruby, S.L., Baldi, A., Ziaie, B., and Siegel, R.A. (2009) An interpenetrating glass-thermosensitive hydrogel construct. gated flow and thermofluidic oscillations. *Sens. Actuators, B. Chem.*, **138**, 631–636.
48. Kim, J.T. and Larter, R. (1991) Simple and complex oscillations in lipid-doped membranes. *J. Phys. Chem.*, **95**, 7948–7955.
49. Urabe, K. and Sakaguchi, H. (1993) Stable self-sustained potential oscillations across a membrane filter impregnated with triolein. *Biophys. Chem.*, **47**, 41–51.
50. Urabe, K. and Sakaguchi, H. (1996) Characteristics of current induced potential oscillations of a triolein impregnated membrane placed between identical salt solutions. *Biophys. Chem.*, **59**, 33–39.
51. Yagisawa, K., Naito, M., Gondaira, K.-I., and Kambara, T. (1993) A model for self-sustained potential oscillation of lipid bilayer membranes induced by the gel-liquid crystal phase transitions. *Biophys. J.*, **64**, 1461–1475.
52. Naparstek, A., Thomas, D., and Caplan, S.R. (1973) An experimental enzyme-membrane oscillator. *Biochim. Biophys. Acta*, **323**, 643–646.
53. Naparstek, A., Romette, J.L., Kernevez, J.P., and Thomas, D. (1975) Memory in enzyme membranes. *Nature*, **249**, 490–491.
54. Goldman, R., Kedem, O., Silman, I.H., Caplan, S.R., and Katchalski, E. (1967) Papain-collodion membranes. I. preparation and properties. *Biochemistry*, **7**, 486–500.
55. Caplan, S.R., Naparstek, A., and Zabusky, N.J. (1973) Chemical oscillations in a membrane. *Nature*, **245**, 364–366.
56. Caplan, S.R. and Naparstek, A. (1977) An enzyme membrane oscillator. *Adv. Biol. Med. Phys.*, **16**, 177–185.
57. Chay, T.R. (1979) pH Oscillations in transport processes. *J. Theor. Biol.*, **90**, 83–99.
58. Ohmori, T. and Yang, R.Y.K. (1996) Self-sustained oscillations in immobilized proteolytic enzyme systems. *Biophys. Chem.*, **59**, 87–94.
59. Ohmori, T., Nakaiwa, M., Yamaguchi, T., Kawamura, M., and Yang, R.Y.K. (1997) Self-sustained pH oscillations in a compartmentalized enzyme reactor system. *Biophys. Chem.*, **67**, 51–57.
60. Hervagault, J.F., Duban, M.C., Kernevez, J.P., and Thomas, D. (1983) Multiple steady states and oscillatory behavior of a compartmentalized phosphofructokinase system. *Proc. Natl. Acad. Sci.*, **80**, 5455–5459.
61. Hervagault, J.-F. and Thomas, D. (1983) Experimental evidence and theoretical discussion for long-term oscillations of phosphofructokinase in

a compartmentalized. *Eur. J. Biochem.*, **131**, 183–187.
62. Chance, B., Ghosh, A.K., Pye, E.K., and Hess, B. (eds) (1973) *Biological and Biochemical Oscillators*, Academic Press, New York.
63. Pettigrew, D.W. and Frieden, C. (1979) Rabbit muscle phosphofructokinase. A model for regulatory kinetic behavior. *J. Biol. Chem.*, **254**, 1896–1901.
64. Hocker, C.G., Epstein, I.R., Kustin, K., and Tornheim, K. (1994) Glycolytic pH oscillations in a flow reactor. *Biophys. Chem.*, **51**, 21–35.
65. Hervagault, J.-F., Friboulet, A., Kernevez, J.P., and Thomas, D. (1980) Spatiotemporal behaviors in immobilized enzyme systems. *Biochimie*, **62**, 367–373.
66. Friboulet, A. and Thomas, D. (1982) Electrical excitability of artificial enzyme membranes. III. Hysteresis and oscillations observed with immobilized acetylcholinesterase membranes. *Biophys. Chem.*, **16**, 153–157.
67. Trubuil, A., Friboulet, A., Joshi, R., Kernevez, J.-P., and Thomas, D. (1988) Electrical excitability of artificial enzyme membranes IV. theoretical approach of the membrane potential of synthetic proteinic films. *Biophys. Chem.*, **31**, 217–224.
68. Chay, T.R. and Zabusky, N.J. (1983) Dual-mode potential oscillations on an immobilized acetylcholinesterase membrane system. *J. Biol. Phys.*, **11**, 27–31.
69. Hideshima, T. and Inoue, T. (1997) Nonlinear oscillatory reaction of catalase induced by gradual entry of substrate. *Biophys. Chem.*, **63**, 81–86.
70. Vincent, J.-C. and Selegny, E. (1982) Bi-enzymatic time oscillations induced by functional structures I. stability analysis. *J. Non-Equilibr. Thermodyn.*, **7**, 259–268.
71. Vincent, J.C., Selegny, E., and Alexandre, S. (1988) Oscillations in open two-michaelian systems. *Biochem. Bioenerg.*, **19**, 247–261.
72. Rapp, P.E. (1976) Analysis of biochemical phase shift oscillators by a harmonic balancing technique. *J. Math. Biol.*, **3**, 203–224.
73. Tyson, J.J. and Othmer, H.G. (1978) The dynamics of feedback control circuits in biochemical pathways. *Prog. Theoret. Biol.*, **5**, 1–62.
74. Schenning, P.H.J., Spelberg, J.H.H., Driessen, M.C.P.F., Hauser, M.J.B., Feiters, M.C., and Nolte, R.J.M. (1995) Enzyme mimic displaying oscillatory behavior. Oscillating reduction on manganese(III) porphyrin in a membrane-bound cytochrome P-450 model system. *J. Am. Chem. Assoc.*, **117**, 12655–12657.
75. Albin, G., Horbett, T.A., Miller, S.R., and Ricker, N.L. (1987) Theoretical and experimental studies of glucose sensitive membranes. *J. Controlled Release*, **7**, 267–291.
76. Klumb, L.A. and Horbett, T.A. (1992) Design of insulin delivery devices based on glucose sensitive membranes. *J. Controlled Release*, **18**, 59–80.
77. Leypoldt, J.K. and Gough, D.A. (1984) Model of a two-substrate enzyme electrode for glucose. *J. Biol. Chem.*, **56**, 2896–2904.
78. Hanazato, Y., Inatomi, K.-I., Shiono, S., and Maeda, M. (1988) Glucose-sensitive field effect transistor with a membrane containing coimmobilized gluconolactonase and glucose oxidase. *Anal. Chim. Acta*, **212**, 49–59.
79. Tse, P.H.S. and Gough, D.A. (1987) Time-dependent inactivation of immobilized glucose oxidase and catalase. *Biotechnol. Bioeng.*, **29**, 705–713.
80. Baker, J.P. and Siegel, R.A. (1996) Hysteresis in the glucose permeability versus pH characteristic for a responsive hydrogel membrane. *Macromol. Rapid. Commun.*, **17**, 409–415.
81. Leroux, J.-C. and Siegel, R.A. (1999) in *Intelligent Materials and Novel Concepts for Controlled Release Technologies* (ed. J. De Nuzzio), American Chemical Society, Washington, DC, pp. 98–112.
82. Flory, P.J. (1993) *Principles of Polymer Chemistry*, Cornell University Press, Ithaca.
83. English, A., Mafe, S., Mazanares, J.A., Yu, X., Grosberg, A.Y., and Tanaka, T. (1996) Equilibrium swelling properties of polyampholytic hydrogels. *J. Chem. Phys.*, **104**, 8713–8720.

84. Hirokawa, Y. and Tanaka, T. (1984) Volume phase transition in a nonionic gel. *J. Chem. Phys.*, **81**, 6379–6380.
85. Hirotsu, S., Hirokawa, Y., and Tanaka, T. (1987) Volume-phase transitions of ionized N-isopropylacrylamide gels. *J. Chem. Phys.*, **87**, 1392–1395.
86. Schild, H.G. (1992) Poly(N-isopropylacrylamide): experiment, theory and application. *Prog. Polym. Sci.*, **17**, 163–249.
87. Sasaki, S. and Maeda, H. (1996) Simple theory for volume phase transition of hydrated gels. *Phys. Rev. E.*, **54**, 2761–2765.
88. Kawasaki, H., Sasaki, S., and Maeda, H. (1997) Effect of pH on the volume phase transition of copolymer gels of N-isopropylacrylamide and sodium acrylate. *J. Phys. Chem. B.*, **101**, 5089–5093.
89. Erman, B. and Flory, P.J. (1986) Critical phenomena and transitions in swollen polymer networks in and linear macromolecules. *Macromolecules*, **19**, 2342–2353.
90. Sato-Matsuo, E. and Tanaka, T. (1988) Kinetics of discontinuous volume-phase transition of gels. *J. Chem. Phys.*, **89**, 1695–1693.
91. Annaka, M., Tokita, M., Tanaka, T., Tanaka, S., and Nakahira, T. (2000) The gel that memorizes phases. *J. Chem. Phys.*, **112**, 471–477.
92. Sekimoto, K. (1993) Temperature hysteresis and morphology of volume phase transition of gels. *Phys. Rev. Lett.*, **70**, 4154–4157.
93. Tanaka, T., Sun, S.T., Hirokawa, Y., Kathayama, S., Kucera, J., Hirose, Y., and Amiya, T. (1987) Mechanical instability of gels at the phase transition. *Nature*, **325**, 796–798.
94. Li, C., Hu, Z., and Li, Y. (1994) Temperature and time dependencies of surface patterns in constrained ionic N-isopropylacrylamide gels. *J. Chem. Phys.*, **100**, 4645–4652.
95. Sultan, E. and Boudaoud, A. (2008) The buckling of a swollen thin gel layer bound to a compliant substrate. *J. Appl. Mech.*, **75**, 051002.
96. Yasuda, H., Lamaze, C.E., and Peterlin, A. (1971) Diffusive and hydraulic permeabilities of water in water-swollen polymer membranes. *J. Poly. Sci. Pt. A-2*, **9**, 1117–1131.
97. Peppas, N.A., Gruny, R., Doelker, E., and Buri, P. (1980) Modelling of drug diffusion through swellable polymeric systems. *J. Membr. Sci.*, **7**, 241–253.
98. Dong, L.-C. and Hoffman, A.S. (1991) A novel approach for preparation of pH-sensitive hydrogels for enteric drug delivery. *J. Controlled Release*, **15**, 141–152.
99. Ogston, A.G., Preston, B.N., and Wells, J.D. (1973) On the transport of compact particles through solutions of chain polymers. *Proc. R. Soc. Lond. A.*, **333**, 297–316.
100. Schnitzer, J.E. (1988) Analysis of steric partition behavior of molecules in membranes using statistical physics. Application to gel chromatography and electrophoresis. *Biophys. J.*, **54**, 1065–1076.
101. Phillips, R.J., Deen, W.M., and Brady, J.F. (1990) Hindered transport in fibrous membranes and gels: effect of solute size and fiber configuration. *J. Colloid Interface Sci.*, **139**, 362–373.
102. Johansson, L., Elvingson, C., and Löfröth, J.E. (1991) Diffusion and interaction in gels and solutions. 3. Theoretical results on the obstruction effect. *Macromolecules*, **24**, 6024–6029.
103. Johansson, L. and Löfroth, J.-E. (1991) Diffusion and interaction in gels and solutions I. method. *J. Colloid Interface Sci.*, **142**, 116–120.
104. Amsden, B. (2001) Diffusion in polyelectrolyte hydrogels: application of an obstruction-scaling model to solute diffusion in calcium alginate. *Macromolecules*, **34**, 1430–1435.
105. Palasis, M. and Gehrke, S. (1992) Permeability of responsive poly(N-isopropylacrylamide) gel to solutes. *J. Controlled Release*, **18**, 1–12.
106. Ishihara, K. and Matsui, K. (1986) Glucose-responsive insulin release from polymer capsule. *J. Polym. Sci. Polym. Lett. Ed.*, **24**, 413–417.
107. Kost, J., Horbett, T.A., Ratner, B.D., and Singh, M. (1985) Glucose-sensitive

membranes containing glucose oxidase: swelling, activity, and permeability studies. *J. Biomed. Mater. Res.*, **19**, 1117–1122.
108. Albin, G., Horbett, T.A., and Ratner, B.D. (1985) Glucose sensitive membranes for controlled delivery of insulin: insulin transport studies. *J. Controlled Release*, **2**, 153–164.
109. Traitel, T., Cohen, Y., and Kost, J. (2000) Characterization of glucose-sensitive insulin release systems in simulated in vivo conditions. *Biomaterials*, **21**, 1679–1687.
110. Podual, K., Doyle, F.J.I., and Peppas, N.A. (2000) Preparation and dynamic response of cationic copolymer hydrogels containing glucose oxidase. *Polymer*, **41**, 3975–3983.
111. Abdekhodaie, M.J. and Wu, X.Y. (2008) Microdomain pH gradient and kinetics inside composite polymeric membranes of pH and glucose sensitivity. *Pharm. Res.*, **25**, 1150–1157.
112. Bhalla, A.S. (2007) Physicochemical investigations of a drug delivery oscillator. PhD thesis, University of Minnesota.
113. Rábai, G. and Hanazaki, I. (1996) pH Oscillations in the bromate-sulfite-marble semibatch and flow systems. *J. Phys. Chem.*, **100**, 10615–10619.
114. Dhanarajan, A. (2004) Mechanistic studies and development of a hydrogel/enzyme drug delivery oscillator. PhD thesis, University of Minnesota.
115. Bhalla, A.S., Mujumdar, S.K., and Siegel, R.A. (2007) Novel hydrogels for rhythmic pulsatile drug delivery. *Macromol. Symp.*, **254**, 338–344.
116. Kang, S.I. and Bae, Y.H. (2001) pH/temperature-sensitive polymer: poly(N-isopropylacrylamide-co-methacryloylated sulfamethoxypyridazine). *Macromol. Chem. Symp.*, **14**, 145–155.
117. Kang, S. and Bae, Y. (2001) pH-induced volume-phase transition of hydrogels containing sulfonamide side group by reversible crystal. *Macromolecules*, **33**, 8173–8178.
118. Kang, S.I. and Bae, Y.H. (2002) pH-induced solubility transition of sulfonamide-based polymers. *J. Controlled Release*, **80**, 145–155.
119. Kang, S.I. and Bae, Y.H. (2003) A sulfonamide based glucose-responsive hydrogel with covalently immobilized glucose oxidase and catalase. *J. Controlled Release*, **86**, 115–122.
120. Mujumdar, S.K. (2007) Stimuli sensitive hydrogels for controlled drug delivery and sensing applications. PhD thesis, University of Minnesota.
121. Thron, C.D. (1991) The secant condition for instability in biochemical feedback control--I. The role of cooperativity and saturability. *Bull. Math. Biol.*, **53**, 383–401.
122. Thron, C.D. (1991) The secant condition for instability in biochemical feedback control--II. models with upper hessenberg jacobian matrices. *Bull. Math. Biol.*, **53**, 403–424.
123. Tsai, T.Y.-C., Choi, Y.S., Ma, W., Pomerening, J.R., Tang, C., and Ferrell, J.E.J. (2008) Robust, tunable biological oscillations from interlinked positive and negative feedback loops. *Science*, **321**, 126–129.

Further Reading

Ohmori, T. and Yang, R.Y.K. (1994) Autonomous pH oscillations in an immobilized papain membrane system. *Biotechnol. Appl. Biochem.*, **20**, 67–78.

11
Structure Formation in Inorganic Precipitation Systems
Oliver Steinbock and Jason Pagano

11.1
Introduction

Throughout nature, simple rules give rise to complex structures and dynamics [1–3]. These complexities often surprise the human observer as they emerge at higher system levels in a seemingly unpredictable fashion. The number of examples for these emergent phenomena is so tremendously large that selecting a few is necessarily misleading. Nonetheless, we remind the reader of some of the classic cases such as the concept of temperature, phase transitions, the stock market, ant piles, biological evolution, and consciousness. Also, chemistry itself provides a beautiful example of emergent phenomena as a few chemical elements combine to an endless number of molecules with microscopic and macroscopic features that are difficult to predict from atomic properties. However, modern chemistry as well as nanotechnology are clearly dominated by reductionism. Accordingly, most chemical approaches are engineering-like and aim to create specific features of a material from molecular building blocks. Despite the tremendous success of this approach, we seem to reach a mysterious boundary that often keeps our devices light-years away from the performance of biological systems.

Clearly, most biological "solutions" to chemical and engineering problems were found by evolution randomly in a tremendously large lab and over time periods that exceed the typical funding cycle in modern science. Hence, it is well possible that such strategies are of no technological value to us. However, it is the underlying hypothesis of our work that the fundamental laws, tools, or tricks of biological ingenuity can also be used by modern chemistry to engineer complex, nonbiological structures, materials, and devices. In our opinion, the two most important aspects are

1) to create emergent complexity from (simple) chemistry with (mild) kinetic nonlinearities and under nonequilibrium conditions and
2) to enhance this complexity and extend it to macroscopic length scales by allowing structure formation, spatial compartmentalization, and transport.

This rather grand manifesto clearly must be substantiated, demonstrated, and tested in reproducible experiments. Furthermore, all results must be analyzed

Nonlinear Dynamics with Polymers: Fundamentals, Methods and Applications.
Edited by John A. Pojman and Qui Tran-Cong-Miyata
Copyright © 2010 WILEY-VCH Verlag GmbH & Co. KGaA, Weinheim
ISBN: 978-3-527-32529-0

in a modern, reductionist fashion. Also suitable experimental models must be found and investigated to establish examples of "lifelike" complexities in nonbiological, chemical systems. Today, we have a small number of such models, many of which are aqueous reaction–diffusion systems. Possibly the best known example is the autocatalytic Belousov–Zhabotinsky (BZ) reaction which self-organizes macroscopic traveling concentration waves, rotating spirals, and more complex three-dimensional vortices [3]. Other well-known examples include the chlorite–iodide malonic acid reaction, known to form stationary Turing patterns, and the Briggs–Rauscher reaction. The spatiotemporal complexities in all of these examples are transient, because they occur in thermodynamically closed systems. Alternatively, they can be maintained in a state far from equilibrium with specialized gel reactors that are continuously fed with reactant solutions. Although all of the resulting phenomena reveal, in a dramatic fashion, how simple reactions coupled by diffusion can provoke intricate spatiotemporal patterns, they unfortunately fail to create permanent, lasting structures.

It is hence not surprising that several groups have explored chemical modifications in which the aforementioned structures are coupled to polymerization or precipitation reactions. For example, Köhler and Müller covered thin layers of BZ gels with silver nitrate solution, which (i) quenches the BZ reaction and (ii) precipitates AgBr along curves tracing the "stopped" chemical wave fronts [4]. Furthermore, Washington *et al.* added acrylonitrile to the stirred BZ reaction and observed periodic polymerization [5]. However, there are no experimental procedures available for transferring the shape of two- or three-dimensional BZ wave to a continuous solid, that is, sufficiently stable to allow its nondestructive removal from the system.

In this chapter, we discuss a different experimental approach that involves inorganic polymerization and precipitation reactions. As we will show, these experiments match the two aforementioned characteristics and could develop into an important model demonstrating the utilization of chemical complexities at the system level in materials science and engineering.

11.2
Permanent Patterns from Inorganic Precipitation and Deposition Processes

There are several precipitation reactions that induce spatial pattern formation, some of which are briefly described in the following to provide an appropriate context for our main topic. Disregarding minerals such as banded agates and zebra spar, the classic example of inorganic precipitation patterns occurs in a type of experiment named after Raphael Liesegang. In the late nineteenth century, Liesegang observed periodic precipitation bands in the reaction between a gel containing potassium dichromate and a solution of silver nitrate [6]. Depending on the geometry of the system, the precipitation bands can form rings or bands. Recently, Grzybowski *et al.* combined this basic experiment with reactive wet stamping techniques and demonstrated the production of intricate microscale designs [7]. Another group

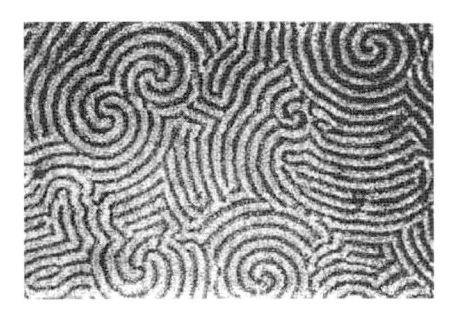

Figure 11.1 Rotating spiral waves during the electrodeposition of silver–antimony alloy. The characteristic wavelength (i.e., the pitch) and the rotation period of the spiral waves are 10 μm and 10 s, respectively. Reprinted with permission from Ref. [11].

showed that information (such as Morse code messages and musical rhythms) can be encoded in Liesegang patterns via intricate control sequences that affect ionic transport through time-dependent electric currents [8].

A qualitatively different type of precipitation patterns concerns structures formed during the electrodeposition of alloys. In 1938, Raub and Schall observed the formation of propagating wave patterns during the electrodeposition of silver–indium alloy [9, 10]. However, their observation was widely ignored, because no systematic theory was available that could classify these patterns as typical features of systems far from thermodynamic equilibrium. In 1986, the phenomenon was studied by Krastev and Nikolova in the possibly related electrodeposition of a silver–antimony alloy (Figure 11.1) [11]. Furthermore, very similar patterns were observed by Saltykova *et al.* during the electrodeposition of iridium–ruthenium alloy from molten salts [12].

The patterns observed during the aforementioned electrochemical growth processes include target patterns and rotating (single- and multiarmed) spiral waves that show all the characteristics expected from wave patterns in excitable reaction–diffusion media. In the silver–antimony and silver–indium systems, the patterns can be seen by the unaided eye as band structures that arise from the simultaneous deposition of separate phases with different metal concentrations. The patterns self-organize on the working electrode under potentiostatic and galvanostatic conditions. Their origins are still controversial but might involve a combination of reaction–diffusion and convective instabilities.

11.3
Tube Formation in Precipitation Systems and Silica Gardens

Hollow tubes are a simple but sufficiently interesting geometric shape that is frequently encountered in nature. They are hence a good target for efforts aiming to create permanent structures via chemical emergence. In nature, tubular structures seem to result from a variety of different mechanisms, but their formation always involves some form of aggregation, precipitation, or mineralization within a chemical gradient field. Examples include iron sulfide chimneys at hydrothermal vents [13], certain forms of stalactites [14], hollow silica fibers in cement [15], corrosion products (S. Thouvenel-Romans and O. Steinbock, unpublished results)

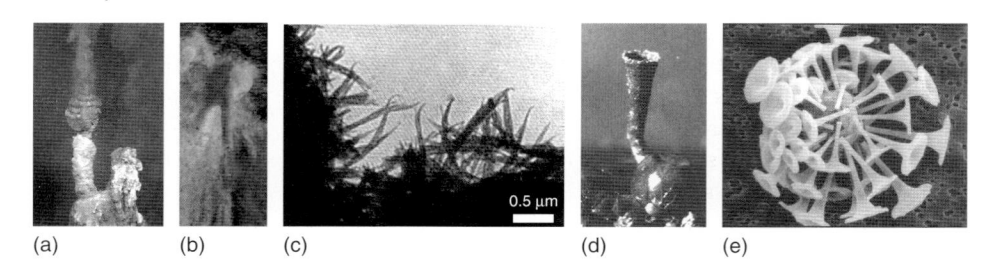

Figure 11.2 Examples of tubular structures. (a) Sulfide mounds with venting black smokers on the East Pacific Rise; (b) 5 cm long soda straw growing on the facade of Florida State University's old chemistry building DLC [28]; (c) silica fibers in Portland cement; (d) rust tube on low-carbon steel undergoing blistering corrosion, as observed by the author [17]; (e) marine algae with an intact calcite shell known as *coccospheres*. (a), (c), and (e) are reprinted with permission from Refs. [13, 15, 20], respectively.

[16, 17], and biomineralization structures [18, 19] such as diatom shells [20] (see Figure 11.2). Some of these structures are closely related to silica tubes forming in precipitation reactions of silicates [21]. The latter example is sometimes referred to as *silica gardens* [22, 23]. It belongs to a larger class of systems that involve the growth of plantlike fibers from aqueous solutions containing anions such as aluminate, borate, carbonate, or silicate [24–26]. The fibers have diameters in the micro- and millimeter range and reach lengths of several decimeters.

Tube growth can be induced by seeding the latter solutions with crystals of various soluble salts excluding group (I) compounds [25]. The qualitative growth mechanism is based on the formation of a semipermeable membrane around the dissolving seed crystal (see, e.g., Ref. [24]). Osmotic pressure gives rise to the inflow of water causing the rupture of the membrane and the ejection of buoyant salt solution into the surrounding medium. In conjunction with polymerization and precipitation reactions [24, 25, 27], this process forms tubular structures that grow upward with speeds of millimeters per day to millimeters per second.

11.4
Historic and Cultural Links

The investigation of silica gardens has a remarkable history dating back to the seventeenth century [29]. In addition, the astonishing self-organized, lifelike shapes in these seemingly simple systems have attracted the imagination of children and adults alike. For example, the German writer and Nobel laureate Thomas Mann discusses these chemical structures in his book "Doktor Faustus" and they are also described in Oliver Sacks' novel "Uncle Tungsten: Memories of a Chemical Boyhood" [30, 31]. It is hence not surprising that silica gardens remain popular demonstration experiments in schools and colleges, and simple "crystal garden kits" are commercially available through various toy and educational companies.

Possibly, the earliest work on silica gardens was carried out by the famous German-Dutch chemist Johann Rudolf Glauber, who also discovered sodium sulfate known to many as *Glauber's salt*. In his 1646 work "Furni Novi Philosophici," he reported experiments involving structures formed from introducing ferrous chloride crystals into a solution of potassium silicate (water glass) [29]. In later years, other well-known scientists such as James Keir and Isaac Newton followed his investigation. The observation of treelike metal structures may have influenced Newton's manuscript "Of Nature's Obvious Laws and Processes in Vegetation" [32].

As suggested by the latter title, the structures in silica gardens resemble similar forms observed in living organisms. As a result, precipitation tubes attracted considerable interest of nineteenth century scientists, who were searching for possible origins of life. During this period, it was commonly believed that silica gardens imitate stems, leaves, twigs, roots, shells, mushrooms and other amoebae, fungi, flowers, and worms [22]. Evidently, advances in biochemistry have dismissed this chemical phenomenon as a paradigm for the genesis of life.

11.5
Some Recent Developments

Modern research on chemical gardens started with, or is at least closely linked to, the work of scientists such as Traube, Pfeffer, and van't Hoff, who carried careful investigations of colloidal membranes and osmosis. However, what is seemingly a modern renaissance of this problem occurred only in the 1990s. For example, Collins *et al.* studied tubular structures in the aluminosilicate system and noted promising catalytic features [33, 34]. Furthermore, precipitation tubes were shown to have intriguing hierarchical nanostructure [35, 36] involving silica nanorods that are organized into cylindrical clusters of about 40 nm within patterned micrometer-scale domains. Also submicron silica tubes have been synthesized using organic crystal templating, solgel methods, and liquid crystal phase transformations (Figure 11.3). These and related inorganic structures formed around organic templates have been reviewed by van Bommel *et al.* [37].

Additional developments, which will not be discussed in detail, include qualitative microgravity studies by Jones and Walter [26]. Cartwright *et al.* used Mach–Zehnder interferometry to investigate changes in the refractive index of the surrounding solutions and noted the importance of buoyancy-driven free convection [24]. Moreover, Pantaleone *et al.* observed rhythmic motion of forming precipitation tubes that are reminiscent of a waving arm [38]. The latter group measured pressure oscillations in the moving tube and proposed a model to account for these relaxation oscillations. The latter work focuses on closed precipitation tubes and allows the estimation of the tube's elastic modulus E and critical stress σ_c. For the specific conditions of their experiments, these values are $E = 1 - 10$ MPa and $\sigma_c = 100 - 1000$ Pa [39].

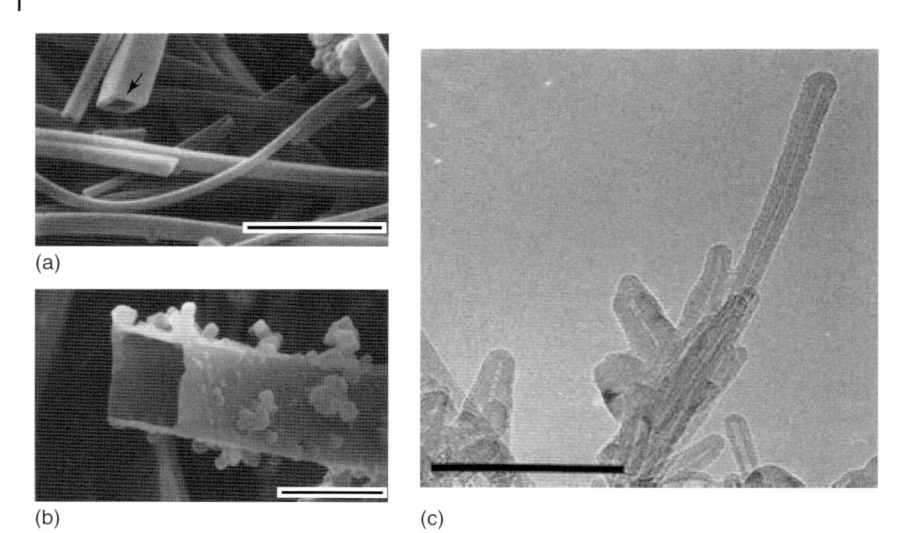

Figure 11.3 (a,b) SEM images of silica tubes with rhomboidal channels. These tubes form in an ethanolic tetraethyl orthosilicate (TEOS) solution containing tartaric acid and ammonium hydroxide. (c) TEM image of a calcined silica nanotube templated with laurylamine hydrochloride. Reproduced with permission from Ref. [37]. According to Ref. [37], the scale bars correspond to 2 μm (a), 5 μm (b), and 100 nm (c).

11.6
Experimental Methods

In the conventional "chemical garden" experiment, small salt particles or crystals are seeded into aqueous solutions containing anions such as silicate, carbonate, borate, or phosphate. Obviously, there are several major disadvantages of such an approach. First, the shape of the seed particle and, typically, also its mass are uncontrolled. Second, the seed is a finite reservoir and the experiment cannot be carried out under continuous conditions. Third, several tubes might simultaneously form from one seed, thus, complicating the overall complex experiments in an undesired fashion. Fourth, chemical reactions begin while the seed is still sinking to the bottom of the reaction vessel or, if solutions are added to the seed, complex fluid flow potentially affects the early growth dynamics.

Some of these problems can be overcome by the use of salt pellets as seeds. This method has been used by Maselko et al., who form pellets from powder under high pressure [38–41]. However, the seed pellets are nonetheless finite sources of metal ion and the resulting experimental conditions are in this sense not truly stationary. The latter problem, however, can be solved using a technique first introduced by Thouvenel-Romans et al. in 2003 [42]. This approach, which is the basis for most of the experimental results presented in this chapter, is based on a very simple idea, namely the salt particle is replaced by an aqueous salt solution. This seed solution is injected through a small glass capillary into a large reservoir of silicate solution. The injection is carried out with a syringe pump and occurs typically at constant pump rates in the range of several milliliters per hour. If the density of the seed solution is smaller than the density of the water glass, then injection is carried out in upward direction and yields upward growing silica tubes.

During the growth optical micrographs are acquired at a rate of typically 10 frames/s with a monochrome charged-coupled device camera connected to a PC via a frame grabber board. The images are then analyzed using in-house software. All measurements are corrected for optical deformations caused by the cylindrical geometry of the main reaction vessel. For electron microscopy as well as all other post-synthesis analyses, the tubes are carefully removed from the silicate solution, rinsed with water, dried at 45 °C overnight, and finally placed in a vacuum desiccator to remove remaining moisture. Then, they are either carbon coated or gold sputtered prior to acquiring micrographs using a JEOL JSM-5900 scanning electron microscope (SEM). Additional characterization experiments include transmission electron microscopy (TEM) and energy dispersive X-ray spectroscopy (EDS).

11.7
Growth Regimes

In 2003, our group reported first results on tube growth in experiments employing hydrodynamic injection of aqueous cupric sulfate into a large volume of sodium silicate solution [42]. These studies allowed us to identify three distinct regimes of tube growth (Figure 11.4). At low concentrations of cupric sulfate (<0.1 M), one observes thin silica tubes (~200 µm) that form along a continuous jet of buoyant copper solution (Figure 11.4i–l). If the concentration is increased to 0.15 M, this steady growth gives way to oscillatory dynamics. As shown in Figure 11.4a–h, a very thin membrane-bound droplet forms at the tip of the structure, inflates, pops off, and rises to the surface of the silicate solution. The period of this rhythmic process is on the order of seconds and the resulting tubes are much wider than those at low concentrations. If the concentration is increased to approximately 0.35 M, one observes a second transition. For these high concentrations, the expanding droplet does not *detach* but bursts and creates a new droplet at the rupture site (Figure 11.4m–p). The repetitive stretching and bursting creates a very wide (here 2–3 mm) bulging tube.

The latter experiments considered a special case of delivery-controlled tube growth, namely the injection of *buoyant* cupric sulfate solution into a *denser* solution of silicate. Clearly, this density difference can be reversed. For the injection of denser cupric sulfate solution into lighter silicate solution, tube growth occurs in downward direction. In addition, we can consider the injection of silicate into cupric sulfate solution. If we again consider different density relations, these combinations total to four distinctly different experimental scenarios.

In the following, we describe the results obtained from the injection of water glass into a large reservoir of cupric sulfate solution [43]. As the density of the injected liquid is selected to be larger than the density of the cupric sulfate solution, tube

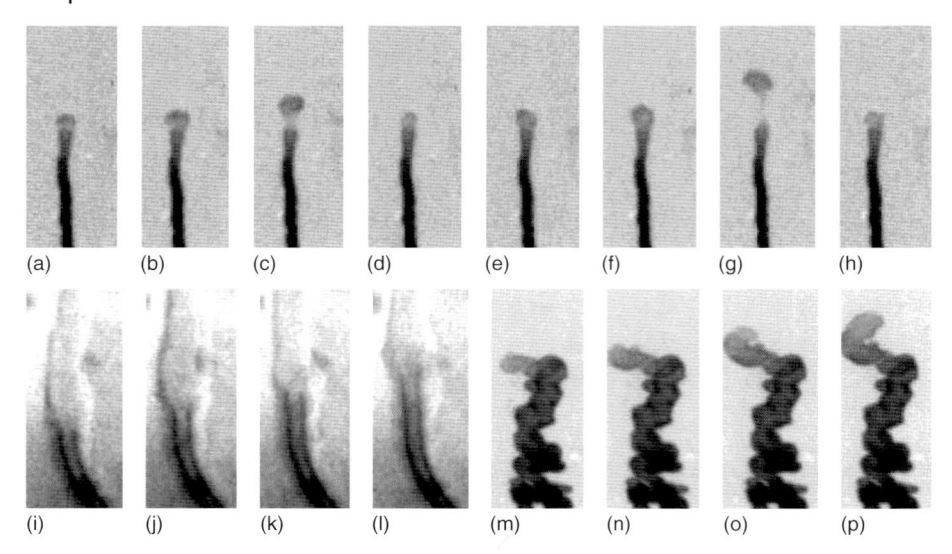

Figure 11.4 Image sequences of tube growth in the popping (a–h), jetting (i–l), and budding (m–p) regimes. In all experiments, aqueous cupric sulfate solution is injected into sodium silicate solution (100 ml, 1 M in Si, 25 °C). Injection is carried out with a syringe pump at constant flow rate (here 7.0 ml h^{-1}) through a vertical glass capillary. The time intervals between frames, and the image areas are 0.5 s, 9.1 × 20.4 mm^2 (a–h, m–p), and 2.0 s, 1.6 × 3.6 mm^2 (i – l), respectively. The concentrations of cupric sulfate are 0.25 M (a–h), 0.05 M (i–l), and 0.50 M (m–p). Reprinted with permission from Ref. [42].

growth occurs in downward direction. As before, we only consider experiments carried out at constant injection rates.

Qualitatively, such experiments differ from the one described above in that the resulting tubes are much softer. Accordingly, attempts to remove the hollow structures from the reaction system after synthesis usually resulted in unsatisfactory results making spatial resolved material characterization difficult. However, one can speculate that any radial composition gradients in the wall should be reversed between these different types of experiments.

Figure 11.5 shows representative sequences of snapshots illustrating four distinct regimes of tube growth under these "reverse" conditions. The sequence in Figure 11.5a illustrates tube formation in a dynamic regime that will be referred to as *reverse jetting* since it is reminiscent of the jetting growth described above. Reverse jetting growth occurs for high flow rates (35 – 200 ml h^{-1}) and requires high density differences (100 – 120 kg m^{-3}) between the cupric sulfate solution in the reservoir and the denser, injected sodium silicate solution. Specifically, we find that solitary tubes form around a continuous, descending jet. Owing to differences in refractive index, this jet can be seen in Figure 11.5a as a very faint line. The smooth and cylindrical precipitation tube appears only slightly darker than the descending fluid jet while the conical base close to the injection nozzle is rather thick and readily discernible.

11.7 Growth Regimes

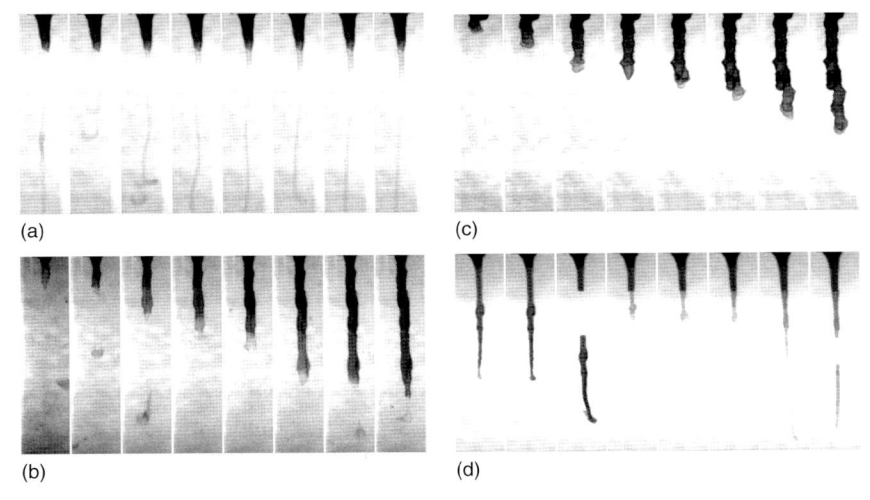

Figure 11.5 Image sequences illustrating four reverse growth regimes observed for the injection of sodium silicate (1.0 M) into cupric sulfate solution. The growth regimes are referred to as (a) *jetting* (b) *popping*, (c) *budding*, and (d) *fracturing*. Flow rates, cupric sulfate concentrations, and density differences are: (a) 50.0 ml h^{-1}, 0.107 M, 110 kg m^{-3}, (b) 30.0 ml h^{-1}, 0.173 M, 100 kg m^{-3}, (c) 5.0 ml h^{-1}, 0.25 M, 90 kg m^{-3}, and (d) 1.1 ml h^{-1}, 0.075 M, and 115 kg m^{-3}. The times elapsed between snapshots are (a) 0.5 s, (b) 12.5 s, (c) 10.0 s, and (d) 15.0 s. All image areas are (1.3 × 5.4) cm^2. Reprinted with permission from Ref. [43].

Reverse jetting behavior ceases and gives way to a qualitatively different form of tube growth if the density difference between the injected and outer solution is smaller than 110 kg m^{-3} and the flow rate is lowered to values between approximately 25 and 185 ml h^{-1}. Figure 11.5b shows a sequence of eight consecutive snapshots that illustrate the dynamics of this reverse popping growth. Most importantly, it involves the periodic release of droplike pieces of colloidal matter that sink to the bottom of the container. For the specific parameters of experiment in Figure 11.5b, we find an average oscillation period of 1.8 s. The average speed of tube growth is 0.30 mm s^{-1} and the average outer radius is 1.2 mm.

The third growth regime, called *reverse budding* in the following discussion, can be observed for even smaller density differences between the two reactant solutions. As illustrated in Figure 11.5c, reverse budding yields bulging, unbranched structures that perhaps are best described as a tubelike chain of hollow nodules. The growth dynamics are similar to that in the reverse popping regime but no detectable fragments of colloidal matter are released. The overall process involves the repetitive generation and expansion of small colloidal droplets. Once these nodules reach a critical size, they burst and nucleate a new nodule thus completing the growth cycle.

The latter three regimes and the resulting macroscopic tube morphologies are similar to the jetting, popping, and budding growth observed earlier for nonreverse conditions. For reverse conditions, however, we succeeded in identifying a fourth distinct regime, which, for the lack of a better term, has been coined *fracturing* growth [43]. The image sequence in Figure 11.5d presents a typical example for this novel type of tube formation. During rather long periods, the tubular structures steadily grow downward although no jet of sodium silicate solution can be detected. The steady growth is interrupted by major break-off events, two of which are captured in the third and eighth snapshots of Figure 11.5d. During these catastrophic events, long, cylindrical segments split off and sink toward the bottom. The break-off events repeat rhythmically but lack a unique period. Also the length of the released fragment can vary significantly. Detailed measurements revealed that the lengths of the broken-off tube segments ΔL and also the times between subsequent break-off events Δt can be described by log-normal distributions. Furthermore, it was found that the maximum of the ΔL distribution is independent of the flow rate Q, while the maximum of the Δt distribution shifts to larger values with decreasing Q. These observations suggest that the stochastic aspects of fracturing growth do not result from external noise but rather from intrinsic features of the precipitation system.

11.8
Wall Composition and Morphology

Qualitative analyses and spectroscopic techniques have shown that the tube walls have a radial composition gradient. The following results exclusively describe tubes formed by injection of buoyant salt solution into a large reservoir of water glass. It was found that the exterior layers of the tube structure are rich in silica while the interior layers are rich in metal hydroxide [44]. Additional information can be obtained from SEM-EDS measurements. The inset of Figure 11.6a shows a micrograph of the exterior surface of a popping tube. This image is accompanied by an EDS spectrum obtained from the small rectangular area marked by the white rectangular box in the inset. The elemental analysis provides a weight percent ratio of $m_{Cu}/m_{Si} = 1.58$ or 57% Cu : 36% Si, thus indicating that the exterior is silica rich. Furthermore, the SEM-EDS of the interior surface of the same tube (Figure 11.6b) yields a ratio of $m_{Cu}/m_{Si} = 12.0$ or 91% Cu : 7% Si. This value indicates that the surface is rich in copper but nearly free of silicon. Both findings provide conclusive evidence that a compositional gradient does indeed exist across the tube wall.

Although some of the internal wall structures can be discerned by optical microscopy of the undried tubes, SEM provides more detailed features [44]. Figure 11.6c shows a SEM image of the tube wall's cross section. The wall extends in the vertical direction and spans the entire frame. The left and the right sides of the wall correspond to the interior and exterior surfaces, respectively. The thickness of the wall is approximately 10 μm. One can clearly distinguish a glassy-looking, bright phase on the right (i.e., the exterior layers) from a coarser, darker one on the left (the interior). It is important to note that the interior surface in Figure 11.6c is extremely rough with attached nodules that are several micrometers in height. Further results from SEM (micrographs not shown) illustrate that these nodules

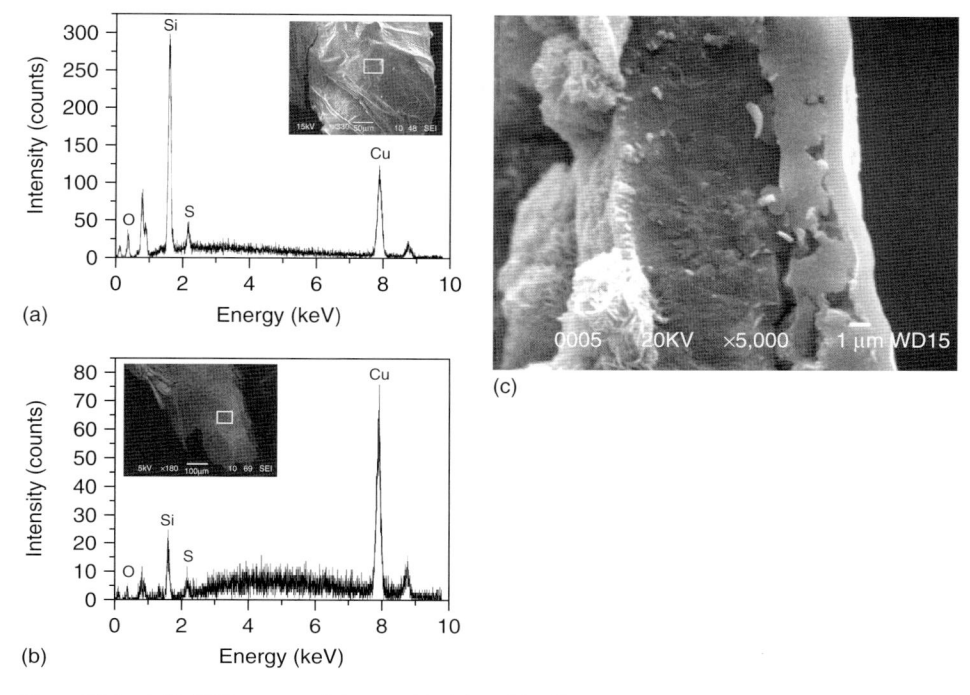

Figure 11.6 (a,b) EDS spectra of the outside (a) and inside (b) surfaces of a popping tube. SEM micrographs of the corresponding surfaces are shown in the insets with white boxes indicating the regions of analysis. (c) SEM micrograph of a tube wall with the exterior surface on the right and the interior on the left. Reprinted with permission from Ref. [44].

are about 10 μm in height and of half-spherical shape. These nodules are either covered by a thin skinlike layer or exposed revealing an interior that is loosely filled with nanorod structures.

The presence of the nodules raises the question whether the tube material contains crystalline regions. Powder X-ray diffraction (XRD) measurements (not shown) reveal that the tube material generated from the popping regime is largely amorphous yet contains crystalline material [44]. Furthermore, by using high-resolution transmission electron microscopy (HRTEM) we can inspect the sample and obtain electron diffraction patterns. Such patterns are used to complement the XRD data for popping tube material. Figure 11.7a shows a small tube fragment with an approximate width of 50 nm. The inset is the corresponding selected area electron diffraction (SAED) pattern. The pattern indicates the presence of a crystalline phase with a d-spacing of 2.45 Å and a cubic crystal structure. Unfortunately, it was not possible to assign this pattern to a specific compound. Consequently, there exists evidence for the presence of small crystalloids provided by HRTEM. An example of this is shown in Figure 11.7b where lattice fringes can be easily discerned. The fringes in Figure 11.7b have a wavelength of 2.5 nm.

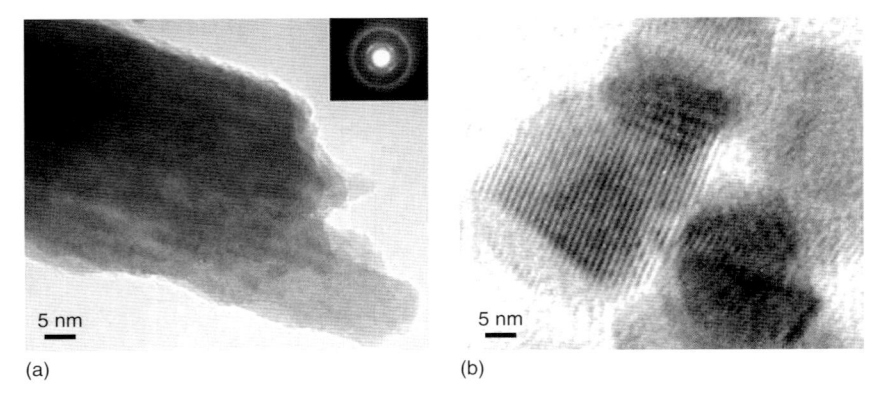

Figure 11.7 HRTEM images of tube fragments from the popping regime. (a) The inset of the HRTEM micrograph is the SAED pattern. (b) The micrograph reveals fringes with a spacing of 2.5 nm. Synthesis parameters are $[CuSO_4] = 0.25$ M, $[Na_2SiO_3] = 1.0$ M, and $Q = 7.0$ ml h^{-1}. Reprinted with permission from Ref. [44].

11.9
Relaxation Oscillations

The oscillatory growth of precipitation tubes is among the most striking, dynamic phenomena in this system. The height evolution of the tube describes sawtoothlike curves [42]. A typical example is shown in Figure 11.8a. In this example, the time between subsequent popping events is approximately 5.6 s. The corresponding height oscillations result in an average growth, which is well described by a constant growth speed. This velocity is proportional to the flow rate and has no strong dependence on the concentration of the injected solution (Figure 11.8b). The average oscillation period T decreases with increasing flow rates Q and the quantities are approximately inversely proportional (Figure 11.8c). Hence, the products vT and QT are nearly constant for a given concentration of cupric sulfate (Figure 11.8d,e). Notice that the product vT is the average growth per period and QT is the volume of solution delivered during one oscillation period. Accordingly, QT approximates the volume of the detaching, critical droplet. This volume increases with increasing concentrations of the cupric sulfate but seems to depend only very weakly on flow rate.

The existence of a critical volume is a key feature of the popping growth as it controls the switching between slow volume expansion of the buoyant droplet and rapid formation of a new membrane at the temporarily uncapped tube. Accordingly, popping silica tubes are typical relaxation oscillators and, in this respect, similar to systems such as the "dripping faucet" [45]. In the latter case, the critical mass of the drop can be approximated by equating the corresponding weight to the surface tension force at the point of detachment. Silica tubes in the oscillatory popping mode can be analyzed analogously by equating the buoyant force of the

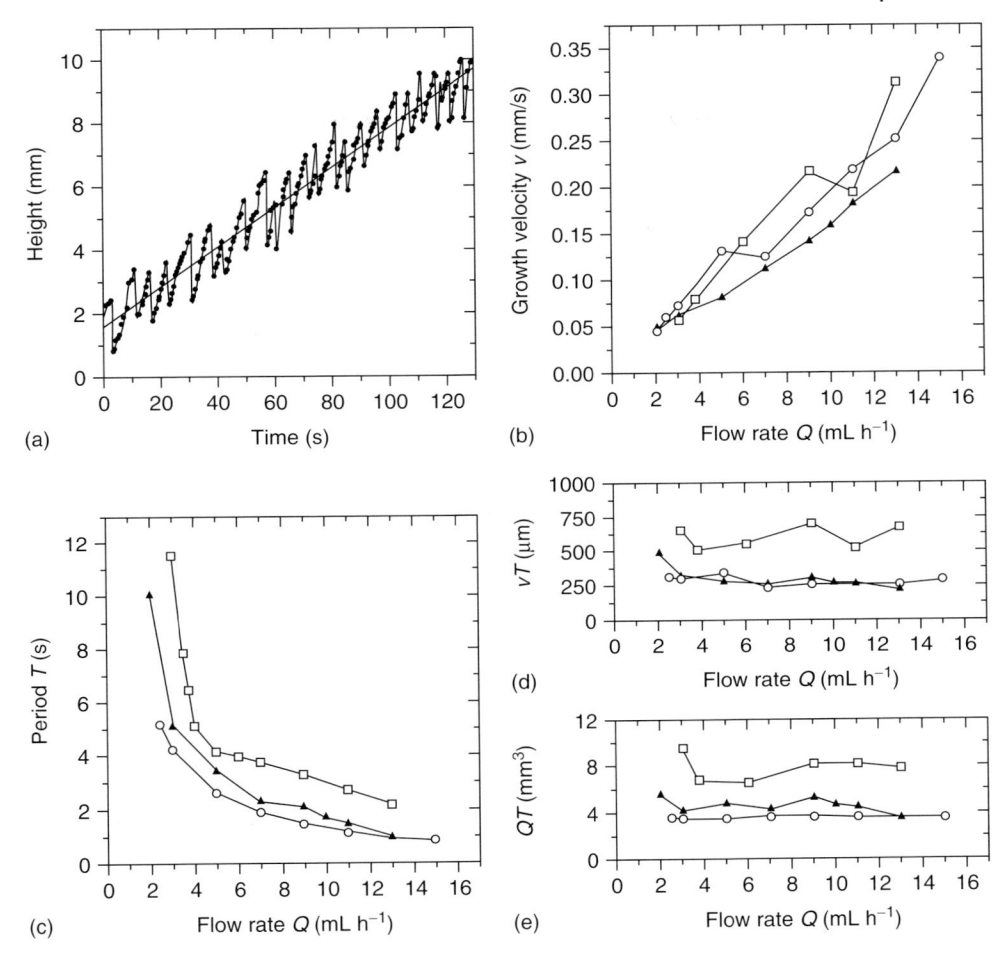

Figure 11.8 Tube growth in the popping mode. (a) Example of the oscillatory height evolution at [CuSO$_4$] = 0.25 M and a flow rate of $Q = 3.0$ ml h^{-1}. In general, this velocity shows a nearly proportional dependence on the flow rate (b). The average oscillation period decreases with increasing flow rates (c) yielding constant values for the growth per period vT (d), and the volume delivered per period QT (e). In (b) through (e), open circles, solid triangles, and open squares indicate data obtained for [CuSO$_4$] = 0.20, 0.25, and 0.30 M, respectively. Reprinted with permission from Ref. [42].

detaching droplet with a constant F_{crit} [42]. We assume that this critical force reflects the physical properties of the droplet's membrane but does not depend on the flow rate. Accordingly, we obtain the equation $gQT\Delta\rho = F_{crit}$, where g denotes earth's gravity (9.8 m s^{-2}) and $\Delta\rho$ is the density difference between the silicate solution, $\rho_2 = 1124$ kg m^{-3}, and the effective density ρ_1 of the droplet. Over the small concentration range for which popping is observed, the density ρ_1 is well described by $\rho_0 + \xi[CuSO_4]$, where ξ is a constant and ρ_0 is the density of pure

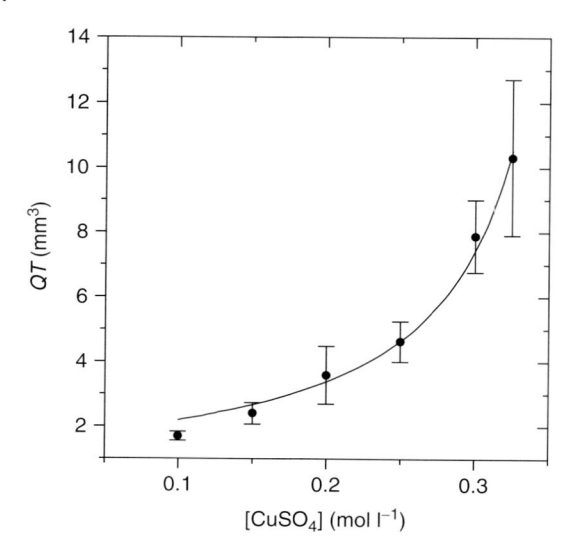

Figure 11.9 The volume of solution delivered during one period of the oscillatory popping events, QT, as a function of the concentration of cupric sulfate. Each data point is the average of several experiments at different flow rates in the range of 2 – 15 ml h^{-1}. QT also measures the maximal volume of the droplets forming at the top of silica tubes. The solid curve represents the fit of the function $QT = 1/(a + b[Cu^{2+}])$ to the experimental data for parameters $a = 0.61$ mm^{-3} and $b = -1.6 \times 10^6$ mol^{-1}. Reprinted with permission from Ref. [42].

water. Thus, we obtain

$$QT = F_{crit}/\{g(\rho_2 - \rho_0 - \xi[CuSO_4])\} \tag{11.1}$$

Least square fitting of the latter expression to our experimental data yields satisfactory results (Figure 11.9). If we further assume that the membrane's contribution to the effective droplet density is small, we find that $F_{crit} = 2$ μN and $\xi = 0.3$ kg mol^{-1}. The latter number differs only slightly from the known value of $\xi' = 0.16$ kg mol^{-1} for pure cupric sulfate solutions. Lastly, the value of F_{crit} might be related to the critical strength of the transient, colloidal membrane. Assuming a wall thickness of $w = 10$ μm, a tube radius of $R = 200$ μm implies a tensile strength of $F_{crit}/(2\pi Rw) = 0.2$ kPa. This value falls within the range of critical membrane stress $\sigma_c(0.1 - 1$ kPa) suggested recently by Pantaleone et al. [39], although the experimental conditions and employed reactants differ significantly.

The latter discussion shows the importance of the interplay between buoyant forces and the physical characteristics of the freshly formed membrane. We note that these factors also control the transition from popping to budding dynamics (see Figure 11.4). However, in the budding regime, the buoyancy of the droplets is too weak to cause their detachment and the continuing expansion is only limited by the occurrence of small ruptures that nucleate new growth buds.

11.10
Radius Selection

In this section, we address the question as to how the precipitation system selects its tube radius. Developing such an understanding is clearly of fundamental interest and also of critical importance for developing a satisfactory understanding of the overall phenomenon. In addition, it might allow engineering microtubes in a controlled fashion for various microfluidic applications.

Figure 11.10 shows the pump-rate dependence of the outer tube radius. Data were obtained from analyses of video images obtained during the growth process [46]. All measurements were carried out at least 1 cm above the injection nozzle to minimize the influence of nozzle-specific artifacts. Furthermore, they reflect the width of the tubes along a range of tube heights to reduce the effect of erratic variations of their radius. The error bars in Figure 11.10 indicate the corresponding variations in terms of the data sets' standard deviations. We emphasize that our experiments show no systematic change in tube radius with height.

The data in Figure 11.10 are obtained for a constant concentration of water glass (1 mol l^{-1}) and for two concentrations of cupric sulfate: 0.075 mol l^{-1} (circles) and 0.15 mol l^{-1} (squares). Over the range of pump rates investigated, these concentrations give rise to jetting and popping growth, respectively.

The measurements show that the tube radius R increases with increasing flow rates Q. Furthermore, we find that popping tubes are always wider than those

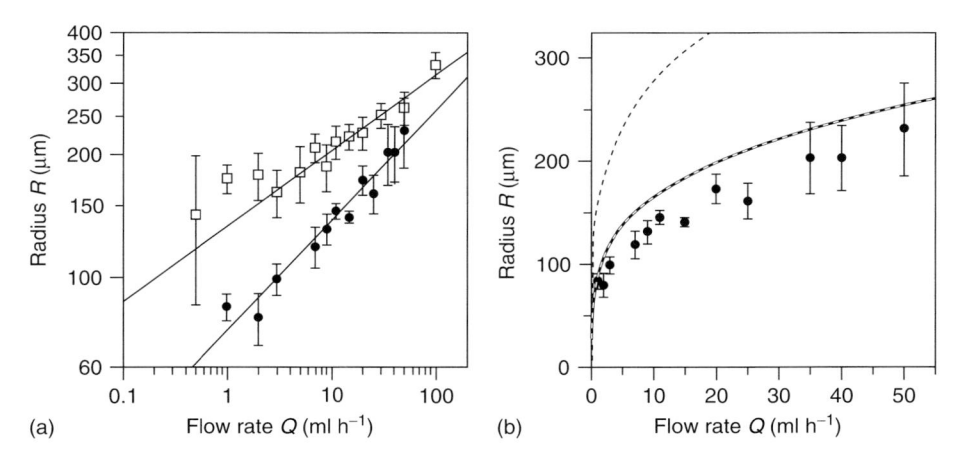

Figure 11.10 Dependence of the outer radius of silica tubes on the employed volume flow rate. Solid circles and open squares indicate seed concentrations of 0.075 and 0.15 mol l^{-1}, respectively. The lines in (a) are the results of linear regressions yielding power law exponents of 0.27 and 0.18 for jetting and popping growth, respectively. The jetting data are shown again in (b) along with the solutions of purely hydrodynamic equations. The solid curve and the nearly identical dashed line correspond to a model without adjustable parameters (cf., Eq. (11.2)) with the latter curve being the approximation specified in Eq. (11.3). The upper dashed curve corresponds to the stiff tube model (i.e., Eq. (11.3) for $\eta_0 \to \infty$). Reprinted with permission from Ref. [46].

grown for jetting conditions. The double-logarithmic plot (Figure 11.10a) suggests simple power laws for both sets of data. The corresponding power law exponents are 0.27 ± 0.02 and 0.18 ± 0.02 for jetting and popping growth, respectively. Up to now there are no measurements for budding tubes mainly because their overall structure hinders accurate characterizations. Qualitatively, however, budding tubes are much wider than those formed under popping and jetting conditions.

We now discuss that in the jetting regime radius selection is dominated by simple hydrodynamics [46]. We assume that the shear stress and the velocity difference across the membrane can be neglected. For the analysis of the hydrodynamic flow profile, we can hence focus on the fluid motion inside and outside of the tube. In the experiments shown in Figure 11.10, the outer silicate solution is confined to a glass cylinder of radius R_{cyl}(1.1 cm). Along its central axis, buoyant cupric sulfate solution ascends as a cylindrical jet of radius R. The cylindrically symmetric velocity fields $v(r)$, which solve the Navier–Stokes equations, are

$$v_o = A_o \ln r + B_o + C_o r^2, \qquad v_i = B_i + C_i r^2 \tag{11.2}$$

where the indices "o" and "i" denote the outer and inner fluids, respectively. Notice that there is no logarithmic term in the expression for v_i, because the velocity in the jet cannot diverge. The five constants in Eq. (11.2) are all fixed by imposing the proper boundary conditions: (i) stick boundary condition at the container wall $v_o(R_{cyl}) = 0$, (ii) continuity of velocity, shear stress, and pressure at the fluid interface $r = R$, (iii) no-mean-flow constraint for the outer fluid (i.e., the overall volume of the silicate solution is constant). Hence, for a given set of radii (R), viscosities ($\eta_{i,o}$), and densities ($\rho_{i,o}$), the flow fields are fully determined. From $v_i(r)$, we can therefore derive the volume flow Q through the jet. Since all relevant fluid properties and the cell size are known, one finds a unique relation between Q and R. This relation can be formulated algebraically but is very cumbersome and, hence, yields little physical insight. However, in the relevant limit $R_{cyl} \gg R$, it simplifies to

$$Q = \pi g(\rho_o - \rho_i)/(8\eta_i) \ R^4 \{1 + (4\eta_i/\eta_o) \ [\ln(R_{cyl}/R) - 1]\} \tag{11.3}$$

This equation expresses simple Poiseuille flow through a pipe but corrects it for jet-driven motion in the outer liquid.

The solid black line in Figure 11.10b shows the result of the full calculation based on Eq. (11.2) and the constraint of no-mean-outer flow. Notice that this description involves *no adjustable parameters*. To the contrary, all quantities can be measured independently or have been reported elsewhere [47–49]. For jetting tubes, the agreement between the experimental data and the theoretical curve is excellent. The dashed line in the figure shows the result of the approximation as stated in Eq. (11.3). Notice that two curves are essentially identical. However, Poiseuille flow without the logarithmic correction (i.e., the expression in Eq. (11.3) up to the term R^4) is too steep and clearly not a satisfactory description. Nonetheless, the latter graph is shown in Figure 11.10b as the upper dashed curve. Lastly, we emphasize that none of the models accounts for the radii of popping tubes.

11.11
Bubbles as Templates

Conventional, crystal-seeded precipitation tubes sometimes have small air bubbles attached to their growth tip. One finds that such tubes tend to be thinner and more linear than their bubble-free counterparts. Thouvenel-Romans et al. [50, 51] took this observation as motivation to investigate the growth of bubble-guided and bubble-templated tubes in controlled experiments that again relied on the aforementioned injection of salt solution into a large reservoir of silicate solution. The experimental setup used for these studies involves a simple stainless steel needle, which is introduced into the solution-delivering capillary with its tip slightly protruding out. Connected to the needle is an air-filled syringe allowing the manual injection of single air bubbles into the flow of solution.

If placed correctly, the small air bubble entrains the formation of an upright growing silica tube. A typical example is shown in Figure 11.11. The growth rate and the tube radius are remarkably constant (here $0.38\,\text{cm}\,\text{s}^{-1}$ and $240\,\mu\text{m}$, respectively). Notice that this bubble-guided tube growth is much faster than that in the bubble-free experiments. However, the velocity is clearly slower than the terminal speed of free rising bubble of comparable radius. Consequently, the gas bubble acts as a buoyant guide for the tube formation and is also pinned to the reaction zone through interfacial tension.

Systematic measurements show that the radius of the precipitation tubes is strongly affected by the size of the bubble. Typically, the ratio of tube to bubble radii varies between about 0.8 and 0.95. Within this range, the ratio increases with increasing pump rate Q, which might indicate some interplay between the

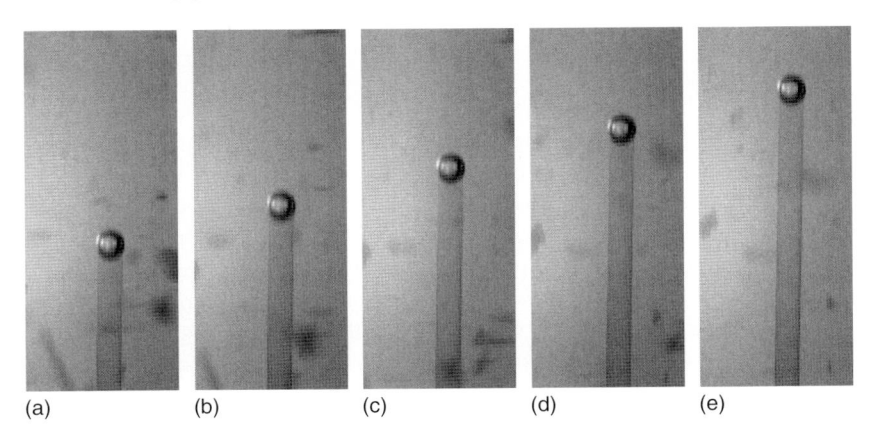

Figure 11.11 Image sequence illustrating the growth of a precipitation tube directed by an air bubble ($R_{bubble} = 0.31$ mm). The concentration of cupric sulfate and water glass are 0.5 and 1.0 M, respectively. Flow rate: 3.0 ml h^{-1}; time between frames: 0.2 s; field of view: 3.0×7.3 mm^2. Reprinted with permission from Ref. [50].

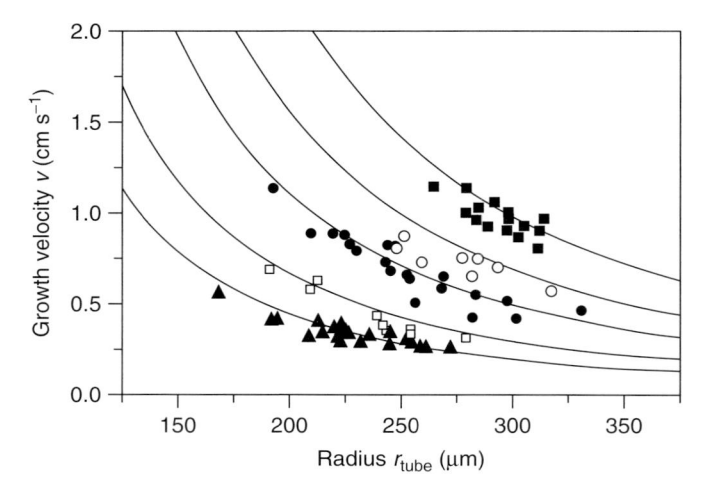

Figure 11.12 Growth speed of bubble-guided tubes versus tube radius. The cupric sulfate solution (0.5 M) is injected at constant flow rates of 10.0 ml h^{-1} (■), 7.0 ml h^{-1} (○), 5.0 ml h^{-1} (●), 3.0 ml h^{-1} (□), and 2.0 ml h^{-1} (▲). The silicate concentration is 1.0 M. The black curves assume volume conservation of the injected liquid in the tubular structure ($v = Q/\pi r_{tube}^2$). Reprinted with permission from Ref. [50].

buoyant force of the bubble and its pinning forces to the forming tube. However, there is currently no detailed understanding of radius selection mechanism in this system.

Another important result of the study by Thouvenel-Romans et al. [50] is that the volume of injected solution per unit time Q is essentially identical to the rate of volume increase of the tube. The latter quantity can be expressed as the product of the tube's cross-sectional area πr^2 and the growth velocity $v = dh/dt$, where h denotes the tube height at time t. As shown in Figure 11.12, the resulting equation, $v = Q/(\pi r^2)$, is in excellent agreement with the experimental data. This finding indicates that there is no or very little leakage of salt solution into the water glass. Consequently, the bubble acts as a sufficiently tight seal between the reacting liquids. However, the literature also notes that sometimes the intricate contact between the bubble and growing tube gets upset [50]. In such cases, the bubble typically shifts its position relative to the reaction zone in upward or downward direction or detaches completely from the tube. Regardless of the details, such a perturbation causes the tube to switch to its bubble-free growth behavior, which we discussed in Section 11.5.

In conclusion, we reiterate that the growth of precipitation tubes is readily affected by buoyant gas bubbles. In such situations, the tube radius is selected by the radius of the bubble and its growth velocity follows from simple volume conservation of the injected solution. Clearly, additional work is needed to unravel the detailed mechanism of bubble pinning to the nonequilibrium but steady reaction zone at the top region of the growing tube.

11.12
Toward Applications

As mentioned in Section 11.1, precipitation tubes have profound potential for applications in microfluidics, materials science, and catalysis. In 2003, our group demonstrated the bridging of two glass capillaries using a self-directing silica tube [42]. This experiment is illustrated in Figure 11.13. The capillaries are submerged in 1 M silicate solution and the lower one injects buoyant cupric sulfate solution into the system. The capillary tube withdraws solution at a rate that matches the delivery rate of the lower one. This arrangement induces tube growth from the lower toward the upper orifice and the tube eventually connects both capillaries by forming a small pluglike structure at the withdrawing port. Subsequent injections of a dye into the continuing flow of cupric sulfate solution reveal no leakage into the silicate solution, thus, demonstrating that the self-organized connection is fully closed to the outside while allowing flow between the two glass capillaries.

More recently, Cronin et al. [52, 53] studied tube formation from small metal-oxide-based inorganic grains. Specifically, they showed that the direction of microtube growth can be controlled by externally applied electric fields. This very interesting approach is illustrated in Figure 11.14. The underlying mechanism is unknown but might involve electromigration of ions within the surrounding solution or possibly large-scale fluid flow.

Combining approaches similar to the ones discussed above could yield powerful means to connect specific locations with intricate networks of hollow microtubes. However, silica tubes might also introduce tailored chemical features to such transport systems. For instance, Pagano et al. [51] showed in 2008 that precipitation tubes obtained from reactions between silicate and zinc sulfate solutions can be converted to silica–ZnO structures using moderate heating. During the heating

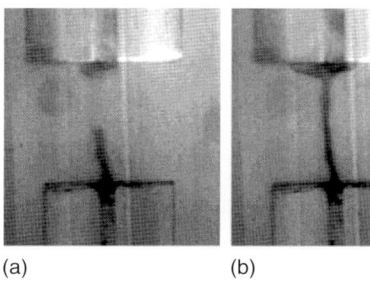

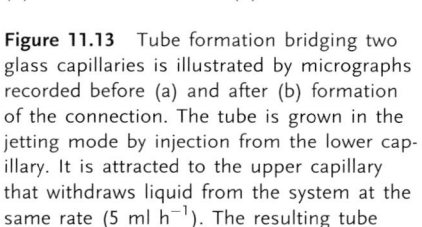

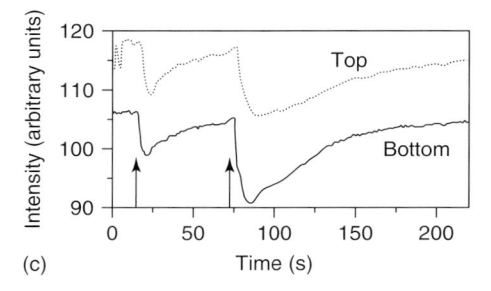

Figure 11.13 Tube formation bridging two glass capillaries is illustrated by micrographs recorded before (a) and after (b) formation of the connection. The tube is grown in the jetting mode by injection from the lower capillary. It is attracted to the upper capillary that withdraws liquid from the system at the same rate (5 ml h^{-1}). The resulting tube has a diameter of 240 µm. Field of view 9.3 × 10.7 mm^2. (c) Injections of rhodamine solution into the flow of copper sulfate (see arrows) lead to temporarily decreasing intensities within the lower and upper glass capillaries, demonstrating that the silica tube forms a closed connection. Reprinted with permission from Ref. [42].

238 | *11 Structure Formation in Inorganic Precipitation Systems*

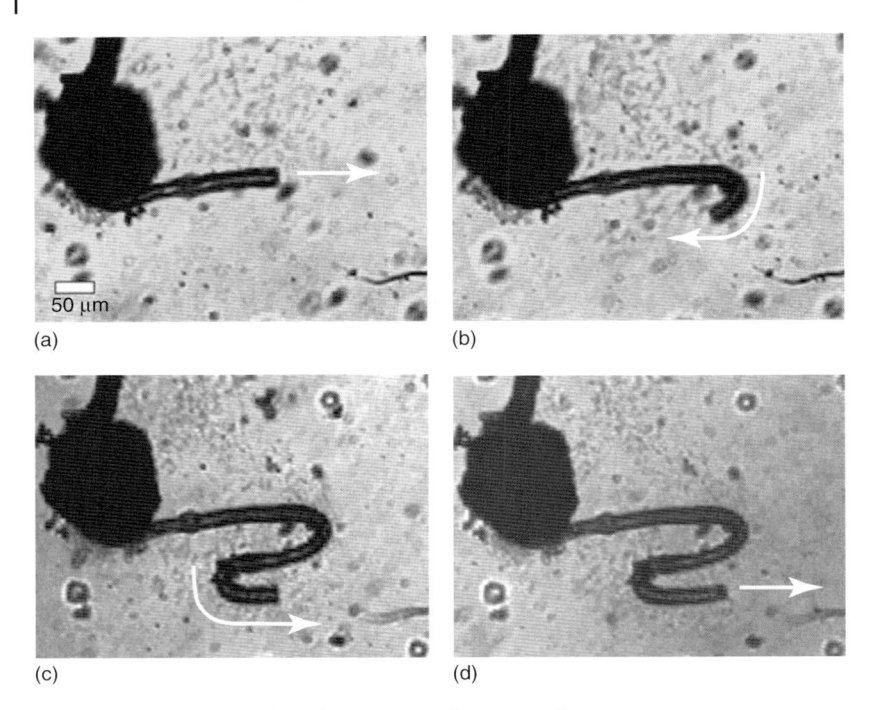

Figure 11.14 Tube growth in the presence of an externally applied electric field. Directional changes are induced by changing the polarity of the employed electrodes. The four subsequent snapshots are taken over a period of approximately 3 min. Reprinted with permission from Ref. [52].

process, the mechanical integrity and overall tubular shape of the precipitation structures are conserved. The resulting samples have photocatalytic activity. This feature was demonstrated for the photodegradation of an organic dye. In addition, ultraviolet light with a wavelength of 325 nm causes the heat-treated tubes to emit band-edge luminescence near 415 nm (3.0 eV), accompanied by a broad green emission around 522 nm (2.4 eV).

11.13
Outlook and Conclusions

The past decade has seen profound progress toward a satisfying and comprehensive understanding of tube formation from inorganic precipitation reactions. In particular, open-tube growth appears to be a solvable problem today as we have already developed a quantitative understanding of radius selection in this system. The two most critical remaining questions regarding open-tube growth relate to the control of growth speeds and the wall thickness. It is likely that both factors are tightly coupled to the diffusion and advection processes in this system, while specific chemical rates are of lesser importance as long as precipitation and polymerization processes are fast.

For closed tube-growth, numerous unsolved problems remain requiring further investigation. Quantitative experiments have shown that these tubes essentially conserve the volume of injected solutions within their expanding structures. In addition, semiquantitative models have been proposed by Thouvenel-Romans et al. [42] and Pantaleone et al. [38] to account for specific features of popping and budding tube growth, respectively. However, there are no quantitative models available today that address tube formation based on detailed reaction and transport processes. A major obstacle toward the study of such models relates to the involved free-boundary problems that affect fluid motion, transport, and possibly reaction rates in a nontrivial fashion.

Another unsolved problem concerns the predictability of tube formation from the underlying reaction mechanisms. In other words, we currently have no theory that would allow us to predict if a certain reaction could potentially induce tube formation. This question is of particular interest as one considers the possibility of forming hollow tubes from organic polymers.

The hope for technologically useful applications has been a driving force of many of the studies that we reviewed in this chapter. These efforts involve a variety of approaches, such as catalysis, microfluidics, and chemical sensors, but clearly work is still in its infancy. It will be interesting to see how these efforts take fruition in the years to come. Explaining and using one of chemistry's oldest experimental systems is definitely an exciting task worth pursuing.

Acknowledgments

This work was supported by the National Science Foundation and the Petroleum Research Fund (ACS-PRF). We thank Stephanie Thouvenel-Romans, Tamás Bánsági, Jr., and Rabih Makki for discussions.

References

1. Ball, P. (1999) *The Self-Made Tapestry: Pattern Formation in Nature*, Oxford University Press, Oxford.
2. Nicolis, G. and Prigogine, I. (1989) *Exploring Complexity*, Freeman, New York.
3. Kapral, R. and Showalter, K. (eds) (1995) *Chemical Waves and Patterns*, Kluwer, Dordrecht.
4. Köhler, J.M. and Müller, S.C. (1995) Frozen chemical waves in the Belousov-Zhabotinsky reaction. *J. Phys. Chem.*, **99**, 980–983.
5. Washington, R.P., West, W.W., Misra, G.P., and Pojman, J.A. (1999) Polymerization coupled to oscillating reactions: (1) a mechanistic investigation of acrylonitrile polymerization in the Belousov-Zhabotinsky reaction in a batch reactor. *J. Am. Chem. Soc.*, **121**, 7373–7380.
6. Liesegang, R.E. (1896) Über einige Eigenschaften von Gallerten. *Naturwiss. Wochenschr.*, **11**(30), 353–362.
7. Grzybowski, B.A., Bishop, K.J.M., Campbell, C.J., Fialkowski, M., and

Smoukov, S.K. (2005) Micro- and nanotechnology via reaction-diffusion. *Soft Matter*, **1**, 114–128.

8. Martens, K., Bena, I., Droz, M., and Racz, Z. (2008) Encoding information into precipitation structures. *J. Stat. Mech.: Theory Exper.*, **P12003**, 1–12.

9. Raub, E. and Schall, A. (1938) Silber-Indium-Legierungen. *Z. Metallkunde*, **5**, 149–151.

10. Raub, E. and Müller, K. (1967) *Fundamentals of Metal Deposition*, Elsevier, Amsterdam.

11. Krastev, I. and Koper, M.T.M. (1986) Pattern formation during the electrodeposition of a silver-antimony alloys. *Phys. A*, **13**, 199–208.

12. Saltykova, N.A. *et al.* (1993) Abstract Book of the 44th Meeting of the International Society of Electrochemistry, Berlin, p. 376.

13. Web Site of the Woods Hole Oceanographic Institution, Deep Submergence Operations Group, Dan Fornari.

14. Andrieux, C. (1965) Morphogenese des helictites monocristallines. *B. Soc. Fr. Mineral Cr.*, **88**, 163.

15. Double, D.D., Hellawell, A., and Perry, S.J. (1978) Hydration of Portland cement. *Proc. R. Soc. Lond.*, **A359**, 435–451.

16. Fontana, M.G. and Greene, N.D. (1967) *Corrosion Engineering*, McGraw-Hill, New York.

17. For related phenomena see: Stone, D.A. and Goldstein, R.E. (2004) Tubular precipitation and redox gradients on a bubbling template. *Proc. Natl. Acad. Sci. U.S.A.*, **101**, 11537–11541.

18. Mann, S. (1993) Molecular tectonics in biomineralization and biomimetic materials chemistry. *Nature*, **365**, 499–505.

19. Young, J.R. and Henriksen, K. (2003) Biomineralization within vesicles: the calcite of coccoliths. *Rev. Mineral. Geochem.*, **54**, 189–215.

20. Mann, S. (2001) *Biomineralization: Principles and Concepts in Bioinorganic Materials Chemistry*, Oxford University Press, Oxford.

21. Coatman, R.D., Thomas, N.L., and Double, D.D. (1980) Studies of the growth of silicate gardens and related phenomena. *J. Mater. Sci.*, **15**, 2017–2026.

22. Leduc, S. (1911) *The Mechanism of Life*, Rebman, London.

23. Roesky, H.W. and Möckel, K. (1996) *Chemical Curiosities*, VCH, Weinheim.

24. Cartwright, J.H.E., García-Ruiz, J.M., Novella, M.L., and Otálora, F. (2002) Formation of chemical gardens. *J. Colloid Interface Sci.*, **256**, 351–359.

25. Balköse, D., Özkan, F., Köktürk, U., Ulutan, S., Ülkü, S., and Níslí, G. (2002) Characterization of hollow chemical garden fibers from metal salts and water glass. *J. Sol-Gel Sci. Technol.*, **23**, 253–263.

26. Jones, D.E.H. and Walter, U.J. (1998) The silicate garden reaction in microgravity: a fluid interfacial instability. *J. Colloid Interface Sci.*, **203**, 286–293.

27. Iler, R.K. (1979) *The Chemistry of Silica: Solubility, Polymerization, Colloid and Surface Properties and Biochemistry of Silica*, John Wiley & Sons, Ltd, Chichester.

28. Photo taken by O. Steinbock (July 10th, 2009) in Tallahassee, Florida.

29. Glauber, J.R. (1646) *Furni Novi Philosophici*, Amsterdam.

30. Mann, T. (1947) *Doktor Faustus*, Bermann-Fischer, Stockholm.

31. Sacks, O. (2001) *Uncle Tungsten: Memories of a Chemical Boyhood*, Vintage Books.

32. Ball, P. (2006) Alchemy Isaac Newton's Curse? New Scientist (Apr. 8), pp. 47–49.

33. Collins, C., Zhou, W., Mackay, A.L., and Klinowski, J. (1998) The silica garden: a hierarchical nanostructure. *Chem. Phys. Lett.*, **286**, 88–92.

34. Collins, C., Mokaya, R., and Klinowski, J. (1999) The silica garden as a bronsted acid catalyst. *Phys. Chem. Chem. Phys.*, **1**, 4669–4672.

35. Collins, C., Mann, G., Hoppe, E., Duggal, T., Barr, T.L., and Klinowski, J. (1999) NMR and ESCA studies of the silica garden bronsted acid catalyst. *Phys. Chem. Chem. Phys.*, **1**, 3685–3687.

36. Collins, C., Zhou, W., and Klinowski, J. (1999) A unique structure of $Cu_2(OH)_3NH_3$ crystals in the 'silica

garden' and their degradation under electron beam irradiation. *Chem. Phys. Lett.*, **306**, 145–148, See also erratum Ibid. 1999, **312**, 346.
37. van Bommel, K.J.C., Friggeri, A., and Shinkai, S. (2003) Organic templates for the generation of inorganic materials. *Angew. Chem. Int. Ed.*, **42**, 980–999.
38. Pantaleone, J. *et al.* (2008) Oscillations of a chemical garden. *Phys. Rev.*, **E77**, 046207.
39. Pantaleone, J., Toth, A., Horvath, D., RoseFigura, L., Morgan, W., and Maselko, J. (2009) Pressure oscillations in a chemical garden. *Phys. Rev.*, **E79**, 056221.
40. Maselko, J., Geldenhuys, A., Miller, J., and Atwood, D. (2003) Self-construction of complex forms in a simple chemical system. *Chem. Phys. Lett.*, **373**, 563–567.
41. Maselko, J., Borisova, P., Carnahan, M., Dreyer, E., Devon, R., Schmoll, M., and Douthat, D. (2005) Spontaneous formation of chemical motors in simple inorganic systems. *J. Mater. Sci.*, **40**, 4671–4673.
42. Thouvenel-Romans, S. and Steinbock, O. (2003) Oscillatory growth of silica tubes in chemical gardens. *J. Am. Chem. Soc.*, **125**, 4338–4341.
43. Pagano, J.J., Bánsági, T., and Steinbock, O. Jr. (2007) Tube formation in reverse in silica gardens. *J. Phys. Chem.*, **C111**, 9324–9329.
44. Pagano, J.J., Thouvenel-Romans, S., and Steinbock, O. (2007) Compositional analysis of copper-silica precipitation tubes. *Phys. Chem. Chem. Phys.*, **9**, 110–116.
45. Wu, X.M. and Schelly, Z.A. (1989) The effects of surface tension and temperature on the nonlinear dynamics of the dripping faucet. *Phys. D*, **40**, 433–443.
46. Thouvenel-Romans, S., van Saarloos, W., and Steinbock, O. (2004) Silica tubes in chemical gardens: radius selection and its hydrodynamic origin. *Europhys. Lett.*, **67**, 42–48.
47. (1958) *Gmelins Handbuch der Anorganischen Chemie*, 8th edn, vol. 60, Verlag Chemie, Weinheim, p. 531.
48. (1967) *Gmelins Handbuch der Anorganischen Chemie*, 8th edn, vol. 21, Verlag Chemie, Weinheim, p. 1526.
49. Roth, W. A. and Scheel, K. (eds) (1943) *Landolt-Bornstein – Physikalisch-chemische Tabellen*, 5th edn, Edwards Brothers, Ann Arbor, MI, p. 136.
50. Thouvenel-Romans, S., Pagano, J.J., and Steinbock, O. (2005) Bubble guidance of tubular growth in reaction–precipitation systems. *Phys. Chem. Chem. Phys.*, **7**, 2610–2615.
51. Pagano, J.J., Bánsági, T., and Steinbock, O. Jr. (2008) Bubble-templated and flow-controlled synthesis of macroscopic silica tubes supporting zinc oxide nanostructures. *Angew. Chem. Int. Ed.*, **47**, 9900–9903.
52. Ritchie, C. *et al.* (2009) Spontaneous assembly and real-time growth of micrometre-scale tubular structure from polyoxometalate-based inorganic solids. *Nat. Chem.*, **1**, 47–52.
53. Cooper, G.J.T. and Cronin, L. (2009) Real-time direction control of self fabricating polyoxometalate-based microtubes. *J. Am. Chem. Soc.*, **131**, 8368–8369.

Index

a

absorption bands, C=O stretching 29
acetate, vinyl 16
acids, *see* malonic acid
– acid core 183
– α-alkylacrylic 210
– gluconic 200–201
– methacrylic 200, 209–210
– nitric 120
acrylamide/bisacrylamide polymerization 53
acrylates
– butyl 54
– ethyl 101
– multifunctional 58
acryloylated sulfamethoxypyridazine (SMPA) 209
actuators 122, 125
adenosine diphosphate (ADP) 198
adhesives 50–51
adiabatic reactors, cylindrical 56
agarose 183
AIBN 84–87
air bubbles 235
"alarm signal", chemical 135
α-alkylacrylic acids 210
alloy, silver–antimony 221
amine hydrochloride, lauryl 224
antagonistic interactions 13
antimony 221
aperiodic behavior 8
applications
– frontal polymerization 49–51, 61–62
– inorganic precipitation systems 237–238
arbitrary symmetry 105–109
arginine ethyl ester, benzyl 196–197
arrays, self-oscillating gel 126–127

aspect ratio 147
autocatalysis 7, 175–176
– hydroxyl ion-reactions 182
autonomous directed motion 154
axial GRINs 72
axisymmetric mode of convection 53

b

BAEE (benzyl arginine ethyl ester) 196–197
beads
– microgel 129–130
– nanogel 174
– spherical 169
Belousov–Zhabotinsky (BZ) reaction 7, 14–15, 116–117
– coupling to 169–171
– homogeneous gels 143–147
– straining heterogeneous BZ gels 147–154
Benard instability, Rayleigh 91
benzoyl peroxide 48
benzyl arginine ethyl, ester (BAEE) 196–197
bifurcation
– Hopf 149, 192
– parameter 53
bilayers 196
binary mixtures, (non)reactive 92–97
binodal line 93
biomimetic mechanochemical response 147
biomimetic micro-/nanoactuators 122
biomimetic soft materials 135–162
bisacrylamide, polymerization 53
bistability
– hydrogel membrane permeability 201–204
– reactions 164
– spatial 174–175
bisulfite 15
black smokers, venting 222
"bottleneck" shape 184

Nonlinear Dynamics with Polymers: Fundamentals, Methods and Applications.
Edited by John A. Pojman and Qui Tran-Cong-Miyata
Copyright © 2010 WILEY-VCH Verlag GmbH & Co. KGaA, Weinheim
ISBN: 978-3-527-32529-0

boundary conditions 74
Brabender internal mixer 38
breakdown of the front 75
Briggs–Rauscher reaction 220
brittle disruption, self-sustained 46
bromate, sodium 120
bromate–sulfite (BS) reaction 164, 184–185
bromide ions, autonomous transport 154
bromide-sensitive electrode 8
bromine dioxide 15
bubbles 58–59
– air 235
– as templates 235–236
– bubble-free growth 236
budding regime 226
buffering capacity 207
building blocks, molecular 219
bulk gel, miniature 117–119
bulk state of polymers 97–98
buoyancy 59
– buoyancy-driven convection 52–55
– cupric sulfate solution 225
– forces 232
butanediol 29
butyl acrylate polymerization 54
BZ (Belousov–Zhabotinsky) reaction 7, 14–15, 116–117
– coupling to 169–171
– homogeneous gels 143–147
– straining heterogeneous BZ gels 147–154

c

C=O stretching absorption bands 29
cables, fiber-optic 73
CAI (computer-assisted irradiation) method 105–106
capillaries, glass 237
capturing effects 143–147
catalase 198–200
catalysts
– auto- 7, 175–176, 182
– concentration 27
– Grubbs 45
– photocatalytic activity 238
– polymer-tethered 147
ceric ion 14
chains
– extenders 33–34
– extension reactions 39
– Gaussian 166
– (non)linear 22–25
– self-oscillating polymer 127–128
channels, rhomboidal 224
chaotic behavior 8

chaotic mixing 31–32
– rotors 37–38
Chapman theory, Gouy– 192
characteristic length scales 106–109
chemical "alarm signal" 135
chemical dynamics, nonlinear 6
chemical oscillations 7
– see also oscillations
chemical potential 171
chemical reactions, see reactions
chemical waves, propagation 119–120, 126–127
chemodynamic oscillations 174–181
chemoelastodynamics, responsive gels 163–188
chemomechanical coupling parameter 151
chemomechanical excitability wave 184
chemomechanical instabilities 174–175, 182
chemoresponsive BZ gels 146
chlorinated polyisoprene 105
chlorite–iodide–malonic acid reaction 9, 220
chlorite–tetrathionate (CT) reaction 164, 182–184
chlorite–thiosulfate reaction 9
clock reaction, formaldehyde–bisulfite–sulfite 15
close-packed colloidal crystal 121
closed tube-growth 239
co-continuous morphology 102–103
coatings 51
– "skinlike" 136
cobalt–hematoporphyrin compound 84
coil 193
colloidal crystal silica 130
color changes, structural 121–122
comonomer 204
complexes 123
– ruthenium 116–117, 170
computer-assisted irradiation (CAI) method 105–106
conical geometry 185
constant shear rate 30
continuous irradiation 108
continuous stirred tank reactor (CSTR) 163, 174–175, 180–182
– membrane oscillators 189
– oscillations in 16
continuous structures, spatially graded 101–105
continuum equations 137–141
control
– hydrophobicity 170
– oscillating behaviors 119

Index

– pH- 131
– propagation of chemical waves 126–127
convective instabilities 52–56
conversion factor 24
copolymerization, pentaerythritol tetra-acrylate 55
copolymers 116, 131
core, acid 183
coupling parameter, chemomechanical 151
cover slips 104
critical solution temperature 92
cross-linked polymers 10
cross-linker reagent 106
crosscut glass wafer 194
crown ether 194
crystals
– close-packed colloidal 121
– cubic structure 229
– organic 223
CT (chlorite–tetrathionate) reaction 164, 182–184
cumene hydroperoxide 60
cupric sulfate 225–235
cure-on-demand putty 49–50
cylindrical adiabatic reactors 56
cylindrically symmetric velocity fields 234

d

1D-gels, symmetric 148
1D-model, gels 137
3D-frontal polymerization 60–61
Darcy law, generalized 165
decay, exponential 5
decomposition 73
– spinodal 96
deflection technique, laser sheet 79
deformation, mechanical 143–154
deformation-sustaining hydrogels 208
degree of ionization 178
degree of polymerization (DP) 23
deposition processes, inorganic 220–221
design
– biomimetic micro-/nanoactuators 122
– mixer 27
– "pared down" 189
– self-oscillating gels 116–117
diacrylate, 1,6-hexanediol 60
dicyclopentadiene 45
diffusion
– double diffusion instability 91
– molecular 36–38
– polymer–solvent interdiffusion 157

– reaction–diffusion equations 168
– reaction–diffusion patterns 163
– side-by-side diffusion cell 202
diisocyanate–poly(ethylene glycol) reactions 28
diluent, neutral 138
dimethacrylate, triethylene glycol 48
diphosphate, adenosine 198
directed motion, autonomous 154
dispersion relation 96
dispersions 15
dispersive mixing 30
dissipation effects 167
dissipative structures 6
distributive mixing 30
"Doktor Faustus" 223
Donnan equilibrium 178–179
Donnan partitioning 207–209
Donnan potential 191–194, 200–203
dopant concentration 72
dopant-IFP 87
double diffusion instability 91
DP (degree of polymerization) 23
"dripping faucet" 230
droplets
– chain extender 34–35
– membranes 231
dynamic inhomogeneity 97
dynamics
– chemoelasto- 163–188
– nonlinear, see nonlinear dynamics
– polymer network 138

e

effects
– capturing 143–147
– complex kinetics 57–58
– dissipation 167
– gel 11
– hydrating 139
– mixing 29–32
– Norrish–Trommsdorff 11
– surface-tension-driven convection 55
– Trommsdorff, see Trommsdorff effect
elastic strain 99–101
elasticity, neo-Hookean 141–142
electrodeposition 221
electrodes
– bromide-sensitive 8
– platinum redox 8–9
electromigration 237
electroneutrality 178–179
electroosmotic properties 199
electrostatic energy 177–178

elements, finite   140
elongation rate   30
energy
– electrostatic   177–178
– Gibbs free   93–95
enzymes, membrane/enzyme oscillators   196–211
epoxy resins   46
equilibrium
– mechanical   165, 178
– thermodynamic   91
ester, benzyl arginine ethyl   196–197
etching, wet   130
ether, crown   194
ethoxylate triacrylate, trimethylolpropane   50
ethyl acrylate   101
ethyl ester, benzyl arginine   196–197
ethylene glycol   28
excitability wave   184
exothermic reactions   78
expanding front, spherically   61
exponential growth/decay   5
extension reactions, chains   33–34, 39
extensional flow   30

f
feedback   6–7
– geometric   174–181
– sources   10–12
fiber-optic cables   73
films
– poly(ethyl acrylate)   101
– thin   46
finite-element approach   140
FKN mechanism   117
"flip-flop" behavior   192
Flory–Huggins parameter   139
Flory–Huggins polymer solution theory   165–169
flow
– extensional   30
– large-scale fluid   237
– Poiseuille   234
flow branch   182
fluctuations
– long-wavelength   94
– pH   205
fluid element   34–35
fluid flow, large-scale   237
fluorescence imaging   105
flux, permeation   167
forces
– buoyant   232
– linear springlike   141–142

– nodal   142
– pinning   236
formaldehyde–bisulfite–sulfite clock reaction   15
formation of structures   219–241
fracturing growth   228
free energy, Gibbs   93–95
free-radical-growth   21
free-radical polymerization   11, 74
free-radical systems   46
free-volume distribution   97
free-radical polymerization, front   47
frequency, oscillation   150–151
friction coefficient   177
fritted glass plate   191
frontal polymerization (FP)   45–67
– applications   61–62
– isothermal, see isothermal FP
– nonlinear dynamics   51–52
– thermal   69
– three-dimensional   60–61
frontal Rayleigh number   52
functional surfaces   128

g
gardens, silica   221–222
Gaussian chains   166
gels
– chemoresponsive   146
– copolymer   116
– 1D-model   137
– gel effect   11
– gel lattice spring model (gLSM)   136, 142–143
– gel point   24
– heterogeneous BZ   147–154
– homogeneous BZ   143–147
– interfacial-gel polymerization   70
– lauryl amine hydrochloride   224
– membrane   122
– micro-   128–129
– miniature bulk   117–119
– nano-   174
– nonchemoresponsive   183
– oscillatory dynamics   166
– oscillatory reactions   115–134
– passive   171–173
– peristaltic motion   119–120
– polyelectrolyte   12
– porous   121–124
– responsive, see responsive gels
– self-oscillating, see self-oscillating gels
– self-walking   122–124
– symmetric 1D   148

generalized Darcy law   165
geometric feedback   174–181
geometry, conical   185
Gibbs free energy   93–95
Gibbs relation   167
glass
– capillaries   237
– cross-cut wafer   194
– fritted plate   191
– ground   84
Glauber's salt   223
gLSM (gel lattice spring model)   136, 142–143
gluconic acid   200–201
glucose   205
glycol   28
glycol dimethacrylate   48
Gouy–Chapman theory   192
graded co-continuous morphology   102–103
gradient refractive index materials (GRINs)   71–74, 85–87
ground glass   84
growth
– bubble-free   236
– closed   239
– exponential   5
– fracturing   228
– free-radical-   21
– (non)linear chains   22–25
– regimes   225–228
– reverse   227
– step-   21
Grubbs catalyst   45
Guth, James and Guth models   166

h

hard polymer product   82
HDDA   58
helical patterns   56
helix–coil transition   193
hematoporphyrin   84
heterogeneous BZ gels, straining   147–154
1,6-hexanediol diacrylate polymerization   60
homogeneous BZ gels   143–147
homogeneous polymerization   82
Hooke's law   5
– neo-Hookean elasticity   141–142
Hopf bifurcation   149, 192
hormones   190
– rhythmic delivery   199–211
horseshoe map   31
hot spot   57

Huggins . . ., see Flory–Huggins . . .
hydrating effect   139
hydrochloride, lauryl amine   224
hydrogels
– deformation-sustaining   208
– hydrogel–enyzme oscillator   199–211
– membrane permeability   201–204
hydrogen bonding interactions   202–203
hydrophobicity
– control   170
– hydrogels   203
– solutes   204
hydrostatic pressure head   194
hydroxyl group   37
hydroxyl ion-autocatalyzed reactions   182
hypogonadotropic hypogonadism   190
hysteresis   149–150, 174–175, 200–204

i

IFP (isothermal FP)   69–90
– mathematical models   74–79
IGP (interfacial-gel polymerization)   70
immiscibility   33
immobilized polymers   128
impermeable wall   181
in-phase modes   152
incompressibility condition   165
inherent nonlinearities   13
inhibitor-IFP   87
inhomogeneity, dynamic   97
inorganic deposition processes   220–221
inorganic precipitation systems   219–241
instabilities
– chemomechanical   174–175, 182
– convective   52–56
– double diffusion   91
– Mullins–Sekerka   91
– Rayleigh–Benard   91
– Rayleigh–Taylor   54, 91
– spatial bistability   174–175
– thermal   56–59
integration, numerical   171–174
intelligent gels   115
intensity, light   157
interdiffusion, polymer–solvent   157
interfacial-gel polymerization (IGP)   70
interferometry, Mach–Zehnder   95, 100–101
internal mixer, Brabender   38
interpatch distance   151, 154
iodide   9, 220
ion mobility   191
ionization, degree of   178
irradiation method, computer-assisted   105–106

isocyanate group   33, 37
N-isopropylacrylamide (NIPA)   200
isothermal FP (IFP)   69–90
– mathematical models   74–79
isothermal processes   166
isotropic pressure   142

j

James and Guth models   166
jet-driven motion   234
jetting regime   226

k

Kapton   51
kinetic equations   74
kinetic nonlinearities   219
kinetics
– non-mean-field   97–99
– phase separation   27, 92–97
– photo-cross-link   98–99
– thermoplastic polyurethane polymerization   21–44
Kohlrausch–Williams–Watts (KWW) function   97

l

Lagrangian mode   180
laminar mixing   30
large-aspect-ratio sample   147
large-scale fluid flow   237
laser sheet deflection technique   79
lattice spring model   136, 142–143
lauryl amine hydrochloride   224
laws and equations
– continuum equations   137–141
– frontal Rayleigh number   52
– generalized Darcy law   165
– Gibbs relation   167
– Hooke's law   5
– kinetic equations   74
– Navier–Stokes equations   234
– Ogston law   177
– reaction–diffusion equations   168
– Snell's law   59–60
– Zeldovich number   57
LCST (lower critical solution temperature)   92
length scales, arbitrary distribution   105–109
Liapunov exponents   31
light, sensitivity to   154–160
light intensity, nonuniform   156
linear chains, growth   22–24
linear polymers   10
linear springlike forces   141–142

lipid membranes   196–211
liquid crystal phase transformations   223
liquid state of polymer mixtures   98–99
local mechanical impact   143–147
long-wavelength fluctuations   94
lower critical solution temperature (LCST)   92
Luperox   51, 55, 61–62

m

MAA (methacrylic acid)   200, 209–210
Mach–Zehnder interferometry   95, 100–101
magnetic stirrer   195
malonic acid   14, 117–120
– chlorite–iodide–malonic acid reaction   9, 220
malonyl radical   15
marble   205–207
mass balances   74
mass transport by peristaltic motion   124
materials
– biomimetic soft   135–162
– GRINs   71–74, 85–87
– processing   91
mathematical models, IFP   74–79
Maxwell–Stefan approach   176
mechanical deformation   143–154
– sensors   153
mechanical equilibrium   165, 178
mechanical impact, local   143–147
mechanochemical response, biomimetic   147
membranes
– droplets   231
– gel   122
– lipid/organic   196–211
– membrane/enzyme oscillators   196–211
– permeability   201–204
– polyelectrolyte membrane-based oscillators   193–194
– polymer   189–217
– resistance   199
– synthetic oscillators   191–199
metal-doped plastic optical fibers (POFs)   84
metathesis polymerization, ring-opening   45
methods and techniques
– computer-assisted irradiation method   105–106
– laser sheet deflection   79
– rheological techniques   166
– solgel methods   223
– Wiener's methods   80–83
methyl methacrylate (MMA)   83–84, 97–99
microactuators, biomimetic   122
microcapsule   60

microdevices   124–126
microfabrication, self-oscillating gels   124–126
microgels   128–129
– microgel beads monolayer   129–130
migration term   179
miniature bulk gel   117–119
mixer design/protocol   27
mixing   21–44
– chaotic   31–32, 37–38
– laminar   30
– timescales   32–35
mixing chamber   37–38
mixing gap   38
mixtures
– binary   92–97
– polymer   98–99
"mode-selector"   94
modulations, time-periodic   169
mold, slime   160
molecular building blocks   219
molecular diffusion   36–38
molecular weight   22–26
– weight-averaged   36
monolayer, microgel beads   129–130
monomer conversion   15
monomer–initiator system   78
monomer solution   70
monomers, chemical structure   25–26
morphology
– arbitrary symmetry   105–109
– graded co-continuous   102–103
– inorganic precipitation systems   228–230
– TPUs   25
motion
– autonomous directed   154
– jet-driven   234
– peristaltic   119–120, 124
mounds, sulfide   222
Mullins–Sekerka instability   91
multifunctional acrylates   58

n

Na, see sodium
nanoactuators, biomimetic   122
nanoconveyor   128
nanogel beads   174
nanooscillators, self-oscillating polymer chains   127–128
nanotechnology   219
Navier–Stokes equations   234
neo-Hookean elasticity   141–142
network formation   21–44
neutral diluent   138

(next-)nearest neighbors   141
NIPA (N-isopropylacrylamide)   200
nitric acid   120
no-dopant IFP   87
nodal forces   142
non-Fickian behavior   11
non-mean-field kinetics   97–99
nonchemoresponsive gel   183
nondecaying oscillations   145
nonequilibrium conditions   219
– stationary   91–113
nonlinear chains   24–25
nonlinear dynamics   5
– chemical   6
– frontal polymerization   51–52
nonlinear rheology   21–44
nonlinearities
– inherent   13
– kinetic   219
nonreactive binary mixtures   92–94
nonuniform conditions   101–109
nonuniform light intensity   156
Norrish–Trommsdorff effect   11
numerical integration   171–174

o

Ogston law   177
one-dimensional gels   148
one-dimensional model   137
operation, oscillators   204–205
optical fibers, plastic   84
optical path length difference (OPLD)   100
optical transmittance   127
Oregonator model   137
organic crystal templating   223
organic membranes   196–211
oscillations
– chemical   7
– chemodynamic   174–181
– frequency   150–151
– in a CSTR   16
– nondecaying   145
– pH   208
– relaxation   171, 230–232
– self-flocculating/dispersing   128–129
– sinusoidal   5
– transient   143
oscillators
– hydrogel–enyzme   199–211
– membrane/enzyme   196–211
– nano-   127–128
– operation   204–205
– pH range tuning   208–211

oscillators (*contd.*)
– polyelectrolyte membrane-based 193–194
– prototype 205–207
– synthetic membrane 191–199
– Teorell 191–193, 199
– thermofluidic 194–196
oscillatory gel dynamics 166
oscillatory reactions 166
– control 119
– gels 115–134
oscillatory systems 189–217
OSFR 181–185
osmotic pressure 176, 179
osmotic stress tensor 170–171
out-of-phase modes 152
oxidant-supplying sites 131

p
papain 196–197
paper, wax 56
"pared down" designs 189
partitioning, Donnan 207
passive gels 171–173
patterns 7–8
– helical 56
– permanent 220–221
– reaction–diffusion 163
– stationary Turing 220
– target 8
– Turing 9
PE (polyethylene) 22–23
pentaerythritol tetraacrylate copolymerization 55
periodic irradiation 108
periodic redox changes 118
peristaltic motion
– gels 119
– mass transport 124
– self-sustaining 124
permanent patterns 220–221
permeability 201–204
permeation flux 167
pH
– control 131
– fluctuating 205
– oscillations 208
– ramp 201
– range tuning 208–211
phase diagram 93
phase separation 21–44
– kinetics 27, 92–97
– nonreactive binary mixtures 92–94
– nonuniform conditions 101–109

– reacting systems 12
– reaction-induced 91–113
– reactive binary mixtures 94–97
phosphofructokinase (PFK) 198
photo-cross-link kinetics 98–99
photocatalytic activity 238
photopolymerization 97–99
Physarum 160
physiological conditions 196
– self-oscillating gels 131–132
pinning forces 236
plasma-grafted poly(*N*-isopropylacrylamide) (pNIPA) 194–195
plastic optical fibers (POFs), metal-doped 84
platinum redox electrode 8–9
PMMA 83–84
POFs (plastic optical fibers), metal-doped 84
Poiseuille flow 234
polyacid hydrogels 203
polybase layer 193
polyelectrolyte gel 12
polyelectrolyte membrane-based oscillators 193–194
polyelectrolyte model 177–181
poly(ethyl acrylate) 101
poly(ethylene glycol) 28
polyethylene (PE) 22–23
polyisoprene, chlorinated 105
polymeric systems 9–10
– chemical reactions 97–99
– nonuniform conditions 101–109
– spatial structures 13
– stationary nonequilibrium conditions 91–113
"polymeric worm" 135, 154
polymerization
– acrylamide/bisacrylamide 53
– butyl acrylate 54
– degree of 23
– free-radical 11, 74
– frontal, *see* frontal polymerization
– 1,6-hexanediol diacrylate 60
– homogeneous 82
– interfacial-gel 70
– photo- 97–99
– polyurethane 21–44
– rheological properties 22–27
– ring-opening metathesis 45
– thermoplastic polyurethane 21–44
– trimethylolpropane triacrylate 61
– vinyl acetate 16
polymers 5
– bulk state 97–98
– co-, *see* copolymers

– Flory–Huggins solution theory 165–169
– hard polymer product 82
– immobilized 128
– liquid state of mixtures 98–99
– membranes 189–217
– network dynamics 138
– polymer–solvent interdiffusion 157
– polymer-tethered BZ catalyst 147
– preformed 69
– self-oscillating 122
– self-oscillating chains 127–128
– spatially graded continuous structures 101–105
– volume fraction 172
poly(methyl methacrylate) 105
polyol 28
polystyrene (PS) 22–23
– photo-cross-link kinetics 98
polyurethane polymerization
– frontal, see frontal polymerization
– thermoplastic 21–44
popping regime 226, 231
popping tube 229
population explosion 7
porous gels 121–124
potential
– chemical 171
– Donnan 191
pre-gel solution 130
precipitation systems, inorganic 219–241
precipitation tube 235
preformed polymer 69
prepolymer 33
pressure
– hydrostatic pressure head 194
– isotropic 142
– osmotic 176, 179
priming 84
processes
– inorganic deposition/precipitation 220–221
– isothermal 166
– transfer 21
propagation
– chemical waves 119–120, 126–127
– polymerization front 47
– velocity 77
protocol, mixer 27
prototype, oscillators 205–207
PS (polystyrene) 22–23
pump rate 233
putty, cure-on-demand 49–50

q
quasi-traveling-wave solutions 75
quaternary copolymers 131

r
radius selection 233–234
ramp pH programs 201
random walk configurations 203
Rauscher reaction, Briggs– 220
Rayleigh–Benard instability 91
Rayleigh number, frontal 52
Rayleigh–Taylor instability 54, 91
reacting systems, phase separation 12
reaction–diffusion equations 168
reaction–diffusion patterns 163
reaction-induced elastic strain 99–101
reaction-induced phase separation 91–113
reaction kinetics 21–44
– see also kinetics
reaction zone 48
reactions
– Belousov–Zhabotinsky, see Belousov–Zhabotinsky reaction
– bistable 164
– Briggs–Rauscher 220
– bromate–sulfite 164, 184–185
– chain extension 39
– chlorite–iodide–malonic acid 9, 220
– chlorite–tetrathionate 164, 182–184
– chlorite–thiosulfate 9
– diisocyanate–poly(ethylene glycol) 28
– effect of mixing 32
– exothermic 78
– FKN mechanism 117
– formaldehyde–bisulfite–sulfite clock 15
– hydroxyl ion-autocatalyzed 182
– oscillating 166
– oscillatory 115–134
– polymeric systems 97–99
– termination 74–75
reactive binary mixtures 94–97
reactor
– CSTR 16, 163, 174–175, 180–182, 189
– cylindrical adiabatic 56
redox changes, periodic 118
redox electrode, platinum 8–9
refractive index, gradient 71–74, 85–87
relaxation 99–101
– dynamics 138
– oscillations 171, 230–232
reorientation 158–159
repair of wood hole 50
resins, epoxy 46
resistance, membrane 199

responsive gels   163–188
"reverse" conditions   226
reverse growth   227
rheological techniques   166
rheology, nonlinear   21–44
rhodamine   104–105, 237
rhomboidal channels   224
rhythmic hormone delivery   199–211
ring-opening metathesis polymerization   45
"ripples"   145
robots, "soft"   160
ROMP   45
rotating spiral waves   221
rotors, chaotic mixer   37–38
ruthenium complexes   116–117, 123, 170

s

salt, Glauber's   223
scales, arbitrary distribution   105–109
seed dissolution   70
segregation coefficient   79
Sekerka instability, Mullins–   91
self-assembly   6
self-flocculating/dispersing oscillations   128–129
self-oscillating gels   116
– arrays   126–127
– biomimetic micro-/nanoactuators   122
– biomimetic soft materials   135–162
– design   116–117
– microfabrication   124–126
– physiological conditions   131–132
self-oscillating polymers   122, 127–128
self-sustained brittle disruption   46
self-sustaining peristaltic motion   124
self-walking gels   122–124
sensitivity
– to light   154–160
– to mechanical deformation   143–154
sensors, mechanical deformation   153
separation, phase, see phase separation
shape changes   158–159
shape functions   140
shear
– constant rate   30
– laminar mixing   30
shear-thickening/-thinning   12
side-by-side diffusion cell   202
silica
– colloidal crystal   130
– gardens   221–222
– tubes   224–225, 233
silicate, sodium   227
silver–antimony alloy   221

sinusoidal oscillations   5
skin layers   204
"skinlike" coating, synthetic   136
slime mold (*Physarum*)   160
smart gels   115
SMPA (sulfamethoxypyridazine), acryloylated   209
Snell's law   59–60
sodium bromate   120
sodium silicate   227
soft materials, biomimetic   135–162
"soft robots"   160
solgel methods   223
solids, viscoelastic   24
solutes, hydrophobic   204
solvents, polymer–solvent interdiffusion   157
sources of feedback   10–12
spatial bistability   174–175
spatial scale   76
spatial structures, polymeric systems   13
spatially graded continuous structures   101–105
spherical beads   169
spherically expanding front   61
spin modes   56
spinodal decomposition   96
spinodal structures   92
spiral waves, rotating   221
spring model, gel lattice   136, 142–143
springlike forces, linear   141–142
stability diagram   53
stationary nonequilibrium conditions   91–113
stationary Turing patterns   220
steady-state assumption (SSA)   76, 143
Stefan approach, Maxwell–   176
step-growth   21
stimuli-responding behavior   116
stirred tank reactor
– continuous   16, 163, 174–175, 180–182, 189
stirrer, magnetic   195
stoichiometric parameter   150
Stokes, Navier–Stokes equations   234
stones, curing   50
strain, elastic   99–101
straining heterogeneous BZ gels   147–154
stress tensor, osmotic   170–171
stretched state   148
stretching   34
– absorption bands   29
striation thickness   30
structural color changes   121–122

structures
- cubic crystal 229
- dissipative 6
- formation 219–241
- spatial 13
- spatially graded continuous 101–105
- spinodal 92
- tubular 222
sulfamethoxypyridazine (SMPA), acryloylated 209
sulfate, cupric 225
sulfide mounds 222
sulfite 15, 164, 184–185
surface chemistry, UHV 186
surface-tension-driven convection 55–56
surfaces
- functional 128
- peristaltic 126
swelling–deswelling changes 116
swelling ratio 202
symmetric 1D gels 148
symmetry, arbitrary 105–109
synchronization 152–153
synthetic membrane oscillators 191–199
synthetic "skinlike" coating 136
systems
- free-radical 46
- inorganic precipitation 219–241
- monomer–initiator 78
- oscillatory 189–217
- polymeric 9–10, 91–113
- reacting 12

t
target patterns 8
Taylor instability, Rayleigh– 54, 91
techniques, *see* methods and techniques
temperature, LCST 92
tension, surface- 55–56
tensor, osmotic stress 170–171
Teorell oscillator 191–193, 199
termination reaction 74–75
test tube wall 79
tetraacrylate, pentaerythritol 55–56
2,2′,6,6′-tetramethyl-1-piperidinyloxy (TEMPO) 82
tetrathionate 164, 182–184
TFP (thermal frontal polymerization) 69
theories and models
- 1D gel model 137
- Flory–Huggins polymer solution theory 165–169

- gel lattice spring model (gLSM) 136, 142–143
- Gouy–Chapman theory 192
- James and Guth models 166
- Oregonator model 137
- polyelectrolyte model 177–181
thermal frontal polymerization (TFP) 69
thermal history 27
thermal instabilities 56–59
thermocouples 83
thermodynamic branch 182
thermodynamic equilibrium 91
thermofluidic oscillator 194–196
thermoplastic polymers 10
thermoplastic polyurethane (TPU) 21–44
- morphology 25
thermosets 10, 24
thermostatic box 83
thickness, striation 30
thin films 46
thin polybase layer 193
thiosulfate 9
three-dimensional frontal polymerization 60–61
time-periodic modulations 169
timescales, mixing 32–35
TMPTA 51
torque 39
TPU (thermoplastic polyurethane) 21–44
- morphology 25
transfer processes 21
transient oscillations 143
transmittance, optical 127
trench walls 195
triacrylate 15, 55
- trimethylolpropane 61
- trimethylolpropane ethoxylate 50
triethylene glycol dimethacrylate 48
trimethylolpropane ethoxylate triacrylate 50
trimethylolpropane triacrylate polymerization 61
trithiol 15
Trommsdorff effect 23, 98, 109
- Norrish– 11
tube formation 221–222
tubes
- closed growth 239
- fragments 230
- popping 229
- precipitation 235
- silica 224–225, 233
tubular structures 222
turbidity 15

Turing patterns 9
– stationary 220

u
UHV surface chemistry 186
UV-induced FP 49

v
velocity, propagation 77
velocity fields, cylindrically symmetric 234
venting black smokers 222
vinyl acetate polymerization 16
viscoelastic solids 24
viscosity, TPUs 21–26
viscous region 86
volume fraction, polymer 172
volume phase transition phenomena 115

w
wafer, glass 194
wall composition, and morphology 228–230
waves 7–8
– chemical 119–120, 126–127
– chemomechanical excitability 184
– rotating spiral 221
wax paper 56
weight, molecular 22–26
weight-averaged molecular weight 36
wet etching 130
Wiener's method 80–83
Williams–Watts function, Kohlrausch– 97
wood hole, repair 50
"worm", polymeric 135, 154

z
Zehnder interferometry, Mach– 95, 100–101
Zeldovich number 57
Zhabotinsky, *see* Belousov–Zhabotinsky reaction